Digital Signal Processing

S. Esakkirajan • T. Veerakumar • Badri N. Subudhi

Digital Signal Processing

Illustration Using Python

 Springer

S. Esakkirajan
Dept of Instrumentation & Control Eng.
PSG College of Technology
Coimbatore, Tamil Nadu, India

T. Veerakumar
Dept of Electronics & Communication Eng.
National Institute of Technology Goa
Ponda, Goa, India

Badri N. Subudhi
Dept of Electrical Engineering
Indian Institute of Technology Jammu
Jammu, Jammu and Kashmir, India

ISBN 978-981-99-6751-3   ISBN 978-981-99-6752-0   (eBook)
https://doi.org/10.1007/978-981-99-6752-0

This Springer imprint is published by the registered company Springer Nature Singapore Pte Ltd.
The registered company address is: 152 Beach Road, #21-01/04 Gateway East, Singapore 189721,
Singapore

Paper in this product is recyclable.

Preface

Motivation

The objective of this book is to implement signal processing algorithms in Python. During COVID-19 pandemic, it was a challenge to conduct the signal processing laboratory session in online mode. It was difficult for the students to install proprietary software as it was costly and hence not affordable. This motivated us to turn our attention toward open-source software. There are many open-source software packages available to implement signal processing algorithms. The reasons for choosing Python are (1) it is a general-purpose programming language that can be used for various tasks beyond scientific computing. (2) Python has an active community of developers who create and maintain a wide range of libraries and frameworks. (3) Python has become the language of choice for many machine learning and deep learning applications with powerful libraries such as TensorFlow, PyTorch, and Keras. The main aim of signal processing is to extract information from the signal. After extracting useful information, further processing, like classification of information, has to be done effectively using machine learning and deep learning libraries in Python.

In this book, Python is used as a tool to implement signal processing algorithms. Teaching Python is not the main aim of this book. Python is used as a vehicle to present concepts related to signal processing. In this book, the signals are generated, manipulated, transformed, and useful information is extracted using libraries available in Python. The Python programs used in this book are purposively made simple and illustrative. The libraries used in this book include (1) Numpy, (2) Scipy, (3) Matplotlib, etc. These libraries provide a wide range of tools and functions for performing operations like filtering, resampling, prediction, etc.

Target Audience

This book is suited for undergraduate students, postgraduate students, research scholars, and faculties working in signal processing. The reader is assumed to be familiar with basic Python programming.

Salient Features of the Book

The salient features of the book are summarized below:

- PreLab questions are included in each chapter. The questions are framed to be concise, clear, and thought-provoking.
- Numerous examples with Python illustrations are provided in each chapter. Python codes that implement signal processing algorithms are explained in step by-step approach. Tasks are given at the end of Python examples. These tasks will help the reader to vary the parameters in the algorithm and realize its impact.
- Exercises are provided in each chapter. These exercises help the reader to develop a deeper understanding of the concepts discussed in the chapter.
- Objective questions are given in each chapter. It helps the reader to prepare for competitive examinations like GATE, IES, etc.

Coimbatore, Tamil Nadu, India   S. Esakkirajan
Ponda, Goa, India   T. Veerakumar
Jammu, Jammu and Kashmir, India   Badri N. Subudhi

Organization of the Book

The book comprises of 12 chapters. Chapter 1 deals with the generation and visualization of continuous-time signals which include periodic signals, non-stationary signals, pulse signals, and standard test signals. Chapter 2 focuses on sampling, quantization, and reconstruction of signals. Both the time domain and frequency domain view of sampling, the effect of undersampling, uniform and non-uniform quantization, and different types of reconstruction like zero-order hold, first-order hold, and sinc interpolation are discussed in this chapter. Chapter 3 is dedicated to the generation of discrete-time signals and mathematical operations that are performed on the discrete-time signals. In this chapter, standard discrete-time signals like unit sample, unit step, unit ramp, exponential, and sinusoidal signals are generated, and mathematical operations like folding, shifting, and scaling are performed on the generated signals. This chapter also discusses two important signal processing operations: convolution and correlation. Different forms of representation of discrete-time system, properties of discrete-time systems, and responses of discrete-time systems are explained with examples in Chap. 4. One of the important topics is signal processing which is analysis of signals and systems using transform. Chapter 5 is devoted to transform domain analysis of signals and systems. Different transforms discussed in this chapter include Z-transform, Fourier transform, Short-Time Fourier transform, and Wavelet transform. Chapter 6 deals with the design of a simple filter using pole-zero placement technique. Different filters discussed in this chapter include moving average filter, digital resonator, notch filter, comb filter, and all-pass filter. Chapter 7 covers the types of FIR filters and the design of FIR filters. Three design approaches covered in this chapter include window-based FIR filter design, frequency sampling-based FIR filter design, and optimal FIR filter design. Chapter 8 deals with the design of IIR filter, mapping from the analog domain to the digital domain. The types of IIR filters discussed in this chapter include Butterworth filter, Chebyshev filter, and Elliptic filter. The mapping techniques discussed in this chapter include the backward difference, impulse invariant, and matched Z-transform techniques. The impact of the finite word length effect in the FIR and IIR filters is discussed in Chap. 9. Concepts like limit cycle oscillation, impact of coefficient quantization, and the nature of coefficient error are discussed in this chapter. Chapter 10 is devoted to multi-rate signal processing. Concepts like multi-rate operators, noble identities, polyphase decomposition, filter bank, and transmultiplexer are covered with detailed examples in this chapter. Design of optimal and adaptive filters and their applications are discussed in Chap. 11. This chapter discusses Wiener filter, LMS algorithm and its variants, RLS algorithm and its applications with necessary examples. Chapter 12 is devoted to case study which discusses the application of signal processing algorithms in analyzing speech signal, ECG signal, and power line signal.

Acknowledgments

The authors are always thankful to the Almighty for guiding them in their persever-ance and blessing them with achievements. The authors wish to thank Shri L. Gopalakrishnan, Managing Trustee, PSG Institutions; Dr. K. Prakasan, Principal, PSG College of Technology, Coimbatore; Prof. Gopal Mugeraya, Director, National Institute of Technology, Goa; and Prof. Manoj Singh Gaur, Director, IIT Jammu for their wholehearted cooperation and constant encouragement given in this successful endeavor.

Dr. S. Esakkirajan would like to express his gratitude to his parents, Mr. G. Sankaralingam, and Mrs. S. Saraswathi, wife Mrs. K. Sornalatha and sons Azhaku Vignesh and Krishnan, for their love and encouragement. He would like to thank his students Mr. Senthil Murugan, Mr. Vijay Bhaskar, Ms. B. Keerthiveena, and Mr. Upendra Vishwanath for their continual support and encouragement.

Dr. T. Veerakumar would like to thank his life guru Dr. S. Esakkirajan for his guidance, motivation, and constant support in completing this work. He also wants to thank his parents, Mr. Thangaraj and Mrs. Muniammal, brothers Mr. Tamilselvan and Mr. Karl Marks, and sister, Mrs. Muniponnu, for their wholehearted support. Finally, he would like to thank his wife, Banupriya, and daughters, Harini and Ishani, for tolerating his late coming home and their support in completing this work on time.

Dr. Badri Narayan Subudhi would like to express his gratitude to his parents, Mr. Ananda Chandra Subudhi and Ms. Subasini Subudhi, wife Ms. Bandanarani Subudhi and children: Aaradhya and Anwit for their unflagging love and support throughout life. He would also like to thank his brother Mr. Rashimi Ranjan Subudhi and Prof. Sarat Kumar Patra for their encouragement and support during his life.

Contents

About the Authors

S. Esakkirajan is a Professor in Instrumentation and Control Engineering Department at PSG College of Technology, where he has been a faculty member since 2004. He did B.Sc. Physics from Sadakathullah Appa College, Palayamkottai, B. Tech. in Instrumentation Engineering from Cochin University of Science and Technology, M.E. in Applied Electronics from PSG College of Technology, and Ph.D. in the area of Image Processing from Anna University. He has successfully guided four research scholars towards their Ph.D. in the area of signal processing. He has published papers in reputed journals and conferences. His research interest includes digital signal processing and digital image processing.

T. Veerakumar is an Associate Professor in the Department of Electronics and Communication Engineering, National Institute of Technology, Goa. He graduated with a B.E. in Electronics and Communication Engineering from RVS College of Engineering Technology, Dindigul. Then, he did an M.E. degree in Applied Electronics from PSG College of Technology, Coimbatore, and a Ph.D. in Image Denoising from Anna University, Chennai. He co-authored the textbook titled *Digital Image Processing* and *Digital Signal Processing*, published by Tata McGraw Hill. In addition, he has published around 60 research articles in reputed Journals and Conferences. His area of interest includes Signal and Image Processing, Biomedical Image Processing, Object Detection, and Tracking.

Badri Narayan Subudhi received M. Tech. in Electronics and System Communication from the National Institute of Technology, Rourkela, India, in 2008–2009. He worked on his Ph.D. from Machine Intelligence Unit, Indian Statistical Institute, Kolkata, India, in 2014 (degree from Jadavpur University). Currently, he is serving as an Associate Professor at the Indian Institute of Technology Jammu, India. Prior to this, he was working as an Assistant Professor at NIT Goa from July 2014 to November 2017. He received CSIR senior research fellowship for the year 2011–2015. He was nominated as the Young Scientist Awardee by Indian Science Congress Association for the year 2011–2012. He was awarded the Young Scientist Travel grant award from DST, Government of India, and Council of Scientific and

Industrial Research, India, in 2011. He received the Bose-Ramagnosi Award for the year 2010 from DST, Government of India, under India-Trento Programme for Advanced Research (ITPAR). He was a visiting scientist at the University of Trento, Italy, during August 2010 to February 2011. His research interests include Video Processing, Image Processing, Medical Image Processing, Machine Learning, Pattern Recognition, and Remote Sensing Image Analysis. He co-authored the textbook titled *Digital Signal Processing*, published by Tata McGraw Hill. He has published around 80 research papers in reputed journals and conferences. He is a senior member of IEEE.

Chapter 1
Generation of Continuous-Time Signals

Learning Objectives

After completing this chapter, the reader should be able to

- Simulate and visualize periodic continuous-time signals.
- Simulate, visualize and interpret non-stationary signals.
- Simulate and visualize standard continuous-time test signals.
- Simulate and visualize continuous-time pulse signals.

Roadmap of the Chapter

This section discusses the flow of contents in this chapter. The objective of this chapter is to generate different types of continuous-time signals, pulse waveforms. The representation of different signals generated in this chapter is given in the form of a flow diagram, which is given below:

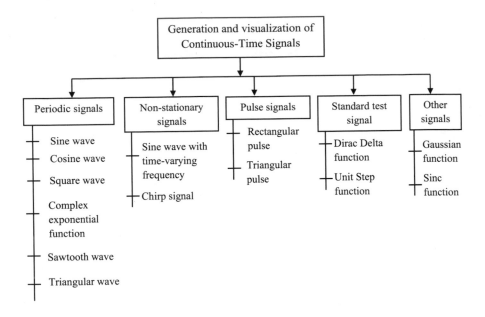

PreLab Questions
1. Give a few examples of real-world signals, which are continuous in nature.
2. Mention the built-in functions available in '*numpy*' library in python to generate data points of specific length to define the independent variable like time.
3. Explain the significance of a sinusoidal signal in signal processing.
4. What do you understand by the term '*phase*' of a signal?
5. Give a few examples of multidimensional signals.
6. Cite an example where the signal or a process can be modelled as a real exponential function.
7. Mention a few significant features of complex exponential signals.
8. Mention the salient features of the '*sinc*' function in signal processing. Is it an even or odd function?
9. Distinguish between stationary and non-stationary signal. Give examples of each category of signal.
10. List a few significant properties of the Gaussian function (signal).

1.1 Continuous-Time Signal

A signal corresponds to a physical quantity that varies with time, space, etc. Signals are represented mathematically as a function of one or more independent variables. The continuous-time signals are defined for a continuum of values of the independent variable. The continuous-time signal is generally represented as $x(t)$. Speech signal as a function of time is an example of continuous-time signal. The signal can be either deterministic or random. Deterministic signals can be described by mathematical functions or expressions. In this chapter, the objective is to generate different types of continuous-time periodic signals, like sinusoidal signal, complex exponential signal, square wave, etc.; non-stationary signals, like chirp signal; standard test signals, like Dirac delta; unit step signal, etc.

1.1.1   Continuous-Time Periodic Signal

A periodic signal is one which repeats itself in an identical manner. Examples of continuous-time periodic signals include sinusoidal signal, complex exponential signal, square wave and sawtooth wave. In this section, python codes are developed to generate a sinusoidal signal, three-phase sinusoidal signal, complex exponential signal, etc. Also, sinusoids are Eigen functions of linear system. Continuous-time sinusoids are described by an amplitude, frequency and phase. Continuous-time sinusoids with distinct frequencies are always distinct.

Experiment 1.1 Generation of Sinusoidal Signal
The aim of this experiment is to generate sinusoidal signal. Sinusoidal signals are periodic functions, which are based on the sine or cosine function. The expression for the sinusoidal signal is given by

$$x(t) = A\sin(2\pi ft + \phi) \tag{1.1}$$

In the above equation, 'A' represents the amplitude of the signal, 'f' denotes the frequency of the signal and 'ϕ' indicates the phase of the signal. To generate sinusoidal signal, one should define three parameters: amplitude, frequency and phase. The independent-variable is '*time (t)*'. In amplitude modulation, the amplitude of the carrier is changed in accordance with the message, while the frequency and phase are kept constant. In frequency modulation, the frequency of the carrier is changed in accordance with the signal, while the amplitude and phase are kept constant. In phase modulation, the phase of the carrier is changed in accordance with the signal, while the amplitude and frequency are kept constant.
　　The steps involved in the generation of sinusoidal signal are summarized below:

Step 1: Defining the independent variable
　　The built-in function '*np.linspace()*' is used to generate the independent variable, which is the time axis.
Step 2: Defining the parameters of the sine wave
　　In this step, the three parameters of sine wave, namely, amplitude, frequency and phase are defined.
Step 3: Generation of sinusoidal signal
　　In this step, the mathematical expression to generate a sine wave is given by

Table 1.1 Built-in functions used in the program

S. No.	Built-in function used	Purpose
1	*np.sin()*	To generate sinusoidal function
2	*np.linspace()*	To generate equally interval data points in an interval
3	*plt.subplot()*	To plot more than one figure in the same plot

```
#Experiment 1: Generation of sinusoidal signal
import numpy as np
import matplotlib.pyplot as plt
#Step 1: Defining the independent variable
t=np.linspace(0,1,1000)
#Step 2: Defining the parameters of sine wave
A=5 #Amplitude of sine wave
f=5 #Frequency of sine wave
ph=0 #Phase of sine wave
#Step 3: Expression of sine wave
x=A*np.sin(2*np.pi*f*t+ph)
#Step 4: Plotting the sine wave
plt.plot(t,x),plt.xlabel('Time'),plt.ylabel('Amplitude')
plt.title('A={}V,F={} Hz,$\phi={}^\circ$'.format(A,f,ph))
```

Fig. 1.1 Python code to generate sinusoidal signal

$$x(t) = A \sin(2\pi f t + \phi)$$

Step 4: Plotting the sinusoidal signal

The built-in function *plt.plot()* is used to plot the generated signal. While plotting the waveform, it is important to mention that the label of x and y axes using *plt.xlabel()* and *plt.ylabel()* command. The command *plt.title()* is used to display the title of the plot.

Built-In Libraries

The built-in libraries used in the program are (1) Numpy and (2) Matplotlib. The '*numpy*' is a general purpose array-processing package. In this program, the numpy library is used to create array (*np.linspace*), and it is used to perform mathematical function (*np.sin*). Matplotlib is a data visualization library used to visualize the generated sinusoidal signal. The built-in functions used in the program is given in Table 1.1.

The python code used to generate sinusoidal waveform is shown in Fig. 1.1, and the corresponding output is shown in Fig. 1.2.

Inference

From Fig. 1.1, the following inferences can be made with respect to python code:

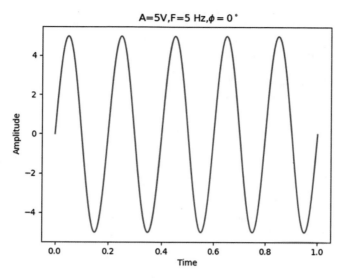

Fig. 1.2 Result of the python code shown in Fig. 1.1

1. The libraries used in the program are (a) Numpy and (b) Matplotlib.
2. The built-in function '*np.linspace()*' is used to generate the independent variable, which is the time axis. In this program, 1000 data points are generated between '0' and '1'.

From Fig. 1.2, it is possible to infer the following:

1. The phase of the signal is '0'; this implies that the waveform starts from the origin.
2. The amplitude of the sine wave is 5 V. The waveform oscillates between -5 and $+5$.
3. The frequency of the generated waveform is 5 Hz. The number of oscillations per second is 5.

Tasks
1. Write a python code to mark the peak of the sinusoidal signal.
2. Write a python code to compute the number of zero crossing of the sine wave.

Experiment 1.2 Sinusoidal Signal with Different Phase
In this experiment, the objective is to generate sine wave of amplitude $= 1$ V, frequency $= 5$ Hz and four different phase angles, namely, 0°, 90°, 180° and 270°. The python code, which does this task, is shown in Fig. 1.3, and the corresponding output is shown in Fig. 1.4. The built-in libraries used in the program are (1) Numpy and (2) Matplotlib.

```
#Generation of sine wave of different phase angles
import numpy as np
import matplotlib.pyplot as plt
t=np.linspace(0,1,100)
#Parameters of sine wave
A=1  #Amplitude
f=5 #Frequency
phi=[0,90,180,270] #Phase
#Generation of sine wave
for i in range(len(phi)):
    x=A*np.sin(2*np.pi*f*t+phi[i]*np.pi/180)
    #Plotting the result
    plt.subplot(2,2,i+1)
    plt.plot(t,x),plt.xlabel('Time (t)'),plt.ylabel('Amplitude (V)')
    plt.title('$\Phi ={}^\circ $'.format(phi[i]))
    plt.tight_layout()
```

Fig. 1.3 Python code to generate sine wave with different phase angle

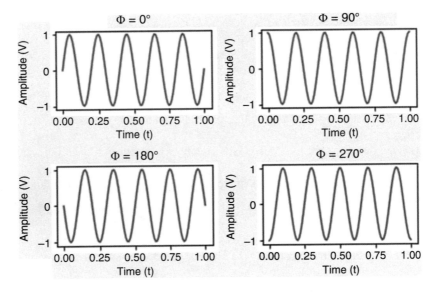

Fig. 1.4 Sine wave with different phase angle

Inference

From the python code shown in Fig. 1.3, the following inferences can be made. The phase angle is varied as 0°, 90°, 180° and 270°. The amplitude of the sine wave is fixed as 1 V and the frequency is fixed as 5 Hz.

Fig. 1.5 Python code to generate three-phase sinusoidal signals

```
#Generation of three phase sine wave
import numpy as np
import matplotlib.pyplot as plt
t=np.linspace(0,1,100)
#Parameters of sine wave
A=1 #Amplitude
f=5 #Frequency
#Three different phases of sine wave
phi_1,phi_2,phi_3=0, 120, 240
x1=A*np.sin(2*np.pi*f*t+phi_1*np.pi/180)
x2=A*np.sin(2*np.pi*f*t+phi_2*np.pi/180)
x3=A*np.sin(2*np.pi*f*t+phi_3*np.pi/180)
#Plotting the result
plt.plot(t,x1,'b',t,x2,'r',t,x3,'g')
plt.xlabel('Time (t)'),plt.ylabel('Amplitude (V)')
plt.title('Three phase sinusoidal signal')
plt.legend(['Phase-1','Phase-2','Phase-3'],loc=1)
```

From Fig. 1.4, it is possible to infer that the starting point of the waveform is different for different phase angle. The phase parameter determines the time locations of the maxima and minima of the sinusoid.

Tasks
1. Write a python code to generate a sinusoidal signal whose phase is varying in a random manner. Assume the phase angle 'Φ' to follow uniform distribution in the range -1 to $+1$.
2. Write a python code to generate a sinusoidal signal, whose frequency is varying in a random manner. Assume the frequency 'f' to follow uniform distribution in the range -1 to $+1$.

Experiment 1.3 Generation of Three-Phase Sinusoidal Signal
The expressions for three-phase sinusoidal signals are given by

$$x_1(t) = A \sin(2\pi ft) \tag{1.2}$$

$$x_2(t) = A \sin(2\pi ft - 120°) \tag{1.3}$$

$$x_3(t) = A \sin(2\pi ft - 240°) \tag{1.4}$$

The amplitude and frequency of the three waveforms are equal. The phase shift between the signals is 120°. The python code, which generates the three-phase sinusoidal waveforms, is shown in Fig. 1.5, and the corresponding output is shown in Fig. 1.6.

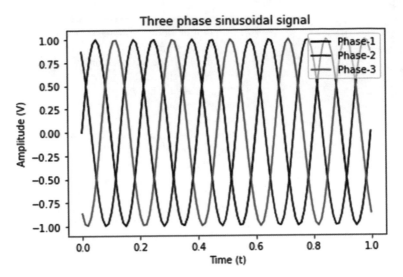

Fig. 1.6 Three-phase sinusoidal signals

Inference

From the python code to generate the three-phase sinusoidal signal, it is possible to observe that the amplitude of each signal is 1 V and frequency is 5 Hz. The phase shift between the signals is 120°.

Task

1. Change the value of amplitude A1, A2 and A3 of three-phase sinusoidal signal in the python code given in Fig. 1.5, and comment on the output waveform.

1.1.2 Exponential Function

Exponential function is of two types: (1) real exponential function and (2) complex exponential function. The real exponential function can be either an increasing function or it could be a decreasing function. The price of petrol is an example of exponentially increasing function. Radioactive decay is an example of exponentially decaying function.

Complex exponential function is of interest in signal processing. Complex exponential function is the basis function of Fourier transform. It is possible to obtain sine wave and cosine wave from complex exponential function. It is an Eigen function for a linear time-invariant system.

Experiment 1.4 Generation of Real Exponential Signal

The general expression for the real exponential signal is given by

Table 1.2 List of built-in functions used in the program

S. No.	Built-in function used	Purpose
1	np.exp()	To generate exponential function
2	np.linspace()	To generate equally interval data points in an interval
3	plt.subplot()	To plot more than one figure in the same plot

```
#Real exponential function
import numpy as np
import matplotlib.pyplot as plt
#Step 1: Defining the time axis
t=np.linspace(-1,1,1000)
#Step 2: Defining the parameter 'alpha'
a,b=2,-2
#Step 3: Generation of function
x1=np.exp(a*t)
x2=np.exp(b*t)
#Step 4: Plotting of the function
plt.subplot(2,1,1),plt.plot(t,x1)
plt.xlabel('Time'),plt.ylabel('Amplitude'),plt.title('Exponentially growing function')
plt.subplot(2,1,2),plt.plot(t,x2)
plt.xlabel('Time'),plt.ylabel('Amplitude'),plt.title('Exponentially decaying function')
plt.tight_layout()
```

Fig. 1.7 Python code to generate real exponential functions

$$x(t) = Ce^{\alpha t} \tag{1.5}$$

where 'C' and 'α' are real. If 'α' is greater than zero, it is exponentially growing function. If 'α' is less than zero, it is exponentially decreasing or decaying function. The built-in functions used in the program are given in Table 1.2.

The python code, which generates exponentially growing and decaying function for $\alpha = 2$ and $\alpha = -2$, is shown in Fig. 1.7, and the corresponding output is shown in Fig. 1.8.

Inference

In exponentially growing function, the value of the function (amplitude of the function) increases with an increase in time. In contrast, in exponentially decaying function, the amplitude of the function decays with respect to time.

Task

1. Write a python code to generate two real exponential functions (one growing and another decaying) with different amplitudes, and add these two functions. Comment on the observed result.

Experiment 1.5 Forward Characteristics of PN Junction Diode

The equation of current through the diode is given by

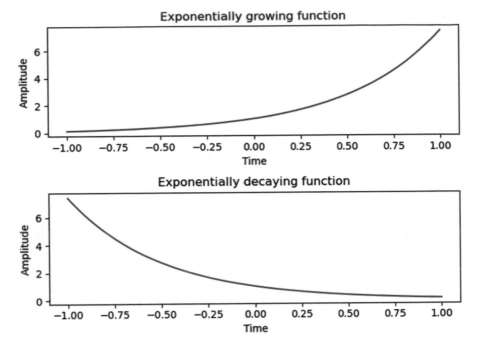

Fig. 1.8 Exponentially growing and decaying functions

$$I_D = I_s \left\{ e^{\frac{V_D}{\eta V_T}} - 1 \right\} \tag{1.6}$$

In Eq. (1.6), I_s represents the reverse saturation current; V_D is the voltage drop across the diode and I_D is the current through the diode; V_T is the volt-equivalent of temperature, which is 26 mV at room temperature; and η is the ideality factor, which is material dependent. The python code, which simulates the V–I characteristics of PN junction diode by assuming $\eta = 1$, $V_T = 26$ mV, and $I_s = 1$ mA is shown in Fig. 1.9, and the corresponding output is shown in Fig. 1.10.

Inference
From the forward characteristics shown in Fig. 1.10, it is possible to observe that the diode current increases after crossing the threshold voltage, generally termed '*knee voltage*'. If one considers the current through the diode as a function, then the function is an exponentially growing function.

Experiment 1.6 Radioactive Decay Function
The equation of radioactive decay is given by $N(t) = N_0 e^{-\lambda t}$. The python code, which implements this equation, is shown in Fig. 1.11, and the corresponding output is shown in Fig. 1.12.

```
#Forward characteristics of PN junction diode
import numpy as np
import matplotlib.pyplot as plt
#Step 1: Defining the voltage axis
v=np.arange(0,1,0.001)
#Step 2: Parameters
vt=0.026  #Volt-equivalent of temp.
i_s=1/1000 #Reverse saturation current
n=1 #ideality factor
#Step 3: Equation of current through diode
i=i_s*(np.exp(v/(n*vt))-1)
#Step 4: Plotting the characteristics
plt.plot(v,i),plt.xlabel('Forward voltage')
plt.ylabel('Forward current')
plt.title('Forward characteristics of PN junction diode')
```

Fig. 1.9 Python code to plot the forward characteristics of PN junction diode

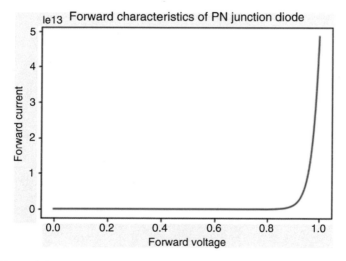

Fig. 1.10 Forward characteristics of PN junction diode

Inference

From Fig. 1.12, it is possible to observe that the radioactive decay activity can be modelled by an exponentially decaying function.

Experiment 1.7 Complex Exponential Function

Generate two complex exponential signals $x_1(t) = e^{j\Omega t}$ and $x_2(t) = e^{-j\Omega t}$. Here the frequency of the signal is fixed as $f = 5$ Hz. After signal generation, extract the magnitude and phase of the two signals, and comment on the observed output.

The built-in functions used in the python program are given in Table 1.3.

Fig. 1.11 Python code for radioactive decay function

```
#Radioactive decay
import numpy as np
import matplotlib.pyplot as plt
#Step 1: Defining the time axis
t=np.linspace(0,80,10)
#Step 2: Parameters
A0=400  #Initial value
T=24
#Step 3: Equation of current through diode
A=A0*np.exp(-t/T)
#Step 4: Plotting the characteristics
plt.plot(t,A),plt.xlabel('Time (Hours)')
plt.ylabel('Counts per second')
plt.title('Radio active decay')
```

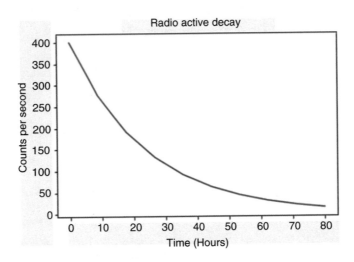

Fig. 1.12 Result of python code shown in Fig. 1.11

Table 1.3 Python built-in functions used in Experiment 1.7

S. No.	Built-in function used	Purpose
1	np.exp()	To generate an exponential function
2	np.abs()	To obtain the magnitude value of a complex number
3	np.angle()	To obtain the phase value of the complex number
4	np.linspace()	To generate equally interval data points in an interval
5	plt.subplot()	To plot more than one figure in the same plot

```
#Complex exponential signals
import numpy as np
import matplotlib.pyplot as plt
#Step 1: Generation of signals x1(t) and x2(t)
t=np.linspace(-1,1,100)
f=5
x1=np.exp(1j*2*np.pi*f*t)
x2=np.exp(-1j*2*np.pi*f*t)
#Step 2: Plotting the signals, its magnitude, and phase
plt.subplot(3,2,1),plt.plot(t,x1)
plt.xlabel('Time (t)'),plt.ylabel('Amplitude (V)')
plt.title('$e^{j\Omega t}$'),plt.subplot(3,2,2),plt.plot(t,x2)
plt.xlabel('Time (t)'),plt.ylabel('Amplitude (V)')
plt.title('$e^{-j\Omega t}$'),plt.subplot(3,2,3),plt.plot(t,np.abs(x1))
plt.xlabel('Time (t)'),plt.ylabel('Magnitude (V)')
plt.title('|$e^{j\Omega t}$|'),plt.subplot(3,2,4),plt.plot(t,np.abs(x2))
plt.xlabel('Time (t)'),plt.ylabel('Magnitude (V)')
plt.title('|$e^{-j\Omega t}$|'), plt.subplot(3,2,5),plt.plot(t,np.angle(x1)*360/(2*np.pi))
plt.xlabel('Time (t)'),plt.ylabel('$Phase(^\circ$)')
plt.title('$\Phi(x_1(t))$'),plt.subplot(3,2,6),plt.plot(t,np.angle(x2)*360/(2*np.pi))
plt.xlabel('Time (t)'),plt.ylabel('$Phase(^\circ$)'),plt.title('$\Phi(x_2(t))$')
plt.tight_layout()
```

Fig. 1.13   Generation of complex exponential signals

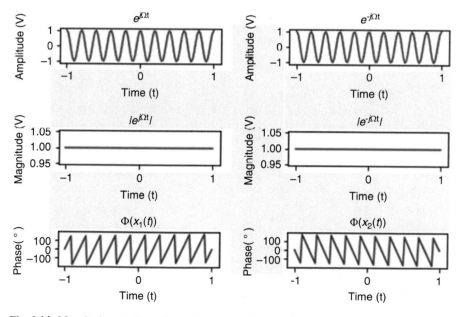

Fig. 1.14   Magnitude and phase of complex exponential signals

```
#Generation of sinusoidal signal
import numpy as np
import matplotlib.pyplot as plt
#Step 1: Generation of rotating phasor
t=np.linspace(0,1,100)
f=5
x1=np.exp(1j*2*np.pi*f*t)
x2=np.exp(-1j*2*np.pi*f*t)
#Step 2: Generation of sine and cosine wave
x_cos=(x1+x2)/2
x_sin=(x1-x2)/(2*1j)
#Step 3: Plotting the result
plt.subplot(2,1,1),plt.plot(t,x_cos),plt.xlabel('Time (t)'),plt.ylabel('Amplitude (V)')
plt.title('Cosine wave'),plt.subplot(2,1,2),plt.plot(t,x_sin,'r')
plt.xlabel('Time (t)'),plt.ylabel('Amplitude (V)'),plt.title('Sine wave')
plt.tight_layout()
```

Fig. 1.15   Generation of sinusoidal signals from complex exponential functions

The python code, which performs the task mentioned above, is shown in Fig. 1.13, and the corresponding output is shown in Fig. 1.14.

Inferences

From Figs. 1.13 and 1.14, the following inferences can be made:

1. Two complex exponential signals $x_1(t) = e^{j2\pi ft}$ $x_2(t) = e^{-j2\pi ft}$ with the frequency value $f = 5$ Hz are generated. The signals $x_1(t)$ and $x_2(t)$ look alike.
2. The magnitude and phase responses of the two signals are plotted. The magnitude of the signal $x_1(t) = e^{j\Omega t}$ is given by $|x_1(t)| = \sqrt{\cos^2(\Omega t) + \sin^2(\Omega t)} = 1$. Similarly, the magnitude of the signal $x_2(t) = e^{-j\Omega t}$ is given by $|x_2(t)| = \sqrt{\cos^2(\Omega t) + \sin^2(\Omega t)} = 1$. Thus, the magnitudes of the two signals are alike.
3. The phase of the signal $x_1(t) = e^{j\Omega t}$ is expressed as $\phi(x_1(t)) = \tan^{-1}\left(\frac{\sin(\Omega t)}{\cos(\Omega t)}\right)$. Upon simplifying the expression, we get $\phi(x_1(t)) = \tan^{-1}(\tan(\Omega t))$, which results in $\phi(x_1(t)) = \Omega t$. The phase of the signal $x_2(t) = e^{-j\Omega t}$ is expressed as $\phi(x_1(t)) = \tan^{-1}\left(-\frac{\sin(\Omega t)}{\cos(\Omega t)}\right)$. Upon simplifying the expression, we get $\phi(x_1(t)) = \tan^{-1}(-\tan(\Omega t))$, which results in $\phi(x_1(t)) = -\Omega t$. The phases of the two signals are different. This implies that the signal $e^{j2\pi ft}$ and $e^{-j2\pi ft}$ represents two phasors, rotating in the opposite direction.

Experiment 1.8 Generation of 'Sine' and 'Cosine' Functions from 'Complex Exponential Function'

This experiment aims to prove that sinusoidal signal can be generated through two phasors rotating in the opposite direction. Mathematically it is expressed as

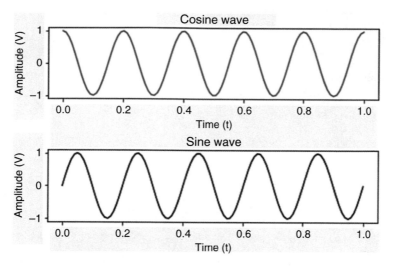

Fig. 1.16 Sine and cosine wave from complex exponential functions

$\cos(\Omega t) = \frac{e^{j\Omega t} + e^{-j\Omega t}}{2}$ and $\sin(\Omega t) = \frac{e^{j\Omega t} - e^{-j\Omega t}}{2j}$. The python code, which generates the cosine and sine wave using a rotating phasor, is shown in Fig. 1.15, and the corresponding output is shown in Fig. 1.16.

Inference
From Fig. 1.16, it is possible to observe that there is a phase difference of 90° between the sine and cosine waveforms.

Task
1. Add the square of the sine and cosine wave obtained in Experiment 1.8, and plot the resultant waveform. Comment on the observed output. [Hint: $\sin^2(\theta) + \cos^2(\theta) = 1$]

Experiment 1.9 Modulating Sinusoidal Signal with an Exponential Signal
This experiment discusses the sinusoidal signal multiplied with a growing and decaying real exponential signal. The python code, which accomplishes this task, is shown in Fig. 1.17, and the corresponding output is shown in Fig. 1.18.

Inference
The following inferences can be made from Fig. 1.18

1. The sinusoidal signal amplitude varies between −1 and +1.
2. Upon multiplying the sinusoidal signal with growing exponential, the amplitude value increases; hence, the plot is shown in the range −5 to +5.
3. Upon multiplying the sinusoidal signal with decaying exponential, the amplitude of the input sinusoidal signal decreases, which is shown between −0.5 and +0.5.

```
#Multiplying sinusoidal signal with an exponential signal
import numpy as np
import matplotlib.pyplot as plt
#Step 1: Defining the time axis
t=np.linspace(0,1,1000)
#Step 2: Defining the parameter
a=2 #Parameter for exponentially growing function
b=-2 #Parameter for exponentially decaying function
#Step 3: Generation of function
x1=np.sin(2*np.pi*5*t)
x2=np.exp(a*t)*np.sin(2*np.pi*5*t)
x3=np.exp(b*t)*np.sin(2*np.pi*5*t)
#Step 4: Plotting of the function
plt.subplot(3,1,1),plt.plot(t,x1)
plt.xlabel('Time'),plt.ylabel('Amplitude'),plt.title('Sinusoidal signal')
plt.subplot(3,1,2),plt.plot(t,x2)
plt.xlabel('Time'),plt.ylabel('Amplitude'),plt.title('Exponentially increasing sinusoidal signal')
plt.subplot(3,1,3),plt.plot(t,x3)
plt.xlabel('Time'),plt.ylabel('Amplitude'),plt.title('Exponentially decaying sinusoidal signal')
plt.tight_layout()
```

Fig. 1.17 Python code to modulate sinusoidal signal by exponential signal

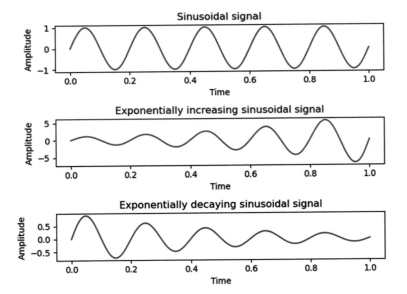

Fig. 1.18 Result of python code shown in Fig. 1.17

1.2   Non-stationary Signal

A stationary signal is one whose statistical characteristics do not change with respect to time. If the signal characteristics change with respect to time, then it is a non-stationary signal. Example of non-stationary signal is a chirp signal whose frequency varies with respect to time. Most of the real-world signals, like the alarm sound from the clock or the sound of an ambulance, are non-stationary. In this section, few stationary and non-stationary signals are generated.

Experiment 1.10 Generation of Stationary and Non-stationary Signal
This experiment deals with the generation of stationary and non-stationary signal. The expression for stationary signal is given by

$$x_1(t) = \sin(2\pi f t) \tag{1.7}$$

The above expression generates a sinusoidal signal of specific frequency. The expression for non-stationary signal is given by

$$x_2(t) = \sin\left(2\pi f t^2\right) \tag{1.8}$$

The frequency of the signal changes with respect to time; hence, it is considered as non-stationary. The python code, which generates the two signals and the corresponding output, is shown in Fig. 1.19, and Fig. 1.20.

```
#Stationary and non-stationary signal
import numpy as np
import matplotlib.pyplot as plt
#Step 1: Generation of signals
t=np.linspace(0,1,1000)
f=5
x1=np.sin(2*np.pi*f*t)
x2=np.sin(2*np.pi*f*t**2)
#Step 2: Plotting of signals
plt.subplot(2,1,1),plt.plot(t,x1),plt.xlabel('Time (t)'),
plt.ylabel('Amplitude'),plt.title('Stationary signal')
plt.subplot(2,1,2),plt.plot(t,x2,'r'),plt.xlabel('Time (t)'),
plt.ylabel('Amplitude'),plt.title('Non-stationary signal')
plt.tight_layout()
```

Fig. 1.19 Python code to generate stationary and non-stationary signal

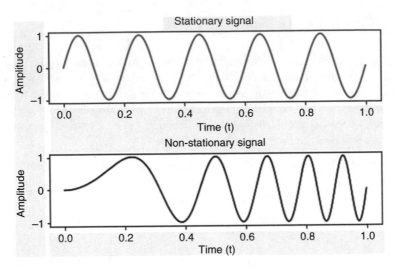

Fig. 1.20 Result of python code shown in Fig. 1.19

Inference
By observing Fig. 1.20, it is possible to conclude that the frequency of the stationary signal does not change with respect to time. On the other hand, the frequency of the non-stationary signal increases with time increases.

Experiment 1.11 Generation of Non-stationary Sinusoidal Signal
The objective of this experiment is to append sinusoidal signals of different frequencies. Signal-1 is generated with 5 Hz frequency appearing first, DC signal next and 10 Hz frequency occurs last. Signal-2 is obtained by interchanging the first and last part of signal-1, which means high frequency occurs first and low frequency occurs next. The python code, which performs this task, is shown in Fig. 1.21, and the corresponding output is shown in Fig. 1.22.

Inference
The signals in Fig. 1.22 are considered as non-stationary, because the signal frequency varies with respect to time. Signal-1 and Signal-2 contain the same frequency components at different instants.

Experiment 1.12 Generation of Chirp Signal
The objective of this experiment is to generate chirp signal. The chirp signal can be considered as a frequency swept sinusoidal signal. Four different methods of frequency sweep are (1) linear, (2) quadratic, (3) logarithmic and (4) hyperbolic. In this experiment, the frequency sweep is from 10 Hz to 1 Hz, as considered. The python code, which generates the chirp signals, are shown in Fig. 1.23, and the corresponding output is shown in Fig. 1.24.

```
import numpy as np
import matplotlib.pyplot as plt
t1=np.linspace(0,1,100)
#Defining signal frequencies
f1,f2,f3=0,5,10
#Generation of signal-1
x1=np.sin(2*np.pi*f2*t1)
#Generation of signal-2
x2=np.sin(2*np.pi*f1*t1)
x3=np.sin(2*np.pi*f3*t1)
x=np.concatenate([x1,x2,x3])
y=np.concatenate([x3,x2,x1])
#Plotting the result
t=np.linspace(0,1,300)
plt.subplot(2,1,1),plt.plot(t,x),plt.xlabel('Time(t)'),plt.ylabel('Amplitude (V)')
plt.title('Signal-1'),plt.subplot(2,1,2),plt.plot(t,y)
plt.xlabel('Time (t)'),plt.ylabel('Amplitude (V)'),plt.title('Signal-2')
plt.tight_layout()
```

Fig. 1.21   Generation of non-stationary signal

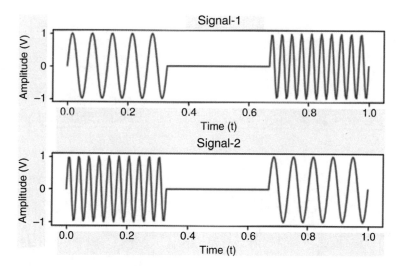

Fig. 1.22   Result of non-stationary signals

Inference

From Fig. 1.24, it is possible to observe that the frequency varies with respect to time in all four types of chirp signals; hence, they are considered non-stationary signals.

```
#Generation of chirp signals
import numpy as np
import matplotlib.pyplot as plt
from scipy.signal import chirp
#Step 1: Generation of chirp signal
t=np.linspace(0,10,10000)
x1= chirp(t, f0=10, f1=1, t1=10, method='linear')
x2= chirp(t, f0=10, f1=1, t1=10, method='quadratic')
x3= chirp(t, f0=10, f1=1, t1=10, method='logarithmic')
x4= chirp(t, f0=10, f1=1, t1=10, method='hyperbolic')
#Step 2: Plotting the signals
plt.subplot(2,2,1),plt.plot(t,x1),plt.xlabel('Time (t)'),plt.ylabel('Amplitude (V)'),
plt.title('Linear chirp'),plt.subplot(2,2,2),plt.plot(t,x2),plt.xlabel('Time (t)'),
plt.ylabel('Amplitude (V)'),plt.title('Quadratic chirp'),plt.subplot(2,2,3),
plt.plot(t,x3),plt.xlabel('Time (t)'),plt.ylabel('Amplitude (V)'),
plt.title('Logarithmic chirp'),plt.subplot(2,2,4),plt.plot(t,x4),plt.xlabel('Time (t)'),
plt.ylabel('Amplitude (V)'),plt.title('Hyperbolic chirp')
plt.tight_layout()
```

Fig. 1.23 Generation of chirp signals

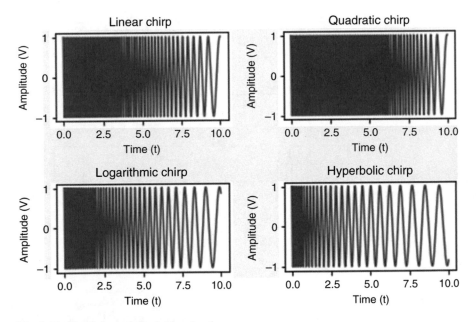

Fig. 1.24 Simulation result of chirp signals

1.3   Non-sinusoidal Waveform

The non-sinusoidal waveform generation considered in this section includes square waveform, triangular waveform, sawtooth waveform, sinc and Gaussian signals.

1.3.1   Square Waveform

A square waveform is a non-sinusoidal periodic waveform. A square wave represents a sudden variation from 'ON' to 'OFF' state and vice versa. Duty cycle is the percentage of time a square wave remains high versus low over one period. Square waves are useful in modelling digital signals. Sine wave contains single frequency, whereas square wave contains a very wide bandwidth of frequencies.

Experiment 1.13 Generation of Square Waveform
The objective is to generate square wave with different duty cycle. The duty option refers to which fraction of the whole duty cycle the signal will be in its 'high' state. The python code, which generates square wave of frequency 5 Hz, is shown in Fig. 1.25, and the corresponding output is shown in Fig. 1.26.

Inference
From Fig. 1.26, it is possible to interpret that the generated waveform is a square waveform of a fundamental frequency of 5 Hz. With increase in the duty, the '*ON time*' of the generated square wave increases. The square waveform takes only binary value, which is either +1 or −1. The state change from +1 to −1 and −1 to +1 occurs immediately.

```
#Square wave with a different duty cycle
import numpy as np
from scipy import signal
import matplotlib.pyplot as plt
t=np.linspace(0,1,100)
f=5
duty=[0.15,0.25,0.5,0.75]
for i in range(len(duty)):
    x=signal.square(2*np.pi*f*t,duty[i])
    plt.subplot(2,2,i+1)
    plt.plot(t,x),plt.xlabel('Time (t)'),plt.ylabel('Amplitude (V)')
    plt.ylim(-2,2),plt.title('Square wave (duty={})'.format(duty[i]))
    plt.tight_layout()
```

Fig. 1.25 Generation of a square wave

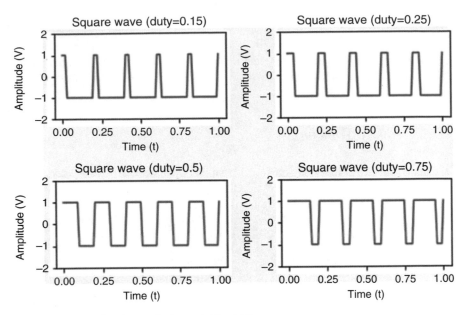

Fig. 1.26 Result of python code shown in Fig. 1.25

1.3.2 Triangle and Sawtooth Waveform

Triangle and sawtooth waveforms are useful for exploring non-linearity in the circuit. A triangle waveform has uniform rise and fall time, whereas in a sawtooth waveform, the rise and fall times are markedly different.

Experiment 1.14 Generation of Sawtooth and Triangular Waveforms
The python code, which generates the sawtooth waveform of frequency 5 Hz, is shown in Fig. 1.27, and the corresponding output is shown in Fig. 1.28.

Inference
When the width is 0.5, the sawtooth waveform turns out to be a triangular waveform. In square waveform, the state change from −1 to +1 and from +1 to −1 occurs instantaneously, whereas, in triangular waveform, the change of state from −1 to +1 and from +1 to −1 occurs gradually.

1.3.3 Sinc Function

A sinc function is represented as

```
#Sawtooth wave with different width
import numpy as np
from scipy import signal
import matplotlib.pyplot as plt
t=np.linspace(0,1,1000)
f=5
width=[0.1,0.2,0.5,0.8]
for i in range(len(width)):
    x=signal.sawtooth(2*np.pi*f*t,width[i])
    plt.subplot(2,2,i+1)
    plt.plot(t,x),plt.xlabel('Time (t)'),plt.ylabel('Amplitude (V)')
    plt.ylim(-2,2),plt.title('Sawtooth wave (width={})'.format(width[i]))
    plt.tight_layout()
```

Fig. 1.27 Python code to generate sawtooth waveform

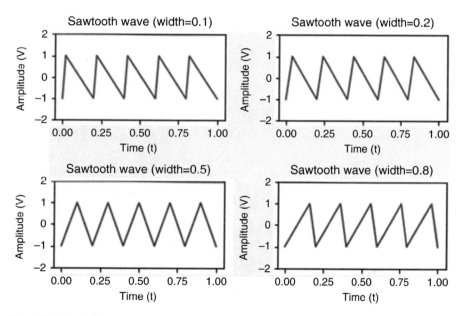

Fig. 1.28 Simulation result of sawtooth waveform

$$x(t) = \frac{\sin \pi t}{\pi t}, \quad -\infty < t < \infty \tag{1.9}$$

A sinc function is an even function with a unit area. It is a symmetric function with respect to the origin. Fourier transform of sinc function will result in rectangular function and vice versa. Thus, sinc function is the impulse response of the ideal lowpass filter.

Fig. 1.29 Python code to
generate *sinc* function

```
#Generation of sinc function
import numpy as np
import matplotlib.pyplot as plt
#Step 1: Defining the independent variable
t=np.linspace(-10,10,1000)
#Step 2: Generating sinc function
x=np.sinc(t)
#Step 3: Plotting the sinc function
plt.plot(t,x),plt.xlabel('Time'),plt.ylabel('Amplitude')
plt.title('Sinc function')
plt.tight_layout()
```

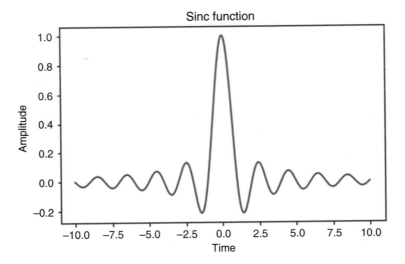

Fig. 1.30 Result of python code shown in Fig. 1.29

Experiment 1.15 Generation of Sinc Function

The expression for sinc function is given by $\sin c(t) = \frac{\sin(\pi t)}{\pi t}$. In this experiment, the sinc function is generated using the built-in function (*sinc*()) available in numpy library. The python code, which generates the sinc function, is shown in Fig. 1.29, and the corresponding output is shown in Fig. 1.30.

Inference

From Fig. 1.30, the following observations can be made:

1. Sinc function has a main lobe and side lobes.
2. The sinc function is symmetric with respect to the origin. It is an even function.
3. The sinc function attains the maximum value at the origin.

Task
1. Write a python code to prove that *sinc* function is an even function.

1.3.4   Pulse Signal

A rectangular pulse can be considered as a positive going edge, followed by negative going one. Convolution of two rectangular pulses results in a triangular pulse.

Experiment 1.16 Generation of Rectangular and Triangular Pulse Signal

The expression for rectangular pulse is given by $x(t) = \begin{cases} 1, & |t| < 1 \\ 0, \text{otherwise} \end{cases}$. It can be considered as a positive going edge followed by negative going one. Rectangular pulse represents a drastic variation from level 0 to 1 and from 1 to 0. The expression for the triangular pulse is $x(t) = \begin{cases} 1 - \dfrac{|t|}{T}, & \text{for } |t| < T \\ 0, & \text{otherwise} \end{cases}$. The triangular pulse represents a gradual variation from level 0 to 1 and from 1 to 0.

The python code, which generates the rectangular and triangular pulse signal, is shown in Fig. 1.31, and the corresponding output is shown in Fig. 1.32.

Inference
Figure 1.32 shows that rectangular pulse exhibits a drastic change in amplitude from 0 to 1 V, whereas triangular pulse exhibits a gradual variation in amplitude from 0 to 1 V. In later section, it will be proved that the convolution of two rectangular pulses will result in a triangular pulse.

```
#Generation of rectangular and triangular pulse signal
import numpy as np
import matplotlib.pyplot as plt
#Step 1: Generation of signals
t=np.linspace(-2,2,100)
rect_pulse=abs(t)<1 #Rectangular pulse
tri_pulse=(1 - abs(t)) * (abs(t) < 1) #Triangular pulse
#Step 2: Plotting of the pulse signals
plt.subplot(2,1,1),plt.plot(t,rect_pulse)
plt.xlabel('Time (t)'),plt.ylabel('Amplitude (V)'),plt.title('Rectangular pulse')
plt.subplot(2,1,2),plt.plot(t,tri_pulse)
plt.xlabel('Time (t)'),plt.ylabel('Amplitude (V)'),plt.title('Triangular pulse')
plt.tight_layout()
```

Fig. 1.31 Generation of rectangular and triangular pulse

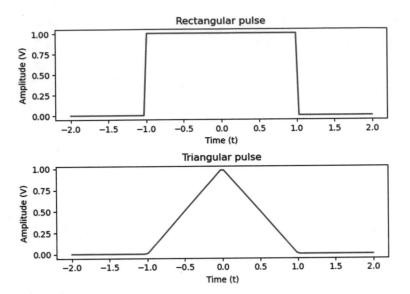

Fig. 1.32 Rectangular and triangular pulse signal

Task
1. Write a python code to illustrate the fact that convolution of two rectangular pulse signals results in a triangular pulse.

1.3.5 Gaussian Function

The Gaussian function is expressed as

$$x(t) = \frac{1}{\sqrt{2\pi}\sigma} e^{-\frac{(t-\mu)^2}{2\sigma^2}} \qquad (1.10)$$

where 'μ' represents the mean and 'σ' represents the standard deviation. Fourier transform of a Gaussian function results in another Gaussian function. The product of two Gaussian functions is a Gaussian function. Gaussian window is an optimal window for time-frequency localization. Smoothening by Gaussian function is widely employed in image processing.

Experiment 1.17 Generation of Gaussian Function
The Gaussian function is widely used in signal processing, image processing and communication fields. The expression for Gaussian function with the mean value 'μ' and standard deviation 'σ' is given by $x(t) = \frac{1}{\sqrt{2\pi}\sigma} e^{-\frac{(t-\mu)^2}{2\sigma^2}}$. This experiment aims to

```
#Generation of Gaussian function
import numpy as np
import matplotlib.pyplot as plt
t=np.linspace(-10,10,1000)
mu=0  #Mean value
sigma=[0.01,0.5,1,10] #Standard deviation
for i in range(len(sigma)):
    k=1/np.sqrt(2*np.pi*sigma[i])
    x=k*np.exp(-np.power(t-mu,2.)/2*np.power(sigma[i],2.))
    plt.subplot(2,2,i+1),plt.plot(t,x),plt.xlabel('Time'),
    plt.ylabel('Amplitude'),plt.title('$\sigma={} $'.format(sigma[i]))
    plt.tight_layout()
```

Fig. 1.33 Gaussian function for different values of standard deviation

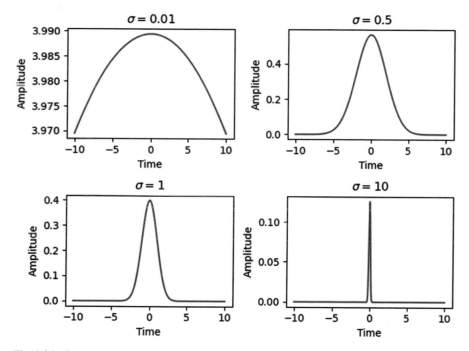

Fig. 1.34 Gaussian function for different values of standard deviation

generate Gaussian function for different values of standard deviation, namely, 0.01, 0.5, 1 and 10. The mean value is taken as zero. The python code, which generates the Gaussian function, is shown in Fig. 1.33, and the corresponding output is shown in Fig. 1.34.

Inference

From Fig. 1.34, it is possible to observe the following facts:

Fig. 1.35 Python code to
generate sinusoidal note

```
#Hearing sinusoidal signal
import numpy as np
import sounddevice as sd
f=1000 #Signal frequency
fs=8000 #Sampling rate
t=np.linspace(0,1,fs)
x=np.sin(2*np.pi*f*t)
sd.play(x,fs)
```

1. The Gaussian function is characterized by two parameters, which are mean and standard deviation.
2. The mean value of the Gaussian function is zero; hence, the maximum value occurs at the origin.
3. With an increase in the value of standard deviation, the narrower the Gaussian function.

Task
1. Write a python code to prove that the multiplication of two Gaussian functions results in a Gaussian function.

Experiment 1.18 Hearing a Sinusoidal Signal
Human ears can hear sound in the frequency range from 20 Hz to 20 kHz. In this experiment, sine wave of particular frequency is heard as a tone. The sampling frequency is chosen as 8000 Hz, and the signal frequency is chosen as 1000 Hz. The library functions used are (1) Numpy and (2) Sounddevice. The built-in function in sound device library (*sd.play*) is used to play the sound. The python code, which generates the sinusoidal tone, is shown in Fig. 1.35. The user can hear the audio using headphone.

Inference
From Fig. 1.35, the following inferences can be made:

1. The signal frequency is 1000 Hz, and the sampling frequency is 8000 Hz.
2. The library used to hear the audio is '*sounddevice*' library.
3. The built-in function (*sd.play*) is used to hear the audio.

Task
1. Human ear can hear an audio signal whose frequency is between 20 Hz and 20 kHz. Generate 10 Hz sinusoidal waveform; try to hear the waveform. It should not be audible. Now increase the frequency of sine wave to 100 Hz; now it should be possible to hear the sinusoid as a single note.

Experiment 1.19 Hearing Amplitude Modulated Sinusoidal Signal
The impact of modulating the amplitude of the sinusoidal signal is observed in this experiment. In this experiment, the amplitude of the sinusoidal signal is modulated by both exponentially decaying and growing functions. The python code, which

Fig. 1.36 Amplitude
modulated sinusoidal signal

```
#Hearing amplitude modulated sinusoidal signal
import numpy as np
import sounddevice as sd
f=1000 #Signal frequency
fs=8000 #Sampling rate
a=-5 #Decaying factor
b=5  #Growing factor
t=np.linspace(0,1,fs)
x1=np.exp(a*t)*np.sin(2*np.pi*f*t)
x2=np.exp(b*t)*np.sin(2*np.pi*f*t)
sd.play(x1,fs)
sd.wait()
sd.play(x2,fs)
```

Table 1.4 Built-in functions used in Experiment 1.19

S. No.	Built-in function used	Purpose
1	np.exp()	To generate an exponential function
2	np.sin()	To generate a sinusoidal function
3	sd.play()	To play the audio signal
4	sd.wait()	To pause the audio signal

performs this task is shown in Fig. 1.36. The built-in functions used in the program are summarized in Table 1.4.

Inference
The following inferences can be made from Fig. 1.36:

1. The signal 'x1' refers to a sine wave modulated by an exponentially decaying function.
2. The signal 'x2' refers to a sine wave modulated by an exponentially growing function.

Experiment 1.20 Generation of Amplitude Modulated Signal
In amplitude modulation, the amplitude of the carrier signal is varied in accordance with the message signal. The expression for amplitude modulated signal is given by

$$x(t) = (1 + m\sin(2\pi f_m t))\sin(2\pi f_c t) \tag{1.11}$$

In the above expression, 'm' denotes the modulation index, f_m represents the frequency of the modulating signal and f_c denotes frequency of the carrier signal. The python code, which generates the amplitude modulated signal for different modulating indices, is shown in Fig. 1.37, and the corresponding output is shown in Fig. 1.38.

Inference
From Figs. 1.37 and 1.38, the following inferences can be drawn:

```
#Amplitude modulation
import numpy as np
import matplotlib.pyplot as plt
t=np.linspace(0,1,1000)
fm=10 #Frequency of modulating signal
fc=100 #Frequency of carrier signal
message=np.sin(2*np.pi*fm*t)
carrier=np.sin(2*np.pi*fc*t)
m=[0.25,0.5,1,1.5]  #modulation index
for i in range(len(m)):
    mod_sig=(1+m[i]*message)*carrier
    plt.subplot(2,2,i+1),plt.plot(t,mod_sig)
    plt.xlabel('Time'),plt.ylabel('Amplitude')
    plt.title('Modulated signal with m={}'.format(m[i]))
    plt.tight_layout()
```

Fig. 1.37 Python code to generate amplitude modulated signal

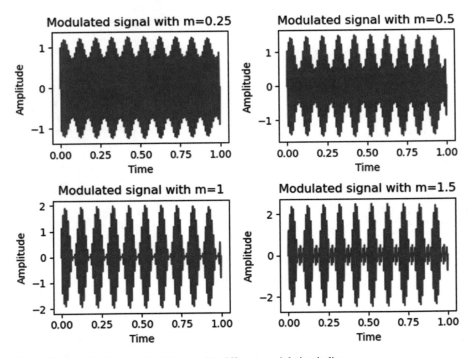

Fig. 1.38 Amplitude modulated signal with different modulation indices

1. The frequency of the message signal is 10 Hz; the frequency of the carrier signal is 100 Hz. The modulation index is varied as 0.25, 0.5, 1.0 and 1.5.
2. It is possible to observe that the amplitude of the carrier is changed in accordance with the message signal.

3. Modulation index less than one corresponds to under modulated signal. Modulation index greater than one corresponds to over modulated signal. Modulation index equal to one corresponds to perfect modulation.

Exercises

1. Generate the following sinusoidal signal $x(t) = A \sin(2\pi ft + \phi)$ with the amplitude $A = 2$ V, frequency $f = 10$ Hz and phase $\phi = 0$. Let the length of the signal be 100 samples. Store this signal in your system in a particular folder along with the time stamp in an Excel sheet. From the Excel sheet, read the data and the time stamp and plot the signal.
2. Write a python code to generate the sinusoidal signal of 1 V amplitude, 5 Hz frequency and phase $\phi = 0$. Mark the positive peak of the waveform. That is the positive peak of the waveform should be marked with 'x' mark.
3. Write a python code to compute the number of zero crossings of sine wave of 2 V amplitude, 5 Hz frequency and phase $\phi = 0$.
4. Write a python code to generate the seven nodes 'sa', 're', 'ga' and 'ma'. Use the sounddevice library to play the seven notes.
5. Generate the Gaussian function, which is given by $x(t) = \frac{1}{\sqrt{2\pi}\sigma} e^{-\frac{(t-\mu)^2}{2\sigma^2}}$ for different mean values $\mu = 0, 1, 2, 4$ with the fixed standard deviation value $\sigma = 1$. Use subplot to plot the generated Gaussian functions.

Objective Questions

1. What will the signal's length be if the following code is executed?

```
t=np.linspace(0,1,100)
x=np.sin(2*np.pi*5*t)
```

   A. 10
   B. 50
   C. 75
   D. 100

2. What will be the magnitude of the variables 'x' and 'y' if the following code segment is executed?

```
t=np.linspace(0,1,100)
f=0
x=np.sin(2*np.pi*f*t)
y=np.cos(2*np.pi*f*t)
```

   A. 1 and 0, respectively
   B. −1 and 1, respectively
   C. 0 and 1, respectively
   D. 1 and −1, respectively

3. What will be the output plot if the following segment of code is executed?

```
t=np.linspace(0,1,100)
x=np.sin(2*np.pi*5*t)
y=np.cos(2*np.pi*5*t)
plt.plot(t,(x**2+y**2))
```

A. DC signal of magnitude 1
B. DC signal of magnitude 5
C. Sine wave of frequency 5 Hz
D. Cosine wave of frequency 5 Hz

4. What will be stored in the variable 'z' if executing the following code segment?

```
t=np.linspace(0,1,100)
x=np.exp(1j*2*np.pi*5*t)
y=np.exp(-1j*2*np.pi*5*t)
z=(x+y)/2
```

A. Sine wave of 5 Hz frequency
B. Cosine wave of 5 Hz frequency
C. Square wave of 5 Hz frequency
D. Sawtooth wave of 5 Hz frequency

5. The phase difference between each signal in a three-phase sinusoidal signal is

A. 45°
B. 90°
C. 120°
D. 240°

6. What will be stored in the variable 'z' if executing the following code segment?

```
t=np.linspace(0,1,100)
x=np.exp(1j*2*np.pi*5*t)
y=np.abs(x)
```

A. Phase angle of the signal 'x'
B. Magnitude of the signal 'x'
C. Frequency of the signal 'x'
D. Number of zero crossings of the signal 'x'

7. The audible frequency range for human beings is

A. 10 Hz to 100 kHz
B. 20 Hz to 20 kHz
C. 1 to 1000 Hz
D. 200 Hz to 2 MHz

8. What will the signal's length be if the following code segment is executed?

```
fs=8000
t=np.linspace(0,1,fs)
```

A. 1000
B. 2000
C. 4000
D. 8000

9. Identify the statement that is WRONG with respect to sinc function

A. Sinc function is an even function.
B. Sinc function is an odd function.
C. Fourier transform of sinc function will result in a rectangular function.
D. Sinc function can be used for signal interpolation.

10. The magnitude of the function $x(t) = e^{-j\Omega t}$ is

A. 1
B. 0
C. -1
D. Infinity

Bibliography

1. Alan V. Oppenheim, and Alan S. Willsky. "Signals and Systems", Prentice Hall, 1996.
2. Simon Haykin, and Bary Van Veen, "Signals and Systems", Wiley, 2005.
3. Hwei P. Hsu, "Signals and Systems", Schaum's outline series, McGraw Hill Education, 2017.
4. Charles L. Phillips, John M. Parr, and Eve A. Riskin, "Signals, Systems, and Transforms", Pearson, 2013.
5. Mark Lutz, "Learning Python", O'Reilly Media, 2013.

Chapter 2
Sampling and Quantization of Signals

Learning Objectives

After reading this chapter, the reader is expected to

- Simulate and visualize standard discrete-time signals.
- Simulate and visualize arbitrary discrete-time signals.
- Perform different mathematical operations on discrete-time signals.
- Implement convolution and correlation operations and interpret the obtained results.

Roadmap of the Chapter

The contents discussed in this chapter are given as a flow diagram. The objective is to convert the continuous-time signal into a discrete-time signal. Two important processes in converting the continuous-time signal into a discrete-time signal are (1) sampling and (2) quantization. Also, reconstructing the original signal from the sampled signal is another important task in signal processing. This chapter explores these three processes in detail.

© The Author(s), under exclusive license to Springer Nature Singapore Pte Ltd. 2024   35
S. Esakkirajan et al., *Digital Signal Processing*,
https://doi.org/10.1007/978-981-99-6752-0_2

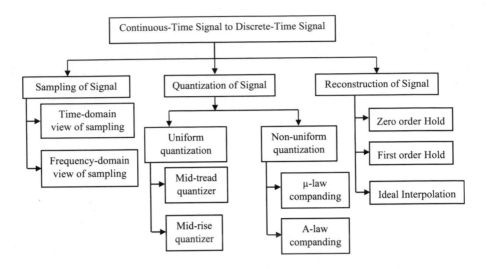

PreLab Questions

1. Mention the steps involved in converting the analogue signal into a digital signal.
2. A real-valued signal is known to be bandlimited. The maximum frequency content in the signal is f_{max}. What is the guideline given by the sampling theorem with respect to the choice of sampling frequency such that from the samples, the signal can be reconstructed without aliasing?
3. What is the impact of sampling a bandlimited signal with too low a sampling frequency?
4. Is it possible to reconstruct a periodic square wave of fundamental frequency 5 Hz from its samples? Explain your answer.
5. Mention the reason for aliasing to occur while sampling the signals?
6. What is the meaning of sampling the signal $x(t)$? What is the meaning of the terms (a) sampling rate and (b) sampling interval?
7. A signal has a bandwidth of 5 kHz. What is the Nyquist rate of the signal?
8. Why quantization is considered as a non-linear phenomenon?
9. Why quantization is considered as irreversible phenomenon?
10. What is signal reconstruction? Mention different types of signal reconstruction strategies.

2.1 Sampling of Signal

Sampling is basically taking a specific instant of the signal. In time domain, it is visualized as passing the signal through a switch. Sampling can be considered as multiplying the continuous-time signal $x(t)$ with train of impulse $c(t)$. The train of impulse will take a value of either one or zero; hence, the multiplication of the signal

```
#Generation of comb function
import numpy as np
import matplotlib.pyplot as plt
Fs=100
t = np.arange(0, 2, 1/Fs)
c=np.zeros(len(t))
T = 0.1
c[::int(Fs*T)]=1
plt.stem(t,c),plt.xlabel('Time'),plt.ylabel('Ampliude'),plt.title('c(t)')
```

Fig. 2.1 Python code to generate comb function

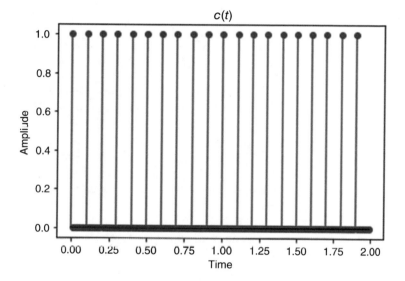

Fig. 2.2 Result of python code shown in Fig. 2.1

$x(t)$ with a train of impulse can be regarded as passing the signal $x(t)$ through a switch. The expression for a train of impulse is given by

$$c(t) = \sum_{n=-\infty}^{\infty} \delta(t - nT) \tag{2.1}$$

The function $c(t)$ takes a value of one whenever $t = nT$; else, it takes a value of zero.

Experiment 2.1 Generation of a Train of Impulse Function

The python code, which generates the train of impulse function or comb function, is given in Fig. 2.1, and the corresponding output is shown in Fig. 2.2.

Inference
1. From Fig. 2.1, it is possible to observe that the variable 'T' (Sampling interval) decides the distance between consecutive samples.
2. From Fig. 2.2, it is possible to confirm that the comb function $c(t)$ takes a value of either '1' or '0'. Whenever $c(t) = 1$, the signal $x(t)$ samples will be collected.

Experiment 2.2 Frequency Domain View of Comb Function
The time-domain expression for the comb function is given by

$$c(t) = \sum_{n=-\infty}^{\infty} \delta(t - nT) \tag{2.2}$$

Upon taking the Fourier transform of the comb function, we get

$$C(\Omega) = \frac{2\pi}{T} \sum_{k=-\infty}^{\infty} \delta(\Omega - k\Omega_s) \tag{2.3}$$

This experiment aims to prove that Fourier transform of a train of impulse will result in a train of impulse function. Here, two comb functions (train of impulse function), namely, $c_1(t)$ and $c_2(t)$ are generated. In the comb function $c_1(t)$, the spacing between consecutive impulses is 0.1 s, whereas in the comb function $c_2(t)$, the spacing between successive impulses is 0.05 s. Upon taking Fourier transform of these two comb functions, the corresponding magnitude spectra $|C_1(f)|$ and $|C_2(f)|$ are obtained. In the magnitude spectrum ($|C_1(f)|$), the spacing between successive peaks is $1/0.1 = 10$, whereas in the magnitude spectrum ($|C_2(f)|$), the spacing between successive peaks is $1/0.05 = 20$. The python code that performs this task is given in Fig. 2.3, and the corresponding output is shown in Fig. 2.4.
    Inferences From Fig. 2.4, the following inferences can be drawn:

1. The spacing between two successive samples in the comb function $c_1(t)$ is 0.1 s.
2. The spacing between two consecutive peaks in $C_1(f)$ is 10 Hz.
3. The spacing between two successive samples in the comb function $c_2(t)$ is 0.5 s.
4. The spacing between two consecutive peaks in $C_2(f)$ is 20 Hz.
5. This experiment illustrates the fact that time and frequency are inversely related to each other. That is, compression in one domain is equivalent to expansion in other domain and vice versa.
6. The Fourier transform of a train of impulse function results in a train of impulse function.

Task

1. Write a python code to generate a function expressed as $x[m] = \frac{1}{M} \times$

$\sum_{k=0}^{M-1} e^{j\frac{2\pi}{M}km}$, $-10 < m < 10$ for $M = 1$ and $M = 2$, and comment on the observed result.

```
#Fourier transform of train of impulse
import numpy as np
import matplotlib.pyplot as plt
from scipy.fft import fft,fftshift
#Step 1: Generation of comb functions
Fs=100
t = np.arange(0, 2, 1/Fs)
f = np.linspace(-Fs/2, Fs/2, len(t), endpoint=False)
T1 = 0.1
c1=np.zeros(len(t))
c1[::int(Fs*T1)]=1
T2=0.05
c2=np.zeros(len(t))
c2[::int(Fs*T2)]=1
#Step 2: Fourier transform of comb function
C1=fftshift(fft(c1))
C2=fftshift(fft(c2))
#Step 3: Plotting the result
plt.subplot(2,2,1),plt.stem(t,c1),plt.xlabel('Time'),plt.ylabel('Ampliude'),
plt.title('$c_1(t)$'),plt.subplot(2,2,2),plt.plot(f, np.abs(C1)/len(C1))
plt.xlabel('Frequency'),plt.ylabel('Magnitude'),plt.title('$|C_1(f)|$')
plt.subplot(2,2,3),plt.stem(t,c2),plt.xlabel('Time'),plt.ylabel('Ampliude'),
plt.title('$c_2(t)$'),plt.subplot(2,2,4),plt.plot(f, np.abs(C2)/len(C2))
plt.xlabel('Frequency'),plt.ylabel('Magnitude'),plt.title('$|C_2(f)|$')
plt.tight_layout()
```

Fig. 2.3 Python code to obtain the spectrum of comb function

2.1.1   Violation of Sampling Theorem

The sampling theorem gives the guideline regarding the choice of the sampling rate. According to the sampling theorem, a continuous-time signal with frequencies no higher than f_{max} (Hz) can be reconstructed exactly from its samples if the samples are taken at a rate greater than $2f_{max}$. That is, $f_s \geq 2f_{max}$. Violation of the sampling theorem results in an aliasing, which can be visualized in both the time and frequency domains.

Experiment 2.3 Illustration of Aliasing in Time Domain
In this experiment, the aliasing is visualized in time domain. The analogue signal to be sampled is represented as $x(t) = \sin(2\pi f t + \phi)$. The frequency of the signal $x(t)$ is 10 Hz, and the phase angle is zero. This signal is sampled at four different sampling frequencies 8, 15, 50 and 100 Hz. Obviously, the first two sampling frequencies ($f_s = 8$ and 15 Hz) are less than the criteria specified by the sampling theorem. This will result in aliasing. The impact of aliasing is visualized in this experiment. The

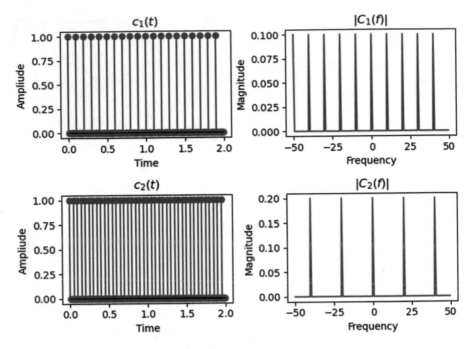

Fig. 2.4 Fourier transform of comb functions

```
#Aliasig in time domain
import numpy as np
import matplotlib.pyplot as plt
f=10  #Signal frequency
fs=[8,15,50,100] #Sampling frequencies
for i in range(len(fs)):
    t=np.arange(0,1,1/fs[i])
    x=np.sin(2*np.pi*f*t)
    plt.subplot(2,2,i+1)
    plt.plot(t,x),plt.xlabel('Time'),plt.ylabel('Amplitude')
    plt.title('$F_s={} $ Hz'.format(fs[i]))
    plt.tight_layout()
```

Fig. 2.5 Python code which illustrates aliasing in time domain

python code that performs this task is shown in Fig. 2.5, and the corresponding output is shown in Fig. 2.6.

Inferences

From Fig. 2.6, the following inferences can be made:

1. The sampling frequency of 8 Hz is insufficient to capture all the information in the signal. The frequency of the sampled signal is given by $f' = f - f_s$. This implies f

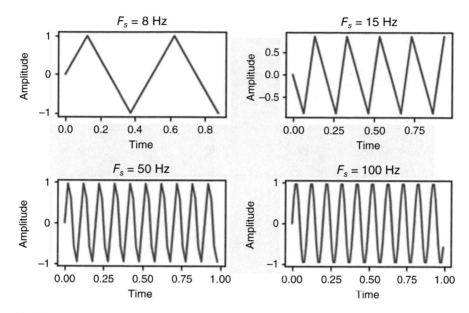

Fig. 2.6 Result of python code shown in Fig. 2.5

$' = 10 - 8 = 2$ Hz. This is the reason that the signal obtained using $f_s = 8$ Hz resembles that of 2 Hz sinusoidal signal.

2. If the sampling frequency is chosen as 15 Hz, then the Nyquist interval is between -7.5 and 7.5 Hz. The signal frequency is not within the Nyquist interval; hence, the frequency of the sampled signal is $f' = f - f_s$. Upon substituting the value, we get $f' = 10 - 15 = -5$ Hz. This is the reason that the signal obtained using $f_s = 15$ Hz resembles that of a 5 Hz sinusoidal signal.

3. For the choice of sampling frequency as 50 and 100 Hz, signal frequency lies well within the Nyquist interval. Hence, no aliasing exists in these cases. As a result, the 10 Hz signal appeared as 10 Hz for $f_s = 50$ and 100 Hz.

Experiment 2.4 Aliasing in the Time Domain

Generate two sinusoidal signals with a frequency of 1 and 6 Hz. Use a sufficiently high sampling frequency to plot the generated signal. Now use the sampling frequency as 5 Hz to plot the 6 Hz frequency component sinusoidal signal, and comment on the observed output. Illustration of this experiment is shown in Fig. 2.7.

The steps involved in the python code implementation of this experiment are as follows:

Step 1: Generation of sine wave of 1 and 6 Hz sinusoidal signals. Let it be represented by the variables '$x1$' and '$x2$'. '$x1$' represents a 1 Hz sine wave, and '$x2$' represents a 6 Hz sine wave. The sampling frequency chosen is 100 Hz ($f_s = 100$ Hz), which is sufficient to represent these two signals without ambiguity.

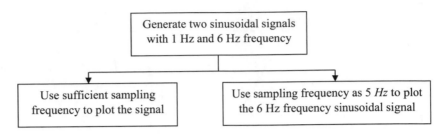

Fig. 2.7 Illustration of Experiment 2.4

```
import numpy as np
import matplotlib.pyplot as plt
#Step 1: To generate x1 and x2
f1=1  #Signal frequency
f2=6
fs=100
t=np.arange(0,1,1/fs)
x1=np.sin(2*np.pi*f1*t)
x2=np.sin(2*np.pi*f2*t)
#Step 2: New sampling frequency is 5 Hz
fs1=5
t1=np.arange(0,1.1,1/fs1)
x3=np.sin(2*np.pi*f1*t1)
#Step 3: Plotting the result
plt.plot(t,x1,'k--',t,x2,'k'),#plt.plot(t,x2,'k')
plt.stem(t1,x3,'r'),plt.xlabel('Time'),plt.ylabel('Amplitude')
plt.legend(['1 Hz Sine wave','6 Hz Sine wave','Sampling with 5 Hz']),
plt.title('Aliasing in Time Domain')
plt.tight_layout()
```

Fig. 2.8 Python code to illustrate aliasing in time domain

Step 2: Now, the new sampling frequency chosen is 5 Hz. That is, $f' = 5$ Hz. This sampling frequency is used to represent a 6 Hz sine wave, which is stored in the variable 'x3'. It is well-known that 5 Hz is insufficient to represent a sine wave of 6 Hz frequency. Because of aliasing, the new frequency will appear at 1 Hz.

Step 3: From the samples taken using $f' = 5$ Hz, it is not possible to distinguish between 1 and 6 Hz sine waves. This phenomenon is termed as 'aliasing'. This occurs due to spectral folding.

The python code used to illustrate this concept is shown in Fig. 2.8, and the corresponding output is shown in Fig. 2.9.

Inferences

From Fig. 2.9, the following inferences can be made:

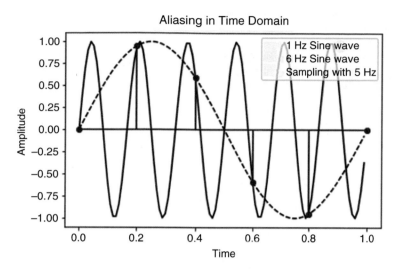

Fig. 2.9 Result of python code shown in Fig. 2.8

1. The solid line shows a sine wave of 6 Hz frequency. The dotted line represents a sine wave of 1 Hz frequency. Since the sampling frequency is 100 Hz, both waveforms appear as desired without ambiguity.
2. The new sampling frequency is chosen as 5 Hz. This sampling frequency is used to represent a 6 Hz sine wave. This sampling frequency is insufficient to represent the 6 Hz. Represent a 6 Hz sine wave; the sampling frequency should be greater than 12 Hz. From the discrete samples, it is not possible to interpret whether the samples are taken from a 6 Hz sine wave or a 1 Hz sine wave. This ambiguity is termed as aliasing, which arises due to spectral folding.

Experiment 2.5 Illustration of Aliasing in Frequency Domain
The python code, which demonstrates the phenomenon of aliasing in the frequency domain, is shown in Fig. 2.10. This experiment generates the signal x-$(t) = \sin(10\pi t) + \sin(30\pi t)$ using two different sampling rates: $f_s = 50$ Hz and $f_s = 25$ Hz.

Inferences
The following inferences can be made from Fig. 2.11.

1. The frequency components present in the signal $x(t)$ are $f_1 = 5$ Hz and $f_2 = 15$ Hz.
2. When the sampling rate is 50 Hz, the peak in the magnitude spectrum appears correctly at $f_1 = 5$ Hz and $f_2 = 15$ Hz.
3. On the other hand, if the sampling rate is chosen as $f_s = 25$ Hz, there is no change with respect to $f_1 = 5$ Hz frequency component, whereas the frequency component $f_2 = 15$ Hz appears as $f_2 = 10$ Hz. Observing a 15 Hz frequency component signal as a 10 Hz frequency component is termed as aliasing.

```
#Sampling theorem
import numpy as np
import matplotlib.pyplot as plt
from scipy.fft import fft,fftfreq
#Step 1: Generate the two signals
f1=5
f2=15
fs=[25,50]
N=256
for i in range(len(fs)):
    T=1/fs[i]
    t=np.linspace(0,N*T,N)
    x=np.sin(2*np.pi*f1*t)+np.sin(2*np.pi*f2*t)
    X=fft(x)
    f_axis=fftfreq(N,T)[0:N//2]
    plt.subplot(2,1,i+1)
    plt.plot(f_axis,2/N*np.abs(X[0:N//2]))
    plt.xlabel('$\omega$-->'),plt.ylabel('|X($\omega$)|'),
    plt.title(r'Spectrum corresponding  to $f_s = {} Hz$'.format(fs[i]))
    plt.tight_layout()
```

Fig. 2.10   Python code to illustrate the concept of aliasing in frequency domain

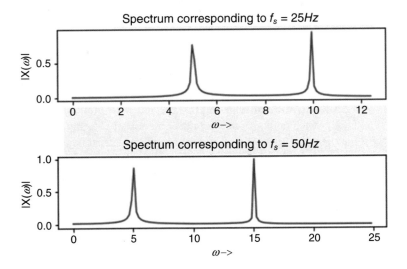

Fig. 2.11   Illustration of aliasing in the frequency domain

Task

1. Change the value of the sampling frequency (f_s) in the python code given in Fig. 2.10, and observe the changes in the output spectrum.

Fig. 2.12 Hearing aliasing effect

```
#Hearing aliasing
import sounddevice as sd
import numpy as np
import matplotlib.pyplot as plt
fs=1500
dur=1
T=1/fs
t=np.linspace(0,1,dur*fs)
x1=np.sin(2*np.pi*500*t)
x2=np.sin(2*np.pi*1000*t)
x=np.concatenate([x1,x2])
sd.play(x,fs)
```

Experiment 2.6 Hearing Aliasing

In this experiment, two sinusoidal tones of frequency $f_1 = 500$ Hz and $f_2 = 1000$ Hz are generated with sampling frequency $f_s = 1500$ Hz. Let $x_1(t)$ and $x_2(t)$ represent the two tones. The maximum signal frequency is 1000 Hz. The minimum sampling rate required is $f_s = 2000$ Hz. Unfortunately, f_s is chosen as 1500 Hz. As a result, 1500 Hz will be heard as 500 Hz. The python code, which illustrates this concept, is given in Fig. 2.12.

Inference

As per the code shown in Fig. 2.12, two sinusoidal tones of frequencies 500 and 1000 Hz are generated. These two tones are appended. Instead of hearing two notes, only one note corresponding to the frequency 500 Hz is heard. This is due to the violation of the sampling theorem. Due to improper sampling, tone of 1000 Hz is heard as a tone of 500 Hz. To overcome the impact of aliasing, the sampling frequency has to be chosen properly.

Task

1. Modify the sampling frequency as 8000 Hz and observe its impact.

2.1.2 Quantization of Signal

Quantization is mapping a large set of values to a smaller set of values. It can be broadly classified into (1) uniform and (2) non-uniform quantization. A uniform quantizer splits the mapped input signal into quantization steps of equal size. The uniform scalar quantization can be broadly classified into (1) mid-tread and (2) mid-rise quantizer.

If 'N' bits are used to represent the value of the signal $x[n]$, then there are 2^N distinct values that $x[n]$ can assume. If the x_{min} and x_{max} are the minimum and maximum values taken by the signal $x[n]$, then the dynamic range of the signal is calculated by

$$\text{Dynamic range} = x_{max} - x_{min} \qquad (2.4)$$

2.1.2.1   Mid-Tread Quantizer

The relationship between the input and output of a mid-tread uniform quantizer is given by

$$y[n] = Q \times \left\lfloor \frac{x[n]}{Q} + \frac{1}{2} \right\rfloor \qquad (2.5)$$

In the above equation, $x[n]$ represents the input signal to be quantized and $y[n]$ represents the quantized signal, 'Q' denotes the quantization step size and the symbol $\lfloor \rfloor$ denotes flooring operation. The expression for quantization step size can be computed by

$$Q = \frac{\text{Dynamic range}}{L} \qquad (2.6)$$

where 'dynamic range' represents the difference between the maximum and minimum value of the signal and 'L' denotes the number of reconstruction levels.

The expression for the number of reconstruction levels is given by

$$L = 2^b \qquad (2.7)$$

In the above expression, 'b' is the number of bits used to represent the signal.

Experiment 2.7 Transfer Characteristics of Mid-Tread Quantizer
The aim of this experiment is to plot the transfer characteristics of mid-tread quantizer for different bit-rate. The bit-rate (b) chosen is $b = 1, 2, 4$ and 8. The python code, which performs this task, is shown in Fig. 2.13, and the corresponding output is shown in Fig. 2.14.

Inferences
The following inferences can be drawn from Figs. 2.13 and 2.14, which are summarized below:

1. From Fig. 2.13, it is possible to observe that the input signal is represented as the variable 'x' and the quantized signal $Q(x)$ is represented as 'y'. The input signal 'x' varies from -20 to $+20$; hence, the dynamic range of 'x' is 40.
2. Figure 2.13 shows that the number of bits used to represent the input signal is varied as 1, 2, 4 and 8. It is represented as the variable 'b' in the code.

```
#Transfer characteristics of mid-tread quantizer
import numpy as np
import matplotlib.pyplot as plt
x=np.linspace(-20,20)
DR=np.max(x)-np.min(x) #Dynamic range
b=[1,2,4,8] #Bits
for i in range(len(b)):
    L=2**b[i] #Reconstruction level
    q=DR/L
    #Mid-tread quantizer
    y=np.sign(x)*q*np.floor((abs(x)/q)+(1/2))
    plt.subplot(2,2,i+1),plt.plot(x,y),plt.xlabel('x'),plt.ylabel('Q(x)')
    plt.title('Quantizer with b={}'.format(b[i]))
    plt.tight_layout()
```

Fig. 2.13 Python code for transfer characteristics of mid-tread quantizer

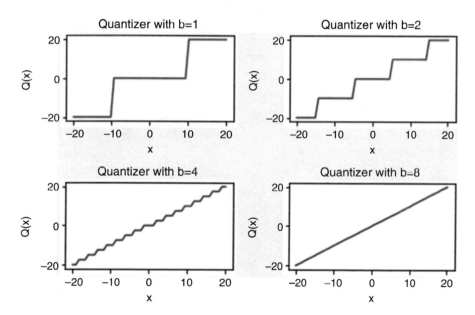

Fig. 2.14 Transfer characteristics of mid-tread quantizer

3. From Fig. 2.14, it is possible to observe that the transfer characteristics of a uniform quantizer is similar to that of a stair-step waveform at low bit rate.
4. At high bit rate ($b = 8$), the relationship between the input signal (x) and the quantized signal ($Q(x)$) is a straight line. This implies that the output follows the input; hence, the error due to quantization will be zero.
5. From Fig. 2.14, it is possible to observe that the number of reconstruction levels depends on the number of bits used to represent the signal.

```
#Uniform Quantization
import numpy as np
import matplotlib.pyplot as plt
from scipy import signal
#Step 1: Generate the input signal
t=np.linspace(0,1,100)
x=signal.sawtooth(2*np.pi*5*t)
#Step 2: Parameters of the quantizer
DR=np.max(x)-np.min(x) #Dynamic range
b=[1,2,4,8] #Number of bits
for i in range(len(b)):
    L=2**b[i] #Quantization level
    q=DR/(L)   #Quantization step size
#Step 3: To obtain the quantized signal
    y=np.sign(x)*q*np.floor((abs(x)/q)+(1/2))
    plt.figure(i+1)
    plt.plot(t,x,'b',t,y,'r'),plt.xlabel('Time'),plt.ylabel('Amplitude')
    plt.legend(['Input signal','Quantized Signal'],loc='upper right')
    plt.title('Quantization with b={}'.format(b[i]))
    plt.tight_layout()
```

Fig. 2.15 Python code to perform uniform mid-tread quantization of the signal

6. The stair tread in a ladder is the horizontal walking surface of an individual step. From Fig. 2.14, it is possible to observe that mid-tread quantizer has a zero-valued reconstruction level.

Tasks
1. Write a python code to plot the error signal. The error signal is the difference between the input and quantized signals. Comment on the observed output.
2. Write a python code to illustrate the fact that quantization error follows a uniform distribution.

Experiment 2.8 Quantization of Input Sawtooth Signal Using Mid-Tread Quantizer

The objective of this python experiment is to perform uniform mid-tread quantization of input sawtooth signal of 5 Hz frequency for different bit rate. The number of bits used to represent the input signal varies as 1, 2, 4 and 8. With an increase in the number of bits used to represent the signal, the quantized signal resembles the input signal. The python code to verify this experiment is shown in Fig. 2.15, and its simulation result is displayed in Fig. 2.16.

Inferences
The following are the inferences can be drawn from Fig. 2.16:

1. The input signal to be quantized is a sawtooth signal whose fundamental frequency is 5 Hz.

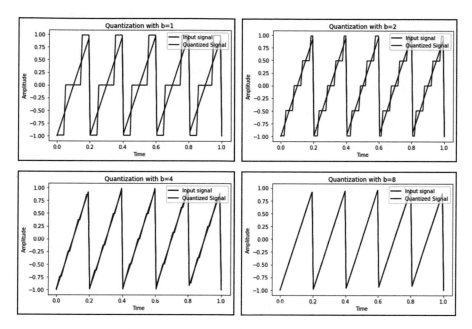

Fig. 2.16 Result of uniform mid-tread quantization

2. The input signal will be uniformly quantized by mid-tread quantizer for different bit rates.
3. It is possible to observe that the quantized signal resembles the input signal with an increase in bit-rate.

2.1.3 Mid-Rise Quantizer

The relationship between the input and output of mid-rise uniform quantizer is given by

$$y[n] = Q \times \left(\left\lfloor \frac{x[n]}{Q} \right\rfloor + \frac{1}{2} \right) \qquad (2.8)$$

In the above equation, $x[n]$ represents the input signal to be quantized and $y[n]$ represents the quantized signal, 'Q' denotes the quantization step size and the symbol $\lfloor \rfloor$ denotes flooring operation.

Experiment 2.9 Transfer Characteristics of Mid-Rise Quantizer
The aim of this experiment is to plot the transfer characteristics of mid-rise quantizer for different bit-rate. The bit-rate (b) chosen is $b = 1, 2, 4$ and 8. The python code,

Fig. 2.17 Python code for transfer characteristics of mid-rise quantizer

```python
#Transfer characteristics of mid-rise quantizer
import numpy as np
import matplotlib.pyplot as plt
x=np.linspace(-20,20)
DR=np.max(x)-np.min(x)  #Dynamic range
b=[1,2,4,8]  #Bits
for i in range(len(b)):
    L=2**b[i]  #Reconstruction level
    q=DR/L
    #Mid-rise quantizer
    y=np.sign(x)*q*(np.floor((abs(x)/q))+(1/2))
    plt.subplot(2,2,i+1)
    plt.plot(x,y),plt.xlabel('x'),plt.ylabel('Q(x)')
    plt.title('Quantizer with b={}' .format(b[i]))
    plt.tight_layout()
```

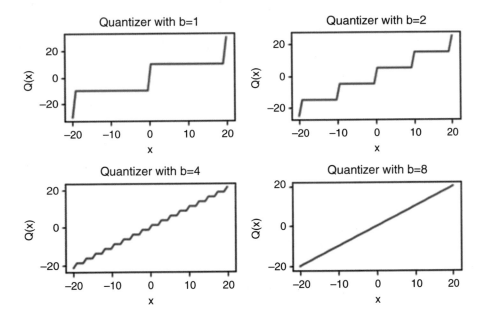

Fig. 2.18 Transfer characteristics of mid-rise quantizer

which performs this task, is shown in Fig. 2.17, and the corresponding output is shown in Fig. 2.18.

Inferences

From Fig. 2.18, it is possible to observe that the reconstruction level rises to the next level at the origin; hence, it is termed as *'mid-rise quantizer'*. It is also possible to observe that with the bit rate increase, the output follows the input. In other words, the quantizer error is minimal with a bit rate increase.

```
#Uniform mid-rise quantizer
import numpy as np
import matplotlib.pyplot as plt
from scipy import signal
#Step 1: Generate the input signal
t=np.linspace(0,1,100)
x=signal.sawtooth(2*np.pi*5*t)
#Step 2: Parameters of the quantizer
DR=np.max(x)-np.min(x) #Dynamic range
b=[1,2,4,8] #Number of bits
for i in range(len(b)):
    L=2**b[i] #Quantization level
    q=DR/(L)   #Quantization step size
#Step 3: To obtain the quantized signal
    y=np.sign(x)*q*(np.floor((abs(x)/q))+(1/2))
    plt.figure(i+1)
    plt.plot(t,x,'b',t,y,'r'),plt.xlabel('Time'),plt.ylabel('Amplitude')
    plt.legend(['Input signal','Quantized Signal'],loc='upper right')
    plt.title('Quantization with b={}'.format(b[i]))
    plt.tight_layout()
```

Fig. 2.19 Python code to perform uniform mid-rise quantization

Task

In the python code given in Fig. 2.17, replace '*np.floor*()' by '*np.ceil*()' function, and comment on the change in the transfer characteristics.

Experiment 2.10 Quantization of Input Sawtooth Signal Using Mid-Rise Quantizer

The objective of this experiment is to perform uniform mid-rise quantization of the input sawtooth signal for different bit rate. The python code, which performs this task, is shown in Fig. 2.19, and the corresponding output is shown in Fig. 2.20.

Inference

From Fig. 2.20, it is possible to interpret that with the increase in the number of bits used to represent the signal, the quantized signal resembles the input signal. In other words, the error due to quantization will be minimum with the increase in the number of bits used to represent the signal.

Experiment 2.11 Quantization of Speech Signal

The objective of this experiment is to analyse the performance of uniform mid-tread quantizer for the speech signal. The experiment consists of two steps. Reading the speech signal from a given location is the first step, and performing uniform midtread-quantization of the input speech signal for different bit rates is the second step. The python code, which does this task, is shown in Fig. 2.21, and the corresponding output is shown in Figs. 2.22 and 2.23.

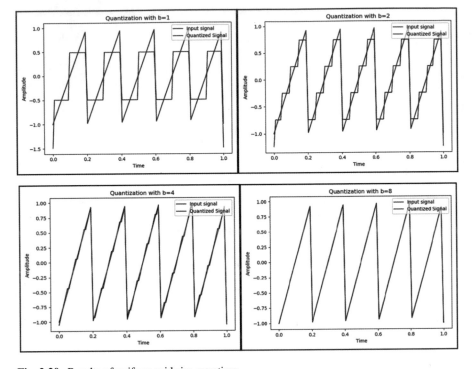

Fig. 2.20 Results of uniform mid-rise quantizer

Inference
The following inference can be made from this experiment:

1. The input speech signal belongs to the uttered word 'Hello'.
2. The quantized signal resembles the original speech signal with the increase in the number of bits of the quantizer.

Experiment 2.12 Uniform Mid-Tread Quantization of Image
In this experiment, a greyscale image, whose intensity varies gradually from black to white, is generated first. This image is subjected to uniform quantization with bit rates 1, 2, 4 and 8. The python code, which performs this task, is shown in Fig. 2.24, and the corresponding output is shown in Fig. 2.25.

Inferences
The following inferences can be drawn from this experiment:

1. The grey level of the input image varies gradually from black to white.
2. The input image is quantized uniformly with a bit rate of $b = 1, 2, 4$ and 8. When $b = 1$, the number of grey levels used to represent the image is minimum. The quantized image is different from the input image.
3. With the increase in the number of bits used to represent the pixel value, the quantized image resembles the input image.

```
#Uniform quantization of speech signal
from scipy.io import wavfile
import numpy as np
import matplotlib.pyplot as plt
#Step 1: Reading of speech waveform
samplerate, x = wavfile.read('C:\\Users\\Admin\\Desktop\\speech1.wav')
duration = x.shape[0] / samplerate
t = np.linspace(0, duration, x.shape[0])
plt.figure(1)
plt.plot(t,x,'k',linewidth=2)
plt.xlabel('Time'),plt.ylabel('Amplitude')
plt.title('Input speech signal')
#Step 2: Performing uniform quantization of the signal
DR=np.max(x)-np.min(x) #Dynamic range
b=[1,2,4,8] #Number of bits
for i in range(len(b)):
    L=2**b[i] #Quantization level
    q=DR/L   #Quantization step size
#Step 3: To obtain the quantized signal
    y=np.floor(x/q)*q-(q/2)
    plt.figure(2)
    plt.subplot(2,2,i+1)
    plt.plot(t,y,'k',linewidth=2),plt.xlabel('Time'),plt.ylabel('Amplitude')
    plt.title('Quantized signal with b={}'.format(b[i]))
    plt.tight_layout()
```

Fig. 2.21 Performing uniform mid-tread quantization of the speech signal

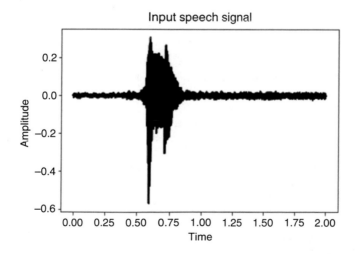

Fig. 2.22 Input speech signal

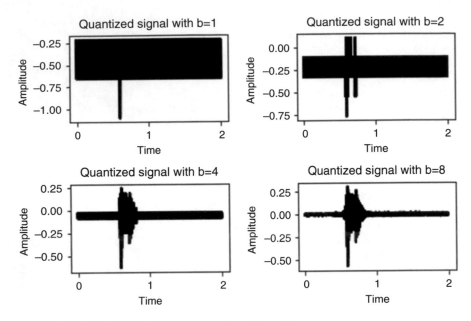

Fig. 2.23 Uniformly quantized speech signal for different bit-rate

```
#Uniform mid-tread quantization of image
import numpy as np
import matplotlib.pyplot as plt
#Step 1: Generation of test image
img=np.zeros([256,256])
img[:,0:256]=np.arange(0,256,1)
plt.figure(1)
plt.imshow(img,cmap='gray')
plt.title('Input image')
#Step 2: Parameters of uniform quantizer
DR=np.max(img)-np.min(img) #Dynamic range
b=[1,2,4,8] #Number of bits
for i in range(len(b)):
    L=2**b[i] #Quantization level
    q=DR/(L)   #Quantization step size
#Step 3: To obtain the quantized signal
    y=np.sign(img)*q*np.floor((abs(img)/q)+(1/2))
    plt.figure(2)
    plt.subplot(2,2,i+1)
    plt.imshow(y,cmap='gray')
    plt.title('b={} '.format(b[i]))
plt.tight_layout()
```

Fig. 2.24 Uniform mid-tread quantization of the image

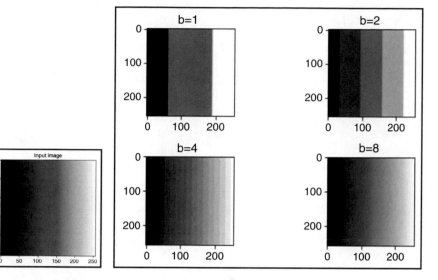

Fig. 2.25 Input and Output of uniform mid-tread quantizer

Task

1. Generate a 256 × 256 image in which half of the pixels are white (grey level 255) and half of the pixels are black (grey level 0). The columns 0 to 127 is white, whereas column 128 to 256 is black. Try to quantize this image for different bit rate and comment on the observed result.

2.2   Non-uniform Quantization

One way to construct non-uniform quantizer is to perform companding.

Companding = Compression + Exp**anding**

The three steps involved in companding are (1) compression, (2) uniform quantization and (3) expanding. In the first step, the input signal is applied to a logarithmic function, and the output of this function is given to a uniform quantizer. Finally, the inverse of the logarithmic function is applied to the output of the quantizer. There are two standards for non-uniform quantizer companding. They are (1) μ-law companding for North America and (2) A-law companding for Europe.

The µ-law compression expression in terms of the input signal $x(t)$ is expressed as

$$x_1(t) = \text{sgn}(x) \ln \frac{(1 + \mu|x|)}{\ln(1 + \mu)} \tag{2.9}$$

```
import numpy as np
import matplotlib.pyplot as plt
from scipy import signal
#Step 1: Generate the input signal
t1=np.linspace(0,1,100)
x=signal.sawtooth(2*np.pi*5*t1)
#Step 2: Mu law Encoding (Non-uniform encoding)
mu=255 # 8 bit Quantization
y1=np.sign(x)*((np.log(1+(mu*abs(x))))/np.log(1+mu))
plt.figure(1)
plt.plot(t1,x,'b',t1,y1,'g'),plt.xlabel('Time'),plt.ylabel('Amplitude')
plt.legend(['Input signal','Encoded'],loc='upper right')
plt.title('Degree of Compression with mu={}'.format(mu))
#Step 3: Parameters of the quantizer
DR=np.max(y1)-np.min(y1) #Dynamic range
b=[1,2,4,8] #Number of bits
for i in range(len(b)):
    L=2**b[i] #Quantization level
    q=DR/(L)   #Quantization step size
#Step 3: To obtain the quantized signal
    y2=np.sign(y1)*q*np.floor((abs(y1)/q)+(1/2))
    y=np.sign(y2)*(((1+mu)**(abs(y2))-1)/mu)
    plt.figure(i+2)
    plt.plot(t1,y2,'r',t1,y),plt.xlabel('Time'),plt.ylabel('Amplitude')
    plt.legend(['Quantized Before decoding','Non-Uniform Quantized'],loc='upper right')
    plt.title('Quantization with b={} and mu={}'.format(b[i],mu))
    plt.tight_layout()
```

Fig. 2.26 Python code for μ-law companding

   In the above expression, 'μ' is the compression parameter, which is 255 for the USA and Japan. During compression, the least significant bits of large amplitude values are discarded.

Experiment 2.13 μ-Law Companding
The python code which performs μ-law companding is shown in Fig. 2.26, and the corresponding output is shown in Figs. 2.27 and 2.28.

Inference
The input signal to be companded is a sawtooth signal. The fundamental frequency of a sawtooth signal is 5 Hz. Figure 2.27 illustrates the signal to be encoded using μ-law companding with $\mu = 255$. Here the signal is basically compressed before passing it to the uniform quantizer. Figure 2.28 shows the uniform quantizer results for different bit-rate values. With increase in bit-rate, the quantized signal resembles the input signal.

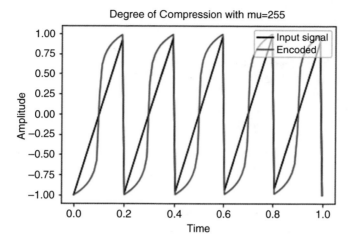

Fig. 2.27   Encoded signal using μ-law companding

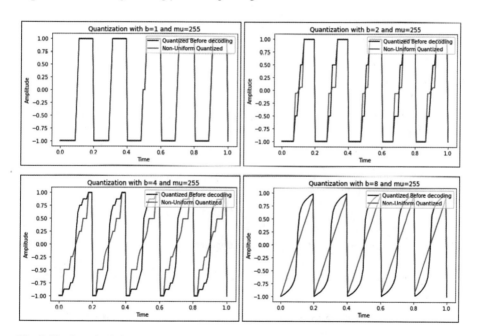

Fig. 2.28   Quantized signal

Experiment 2.14 Error Due to Quantization

Quantization is basically mapping a large set of values to a smaller set of values. It is a non-linear and irreversible process. Quantization leads to loss of information. The loss of information due to quantization can be considered as an error. The error signal is considered as the difference between the quantized signal ($y[n]$) and the input signal ($x[n]$). The objective of this experiment is to quantize the input

```
#Error due to quantization
import numpy as np
import matplotlib.pyplot as plt
#Step 1: Generate the input signal
t=np.linspace(0,1,100)
x=np.sin(2*np.pi*5*t)
#Step 2: Parameters of the quantizer
DR=np.max(x)-np.min(x) #Dynamic range
b=[1,2,4,8] #Number of bits
for i in range(len(b)):
    L=2**b[i] #Quantization level
    q=DR/(L)   #Quantization step size
#Step 3: To obtain the quantized signal
    y=np.sign(x)*q*(np.floor((abs(x)/q))+(1/2))
#Step 4: Obtain the error signal
    e=y-x
#Plot the error signal
    plt.subplot(2,2,i+1), plt.plot(e),plt.xlabel('Time'), plt.ylabel('Amplitude'),
    plt.title('Error signal for b={}'.format(b[i]))
    plt.tight_layout()
```

Fig. 2.29 Error due to quantization

sinusoidal signal of 5 Hz frequency for different bit rate. Then, plot the error signal for different bit-rate. The python code, which performs this task, is shown in Fig. 2.29, and the corresponding output is shown in Fig. 2.30.

Inferences
From Fig. 2.30, the following inferences can be made:

1. The error signal is oscillatory in nature. The magnitude of the error signal varies between positive and negative values.
2. The magnitude of the error signal decreases with increase in bit-rate of the quantizer.
3. Error due to quantization is inevitable; hence, quantization is considered as irreversible phenomenon.

Experiment 2.15 Probability Density Function of Quantization Error
From the previous experiment, it is possible to confirm that error is inevitable in quantization process. The objective of this experiment is to prove that quantization error follows a uniform distribution. The steps followed in this experiment are displayed in Fig. 2.31.

The python code which performs the task mentioned above is shown in Fig. 2.32, and the corresponding output is shown in Fig. 2.33.

Inferences
The following inferences can be drawn from Fig. 2.33:

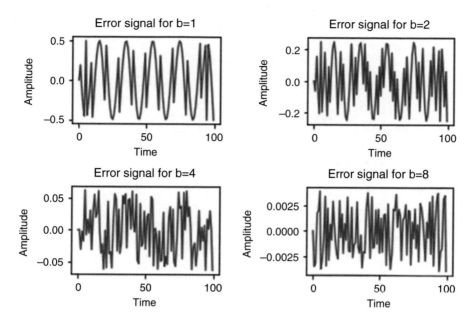

Fig. 2.30 Error signal for different bit-rate of the quantizer

Fig. 2.31 Flow diagram of
Experiment 2.13

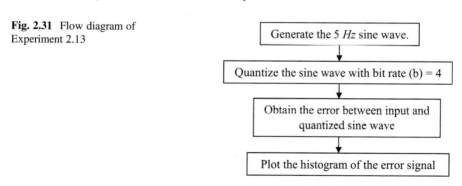

1. The quantization error follows uniform distribution in the range $(-\Delta/2, \Delta/2)$, where 'Δ' is the quantization step size.
2. In this example, the value of 'Δ' is 0.125; hence, $\Delta/2$ value is 0.0625.

2.3   Signal Reconstruction

Signal reconstruction is an attempt to obtain the continuous-time signal from the samples. This is also termed as interpolation. Different types of interpolation schemes include (1) zero-order hold interpolation, (2) first-order hold or linear interpolation and (3) ideal interpolation.

```
#PDF of quantized error signal
import numpy as np
import matplotlib.pyplot as plt
#Step 1: Generate the input signal
t=np.linspace(0,1,1000)
x=np.sin(2*np.pi*5*t)
#Step 2: Parameters of the quantizer
DR=np.max(x)-np.min(x) #Dynamic range
b=4
L=2**b #Quantization level
q=DR/(L) #Quantization step size
#Step 3: Quantize the input signal
y=np.sign(x)*q*(np.floor((abs(x)/q))+(1/2))
#Step 3: Obtain the error signal
e=y-x
#Step 4: Plot the histogram of the error signal
plt.hist(e,10),plt.xlabel('e'),plt.ylabel('$P_e(e)$')
plt.title('PDF of error signal')
```

Fig. 2.32 Python code for PDF of quantization error

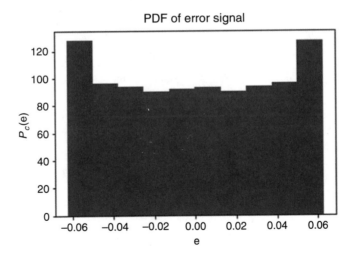

Fig. 2.33 Histogram plot quantization error

2.3.1 Zero-Order Hold Interpolation

A zero-order hold (ZoH) system is a form of simple interpolation, where a line of zero-slope connects discrete samples. The zero-order hold maintains the signal level of the previous pulse until the next pulse arrives. The reconstructed signal will resemble a staircase curve. This is depicted in Fig. 2.34.

Fig. 2.34 Zero-order hold interpolation

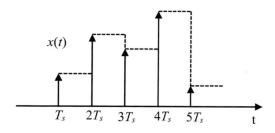

Fig. 2.35 Impulse response of ZoH interpolation function

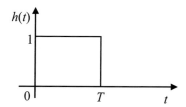

```
#Zero-order hold interpolation
import numpy as np
import matplotlib.pyplot as plt
from scipy.interpolate import interp1d
#Step 1: Generation of sine wave
t=np.linspace(0,2*np.pi,10)
x=np.sin(t)
#Step 2: Performing zero-order hold interpolation
f=interp1d (t,x,kind='previous')
#Step 3: Plotting the results
t1=np.linspace(0,2*np.pi,500)
plt.plot(t1,f(t1),'k--'),plt.stem(t,x,'r'),plt.xlabel('Time'),
plt.ylabel('Amplitude'),plt.title('Sine wave')
plt.legend(['ZOH interpolation','Sine wave samples'],loc=1)
```

Fig. 2.36 Python code of zero-order hold interpolation

The impulse response of a zero-order hold is shown in Fig. 2.35. The transfer function of zero-order hold function is given by

$$H(s) = \frac{1 - e^{-Ts}}{s} \tag{2.10}$$

Experiment 2.16 Zero-Order Hold Interpolation

The python example, which performs zero-order hold interpolation of the sinusoidal signal, is shown in Fig. 2.36, and the corresponding output is shown in Fig. 2.37. In the scipy package, the built-in function 'interp1d' performs the zero-order hold interpolation.

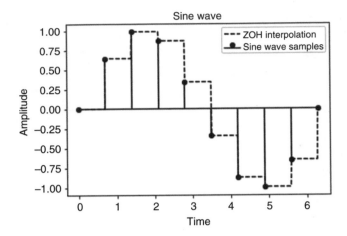

Fig. 2.37 Result of python code shown in Fig. 2.36

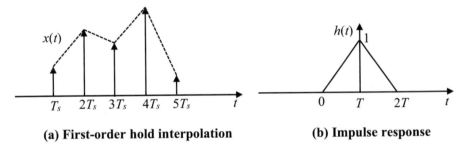

(a) First-order hold interpolation   **(b) Impulse response**

Fig. 2.38 First-order hold interpolation. (**a**) First-order hold interpolation. (**b**) Impulse response

Inference

From Fig. 2.37, it is possible to interpret that zero-order hold interpolation converts the input signal into a piece-wise constant signal. It is possible to observe discontinuity in the zero-order hold interpolated signal.

2.3.2   First-Order Hold Interpolation

In first-order hold (FoH) interpolation, the signal samples are connected by a straight line. This idea is illustrated in Fig. 2.38a.

The first-order hold performs linear interpolation between samples. The impulse response of first-order hold is shown in Fig. 2.38b.

The transfer function of first-order hold is expressed as

```
#First-order hold interpolation
import numpy as np
import matplotlib.pyplot as plt
from scipy.interpolate import interp1d
#Step 1: Generation of sine wave
t=np.linspace(0,2*np.pi,10)
x=np.sin(t)
#Step 2: Performing zero-order hold interpolation
f=interp1d(t,x,kind='linear')
#Step 3: Plotting the results
t1=np.linspace(0,2*np.pi,10)
plt.plot(t1,f(t1),'k--'),plt.stem(t,x,'r'),plt.xlabel('Time'),
plt.ylabel('Amplitude'),plt.title('Sine wave')
plt.legend(['FOH interpolation','Sine wave samples'],loc=1)
plt.tight_layout()
```

Fig. 2.39 Python code to perform first-order hold interpolation

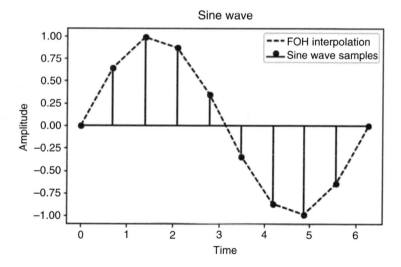

Fig. 2.40 Result of python code shown in Fig. 2.39

$$H(s) = \left(\frac{1 - e^{-sT}}{s}\right)^2 \qquad (2.11)$$

Experiment 2.17 First-Order Hold Interpolation

The python code to illustrate first-order hold interpolation is shown in Fig. 2.39, and the corresponding output is shown in Fig. 2.40.

From Fig. 2.40, it is possible to interpret that first-order hold interpolation attempts to connect the sample points through a straight line.

Inferences
The following inference can be drawn from Fig. 2.40:

1. The zero-order hold yields a staircase approximation of the signal.
2. The first-order hold yields a linear approximation of the signal.
3. The first-order hold connects the samples with straight lines.

2.3.3   Ideal or Sinc Interpolation

The expression for continuous-time signal obtained using sinc interpolation is expressed as

$$x(t) = \sum_{n=-\infty}^{\infty} x[n] \sin c\left(\frac{t - nT_s}{T_s}\right) \tag{2.12}$$

The sinc function is a symmetric function which is square integrable. The decay of the sinc function is slow. The sinc function has infinite support; hence, it is termed as ideal interpolation. The sinc interpolation produces the smoothest possible interpolation of the samples.

Experiment 2.18 Ideal or Sinc Interpolation of a Sinusoidal Signal
The python code, which performs the ideal interpolation of the sine waveform, is shown in Fig. 2.41, and the corresponding output is shown in Fig. 2.42.

Inference
The sinc interpolation produces the smoothest possible interpolation of the samples.

Experiment 2.19 Comparison of Zero-Order Hold and Sinc Interpolation
The python code, which performs the zero-order hold and sinc interpolation of a given sinusoidal signal, is shown in Fig. 2.43, and the corresponding output is in Fig. 2.44.

Inference
By observing Fig. 2.44, it is possible to infer that sinc interpolation smooths the successive samples in the sine wave when compared to zero-order hold interpolation method.

Exercises
1. Write a python code to demonstrate the phenomenon of aliasing in the frequency domain for which the signal $x(t) = \sin(20\pi t) + \sin(50\pi t)$ is generated using two different sampling rates: $f_s = 100$ Hz and $f_s = 25$ Hz. Plot the corresponding spectrum and comment on the observed result.

```
#Ideal or sinc interpolation
import numpy as np
import matplotlib.pyplot as plt
t=np.linspace(0,2*np.pi,10)
t1=np.linspace(0,2*np.pi,100)
x=np.sin(t)
def sinc_interp(x, s, u):
    if len(x) != len(s):
        raise ValueError('x and s must be the same length')
    T = s[1] - s[0]
    sincM = np.tile(u, (len(s), 1)) - np.tile(s[:, np.newaxis], (1, len(u)))
    y = np.dot(x, np.sinc(sincM/T))
    return y
y=sinc_interp(x,t,t1)
plt.plot(t1,y,'r--'),plt.stem(t,x,'k'),plt.xlabel('Time'),
plt.ylabel('Amplitude'),plt.title('Sinc interpolation')
plt.legend(['Ideal interpolation','Sine wave samples'],loc=1)
plt.tight_layout()
```

Fig. 2.41 Python code to perform sinc interpolation

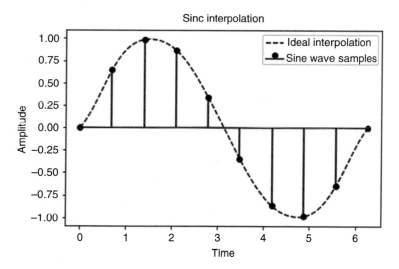

Fig. 2.42 Result of python code shown in Fig. 2.41

2. Generate a sinusoidal signal of 5 Hz frequency. Quantize this signal using uniform mid-rise quantizer with bit-rate, $b = 1$, 2 and 4. Use a subplot to plot the input signal and the quantized signal.

3. Consider an analogue signal $x(t) = \cos(2\pi t) + \cos(14\pi t) + \cos(18\pi t)$, where '$t$' is in seconds. If this signal is sampled at $f_s = 8$ Hz, then it will be aliased with the

```
#Ideal and sinc interpolation
import numpy as np
import matplotlib.pyplot as plt
from scipy.interpolate import interp1d
t=np.linspace(0,2*np.pi,10)
t1=np.linspace(0,2*np.pi,500)
x=np.sin(t)
#Zero-order hold interpolation
f=interp1d(t,x,kind='previous')
#Sinc interpolation
def sinc_interp(x, s, u):
  if len(x) != len(s):
    raise ValueError('x and s must be the same length')
  T = s[1] - s[0]
  sincM = np.tile(u, (len(s), 1)) - np.tile(s[:, np.newaxis], (1, len(u)))
  y = np.dot(x, np.sinc(sincM/T))
  return y
y=sinc_interp(x,t,t1)
plt.plot(t1,f(t1),'b:'),plt.plot(t1,y,'k--'),plt.stem(t,x,'r'),plt.xlabel('Time'),
plt.ylabel('Amplitude'),plt.title('Comparison of interpolation methods')
plt.legend(['ZOH interpolation','Sinc interpolation','Sine wave samples',],loc=1)
```

Fig. 2.43 Comparison of zero-order hold and sinc interpolation

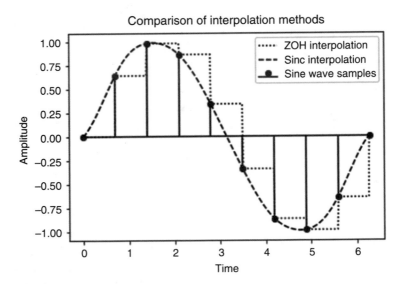

Fig. 2.44 Result of ZOH and sinc interpolation

signal, which is expressed as $x'(t) = 3 \cos(2\pi t)$. Plot $x(t)$ and $x'(t)$ on the same graph to verify the signals inter at the sampling instants.

4. Write a python code to generate a sinusoidal signal of fundamental frequency 1300 Hz and sampling frequency $f_s = 8\,kHz$. Hear this tone. Now downsample this signal by a factor of 2 and hear the tone. Comment on the heard tones.

5. Write a python code to generate a sinusoidal signal of 10 Hz frequency. Quantize this signal using 4-bit uniform mid-tread quantizer. Use a subplot to plot the input, quantized and error signals. Comment on the observed output.

Objective Questions

1. What will be the output if the following code is executed?

```
import numpy as np
x=4.25
print(np.floor(x))
```

A. 4.5
B. 4.0
C. 4.25
D. 5.0

2. What will be the output if the following code is executed?

```
import numpy as np
x=-4.25
print(np.ceil(x))
```

A. −4.0
B. −5.0
C. −4.25
D. −5.25

3. The following python code segment produces

```
t=np.linspace(0,2*np.pi,20)
x=np.sin(t)
f=interp1d(t,x,kind='nearest')
```

A. Zero-order hold interpolation
B. Linear interpolation
C. Polynomial interpolation
D. Sinc interpolation

4. Fourier transform of train of impulse function results in

A. Train of step function
B. Train of impulse
C. Sinc function
D. Triangular function

5. A sinusoidal signal of the form $x(t) = \sin(2\pi ft)$, where '$f = 5$ Hz' is sampled at the rate $f_s = 100$ Hz to obtain the discrete-time sequence $x[n]$. The expression for the signal $x[n]$ is

$$x[n] = \sin\left(\frac{\pi}{2}n\right)$$

$$x[n] = \sin\left(\frac{\pi}{4}n\right)$$

$$x[n] = \sin\left(\frac{\pi}{5}n\right)$$

$$x[n] = \sin\left(\frac{\pi}{10}n\right)$$

6. Assertion: Quantization is an irreversible process.
   Reason: Quantization is many-to-one mapping:

   A. Assertion and reason are true.
   B. Assertion is wrong; reason is true.
   C. Assertion is true; reason is wrong.
   D. Assertion and reason are wrong.

7. Statement 1: Quantization is a non-linear phenomenon
   Statement 2: Quantization is an irreversible phenomenon

   A. Statement 1 and 2 are false.
   B. Statement 1 and 2 are true.
   C. Statement 1 is true; statement 2 is false.
   D. Statement 1 is false; statement 2 is true.

8. The transfer function of zero-order hold is

$$H(s) = 1$$

$$H(s) = \frac{1}{s}$$

$$H(s) = \frac{1 - e^{-sT}}{s}$$

$$H(s) = 1 - e^{-sT}$$

9. An analogue voltage in the range 0–4 V is divided into 32 equal intervals. The quantization step size of this uniform quantizer is

   A. 0.0625
   B. 0.125
   C. 0.25
   D. 0.5

10. If 'Δ' represents the quantization step size of a uniform quantizer, the expression for mean square quantization error is

A. $\dfrac{\Delta^2}{2}$

B. $\dfrac{\Delta^2}{4}$

C. $\dfrac{\Delta^2}{8}$

D. $\dfrac{\Delta^2}{12}$

11. The quantization error follows

A. Normal distribution
B. Uniform distribution
C. Chi-square distribution
D. Exponential distribution

12. The transfer function of first-order hold is

A. $H(s) = \dfrac{1 - e^{-sT}}{s}$

B. $H(s) = \dfrac{1}{s}$

C. $H(s) = \left(\dfrac{1 - e^{-sT}}{s}\right)^2$

D. $H(s) = 1 - e^{-sT}$

13. The signal to be quantized takes the value in the range $(-1,1)$. The dynamic range of the signal is

A. 1
B. −1
C. 0
D. 2

14. If f_s represents the sampling frequency, then the expression for Nyquist frequency is

A. f_s
B. $f_s/2$
C. $f_s/4$
D. $f_s/8$

15. The quantization step size of a two-bit quantizer which accepts the input signal, which varies from 0 to 2 V, is

 A. 0.125
 B. 0.25
 C. 0.5
 D. 0.75

Bibliography

1. Alan V. Oppenheim, and Ronald W. Schafer, "Discrete-Time Signal Processing", Pearson, 2009.
2. Michael Roberts, and Govind Sharma, "Fundamentals of Signals and Systems", McGraw Hill Education, 2017.
3. John G. Proakis, and Dmitris G. Manolakis, "Digital Signal Processing: Principles, Algorithms and Applications", Pearson Education, 2007.
4. Barrie Jervis, Emmanuel Ifeachor, "Digital Signal Processing: A Practical Approach", Pearson, 2001.
5. Allen B. Downey, "Think DSP: Digital Signal Processing in Python", O' Reilly Media, 2016.

Chapter 3
Generation and Operation on Discrete-Time Sequence

Learning Objectives

After completing this chapter, the reader is expected to

- Generate standard discrete-time sequences like unit sample, unit step, unit ramp sequences, etc.
- Perform operations like folding, shifting and scaling on the discrete-time sequence.
- Perform linear convolution and circular convolution between discrete-time sequences.
- Perform autocorrelation and cross-correlation between discrete-time sequences.

Road Map of the Chapter

This chapter aims to generate different discrete-time signals or sequences and perform various mathematical operations on the discrete-time signal. The flow of the concept in this chapter is illustrated in the form of a block diagram, which is given below:

S. Esakkirajan et al., *Digital Signal Processing*,
https://doi.org/10.1007/978-981-99-6752-0_3

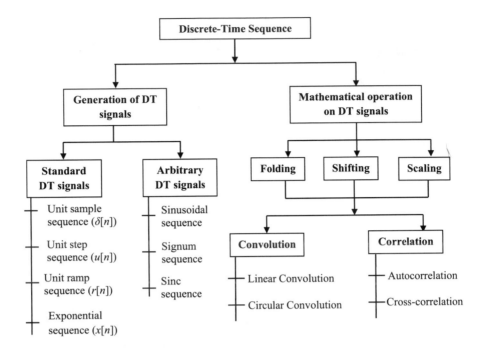

PreLab Questions

1. What are the steps involved in converting the continuous-time signal into a discrete-time signal?
2. Mention different forms of representations of discrete-time signals?
3. Mention a few standard discrete-time sequences.
4. Mention the significant features of the unit sample sequence ($\delta[n]$).
5. State the condition for the discrete-time signal to be periodic.
6. Distinguish between energy and power signal.
7. What are the various mathematical operations that can be performed on discrete-time signals?
8. When a discrete-time signal is said to be (a) an even signal (b) an odd signal? Give an example for each class of signal. Also, give an example of a signal which is neither even nor odd.
9. Give an example of an energy and power signal. Also, give an example of a discrete-time signal which is neither energy nor power signal.
10. Explain in your own word regarding the significance of convolution operation in signal processing.
11. What is the relationship between convolution and correlation? Mention two applications of correlation.

3.1   Generation of Discrete-Time Signals

This section deals with the generation of different types of discrete-time signals like unit sample signal, unit step signal, unit ramp signal, real and complex exponential signals. The following section discusses about different mathematical operations that could be performed on discrete-time signals.

Experiment 3.1 Generation of the Unit Sample Sequence
The mathematical expression of the unit sample sequence ($\delta[n]$) is given by

$$\delta[n] = \begin{cases} 1, & \text{if } n = 0 \\ 0, & \text{Otherwise} \end{cases} \tag{3.1}$$

This experiment discusses the generation of unit sample sequence using '*if*' and '*else*' conditions in python platform. The python code to generate unit sample sequence using '*if*' and '*else*' conditions is shown in Fig. 3.1, and the corresponding output is shown in Fig. 3.2. The built-in functions used in the program are given in Table 3.1.

Inference
It is possible to observe that unit sample sequence takes a value of '1' at 'n' equal to zero and zero at other instances of 'n'.

Experiment 3.2 Generation of Unit Sample Sequence Using the Logical Operation
This experiment deals with the logical operation used to generate unit sample sequence, and the python code for this experiment is shown in Fig. 3.3, and the corresponding output is shown in Fig. 3.4.

Inference
The statement ($x = (n == 0)$) given in Fig. 3.3 implies that the variable 'x' takes a value of '1' if $n = 0$, and it takes a value of '0' for all the other values of 'n'.

Fig. 3.1 Python code to generate unit sample sequence

```
#Python code to generate unit sample sequence
import numpy as np
import matplotlib.pyplot as plt
#Step 1: Generating the sequence
n=np.arange(-10,11) #Define the x-axis
x=[1 if i==0 else 0 for i in n] #Unit sample sequence
#Step 2: Plotting the sequence
plt.stem(n,x),plt.xlabel('n-->'),plt.ylabel('Amplitude'),
plt.title('$\delta[n]$')
plt.xticks(n)
```

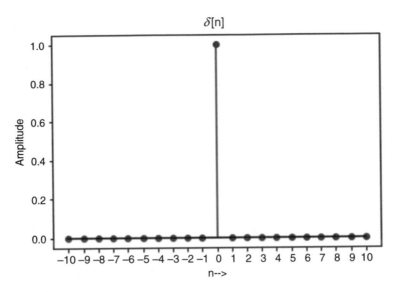

Fig. 3.2   Unit sample sequence

Table 3.1   Built-in functions used in unit sample signal generation

S. No.	Built-in function used	Purpose
1	np.arange()	To generate evenly spaced values within a given interval
2	plt.stem()	To plot the discrete-time signal
3	plt.xticks()	To get or set the current tick locations and labels of the x-axis

Fig. 3.3   Logical operation to generate unit sample sequence

```
#Python code to generate unit sample sequence
import numpy as np
import matplotlib.pyplot as plt
#Step 1: Generating the sequence
n=np.arange(-10,11,1)  #Define the x-axis
x=(n==0) #Unit sample sequence
#Step 2: Plotting the sequence
plt.stem(n,x),plt.xlabel('n-->'),plt.ylabel('Amplitude'),
plt.title('$\delta[n]$'),plt.xticks(n)
```

Experiment 3.3 Generation of Unit Sample Sequence Using the Built-In Function from *the Scipy* Library

The built-in function in *scipy* library '*unit_impulse*' can be used to generate unit sample sequence. The python code, which generates unit sample sequence using the built-in function from *the scipy* library, is shown in Fig. 3.5, and the corresponding output is shown in Fig. 3.6.

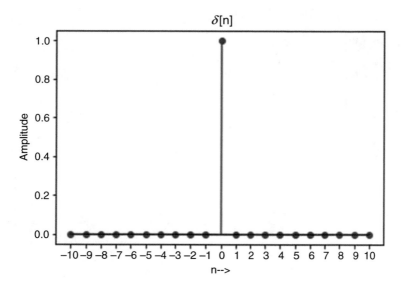

Fig. 3.4   Output of python code shown in Fig. 3.3

Fig. 3.5 Unit sample sequence generation using *scipy* library

```
import matplotlib.pyplot as plt
import numpy as np
from scipy import signal
n=np.arange(-5,6)
x=signal.unit_impulse(len(n), 'mid')
plt.stem(n, x),plt.xlabel('n-->'),plt.ylabel('Amplitude'),
plt.title('$\delta[n]$'),plt.xticks(n)
```

Inference

From Figs. 3.5 and 3.6, it is possible to confirm that unit sample sequence can be generated using *the scipy* library with the built-in command of 'signal.unit_impulse'.

Experiment 3.4 Generation of Unit Step Sequence

The mathematical expression of the unit step sequence is written as

$$u[n] = \begin{cases} 1, & \text{if } n \geq 0 \\ 0, & \text{Otherwise} \end{cases} \tag{3.2}$$

In this experiment, the unit step sequence is generated using two methods. In the first method, '*if*' and '*else*' conditions are used to generate unit step sequence. The second method uses logical operation to generate unit step signal. The python code, which generates unit step signal using two different methods, is shown in Fig. 3.7, and the corresponding output is shown in Fig. 3.8.

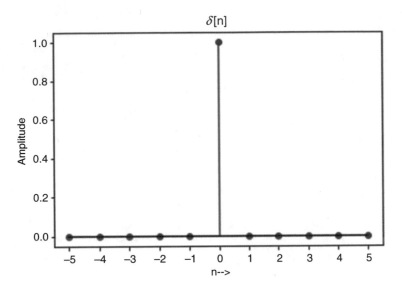

Fig. 3.6 Result of python code shown in Fig. 3.5

```
#Genertion of unit step signal
import numpy as np
import matplotlib.pyplot as plt
#Step 1: Generating the sequence
n=np.arange(-10,11,1)  #Define the x-axis
#Method 1
x1=[1 if i>=0 else 0 for i in n] # if and else
#Method 2
x2=(n>=0) #Logical operation
#Plotting the result
plt.subplot(2,1,1),
plt.stem(n,x1),plt.xlabel('n-->'),plt.ylabel('Amplitude'),
plt.title('u[n]'),plt.xticks(n)
plt.subplot(2,1,2),plt.stem(n,x2),plt.xlabel('n-->'),
plt.ylabel('Amplitude'),plt.title('u[n]'),plt.xticks(n)
plt.tight_layout()
```

Fig. 3.7 Python code to generate unit step signal

Inference

From Fig. 3.8, it is possible to interpret that both methods yield the same result, which is a unit step signal. The unit step signal exhibits a sudden change in state from logic 0 to logic 1 instantaneously.

Experiment 3.5 Generation of the Unit Ramp Signal

The mathematical expression of the unit ramp sequence ($r[n]$) is written as

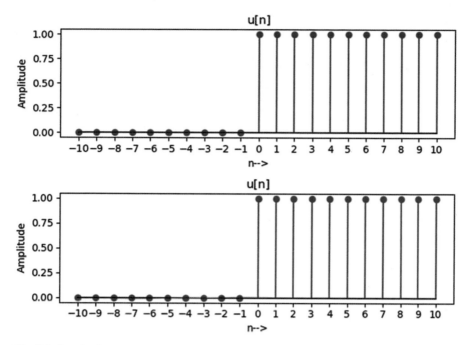

Fig. 3.8 Result of python code shown in Fig. 3.7

$$r[n] = \begin{cases} n, & \text{if } n \geq 0 \\ 0, & \text{Otherwise} \end{cases} \tag{3.3}$$

The python code, which generates unit ramp signal using two methods, is discussed in this experiment. In method 1, '*if*' and '*else*' conditions generate unit ramp signals, whereas in method 2, logical operation is used to generate unit ramp signals. The python code, which generates unit ramp signal using the two methods, is shown in Fig. 3.9, and the corresponding output is shown in Fig. 3.10.

Inference

From Fig. 3.10, it is possible to observe that the ramp signal generated using '*if*' and '*else*' condition and '*logical operation*' are alike. Unlike step signal, the ramp signal gradually increases from low to high value.

Task

1. Write a python code to generate unit ramp signal from unit step signal.

Experiment 3.6

From unit sample signal generates unit step signal, and from unit step signal generates unit ramp signal.

The relationship between unit sample ($\delta[n]$) and unit step ($u[n]$) sequence is given by

Fig. 3.9 Python code to
generate unit ramp signal

```
#Generation of unit ramp signal
import numpy as np
import matplotlib.pyplot as plt
#Step 1: Generating the sequence
n=np.arange(-10,11,1)  #Define the x-axis
#Two methods to generate unit ramp signal
x1=[i if i>=0 else 0 for i in n] #Unit ramp sequence
x2=n*(n>=0) #Logical operation
#Plotting the result
plt.subplot(2,1,1),plt.stem(n,x1),plt.xlabel('n-->'),
plt.ylabel('Amplitude'),plt.title('r[n]'),plt.xticks(n)
plt.subplot(2,1,2),plt.stem(n,x2),plt.xlabel('n-->'),
plt.ylabel('Amplitude'),plt.title('r[n]'),plt.xticks(n)
plt.tight_layout()
```

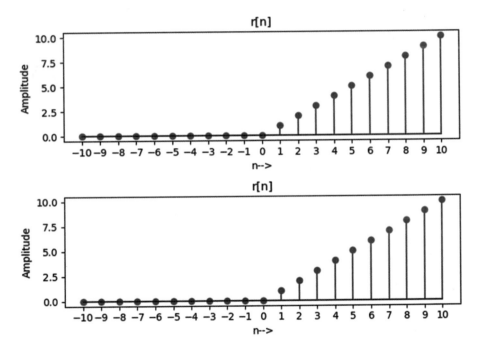

Fig. 3.10 Result of python code shown in Fig. 3.9

$$u[n] = \sum_{k=-\infty}^{n} \delta[k] \qquad (3.4)$$

and

Fig. 3.11 Flow chart depicting the problem statement of Experiment 3.6

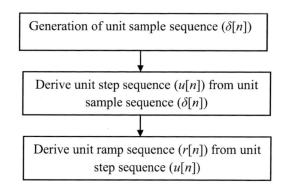

$$\delta[n] = u[n] - u[n-1] \tag{3.5}$$

The relationship between the unit ramp and unit step sequence is given by

$$r[n] = nu[n] \tag{3.6}$$

The flow chart, which depicts the objective of this experiment, is shown in Fig. 3.11.

From the flow chart, the unit sample sequence is generated first. From unit sample sequence, unit step sequence is obtained by repeated addition. From unit step sequence, unit ramp sequence is derived. The python code, which performs the above-mentioned task, is shown in Fig. 3.12, and the corresponding output is shown in Fig. 3.13.

Inferences

From the python code shown in Fig. 3.12, it is possible to infer that unit step sequence is obtained by repeatedly adding the unit sample sequence. The unit ramp sequence is obtained by weighting the unit step signal by a factor of 'n'. From this example, it is possible to infer that any arbitrary signal $x[n]$ can be obtained from the unit sample sequence by scaling and shifting operations.

Task

1. Write a python code to generate a unit sample signal from the unit step signal.

Experiment 3.7 Generation of Real Exponential Sequence

The expression for a real exponential signal is given by

$$x[n] = \alpha^n \tag{3.7}$$

where α must be a real value. The aim of this experiment is to generate real exponential sequence for four different values of 'α', namely, $\alpha = 0.5$, $\alpha = -0.5$, $\alpha = 1.0$ and $\alpha = -1.0$. The python code, which performs this task, is shown in Fig. 3.14, and the corresponding output is shown in Fig. 3.15.

```
#Generation of test sequences from unit sample sequence
import numpy as np
import matplotlib.pyplot as plt
#Step 1: Generation of unit sample sequence
n=np.arange(-10,11)
x=[1 if i==0 else 0 for i in n]  #delta[n]
#Step 2: Unit step sequence from unit sample sequence
y=np.zeros_like(n)
for k in range(len(x)):
    y[k]=np.sum(x[:k+1])
#Step 3: Unit ramp sequence from unit step sequence
z=n*y
#Step 4: Plotting the result
plt.subplot(3,1,1),plt.stem(n,x),plt.xlabel('n-->'),plt.ylabel('Amplitude'),
plt.title('$\delta[n]$'),plt.xticks(n),plt.yticks(x),
plt.subplot(3,1,2),plt.stem(n,y),plt.xlabel('n-->'),plt.ylabel('Amplitude'),
plt.title('u[n]'),plt.xticks(n),plt.yticks(y),
plt.subplot(3,1,3),plt.stem(n,z),plt.xlabel('n-->'),plt.ylabel('Amplitude'),
plt.title('r[n]'),plt.xticks(n),
plt.tight_layout()
```

Fig. 3.12   Python code to generate test signals from unit sample sequence

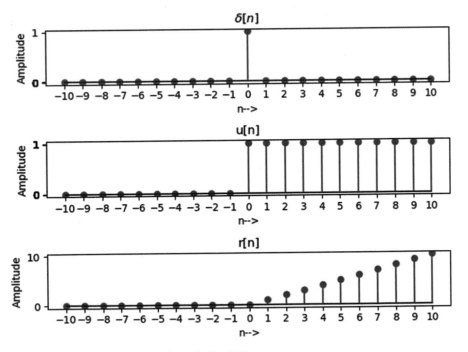

Fig. 3.13   Result of python code shown in Fig. 3.12

```
#Python code to generate real exponential sequences
import numpy as np
import matplotlib.pyplot as plt
n=np.arange(-5,6,1) #Define the x-axis
alpha=[0.5, -0.5, 1.0, -1.0]
for i in range(len(alpha)):
    x=alpha[i]**n #Real exponential sequence
    plt.subplot(2,2,i+1)
    plt.stem(n,x),plt.xlabel('n-->'),plt.ylabel('Amplitude')
    plt.title(r'$\alpha$={}'.format(alpha[i]))
    plt.xticks(n)
plt.tight_layout()
```

Fig. 3.14 Python code to generate real exponential signal

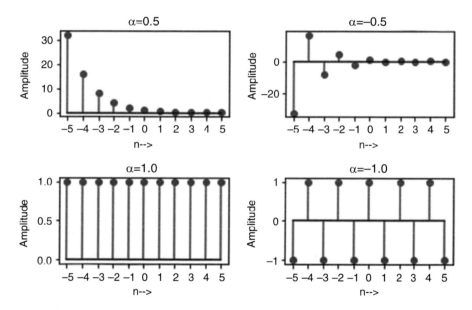

Fig. 3.15 Result of python code shown in Fig. 3.14

Inferences

The following inference can be made from this experiment:

1. If the value of α is $0 < \alpha < 1$, then the signal $x[n]$ decreases in magnitude. This is evident by observing the first subplot for $\alpha = 0.5$.
2. If the value of α is $-1 < \alpha < 0$, then the signal $x[n]$ alternates in sign but decreases in magnitude. This is evident by viewing the second subplot in Fig. 3.15 for $\alpha = -0.5$.
3. For $\alpha = 1.0$, there is no oscillation and the amplitude is always one.

```
#Python code to generate complex exponential sequences
import numpy as np
import matplotlib.pyplot as plt
n=np.arange(-5,6,1)  #Define the x-axis
omega_degree=[0, 90, 180, 270]
omega_radians=np.deg2rad(omega_degree)
for i in range(len(omega_radians)):
    x=np.exp(1j*omega_radians[i]*n) #Complex exponential sequences
    plt.subplot(2,2,i+1)
    plt.stem(n,x),plt.xlabel('n-->'),plt.ylabel('Amplitude')
    plt.title(r'$\omega={}^\circ$'.format(omega_degree[i]))
    plt.xticks(n)
plt.tight_layout()
```

Fig. 3.16 Python code to generate complex exponential sequences

4. For $\alpha = -1.0$, the signal $x[n]$ toggles. This is the highest frequency in digital sequence.

Task

1. Obtain the real exponential sequence for $\alpha = 2$ and comment on the nature of the signal. Here the term 'nature' refers to whether the signal is a bounded or not.

Experiment 3.8 Generation of Complex Exponential Signal

The general form of complex exponential signal is given by

$$x[n] = e^{j\omega n} \tag{3.8}$$

where 'ω' represents the angular frequency in radians. The python code to generate complex exponential sequences for four different values of 'ω' such as $\omega = \left[0, \frac{\pi}{2}, \pi, \frac{3\pi}{2}\right]$ is given in Fig. 3.16, and the corresponding output is shown in Fig. 3.17.

Inferences

The following inference can be drawn from this experiment:

1. When $\omega = 0$, the frequency is zero, the amplitude of the signal is constant and there is no variation in the signal. This is termed as DC signal. For a DC signal, the frequency is zero.
2. With increase in the value of 'ω', the oscillation exhibited by the signal increases. At $\omega = \pi$, the signal takes alternate values of +1 and −1. It is the highest frequency in the digital signal.

Task

1. Write a python code to prove the fact that digital frequency 'ω' is unique in the range 0 to 2π or from $-\pi$ to π.

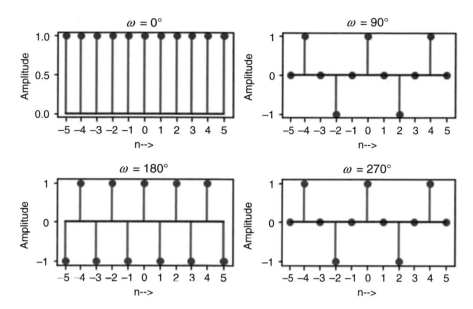

Fig. 3.17 Complex exponential sequences for different values of 'ω'

```
#Generation of signum function
import numpy as np
import matplotlib.pyplot as plt
n=np.arange(-5,6)
x=np.sign(n)
plt.stem(n,x)
plt.xlabel('n-->'),plt.ylabel('Amplitude'),plt.xticks(n)
plt.yticks(x),plt.title('Signum function')
```

Fig. 3.18 Python code to generate signum function

Experiment 3.9 Generation of Signum Function

Signum function is defined as a mathematical function that gives the sign of a real number. The signum function $f : R \rightarrow R$ is defined as

$$\text{sgn}[n] = \begin{cases} 1, \text{if } n > 0 \\ 0, \text{if } n = 0 \\ -1, \text{if } n < 0 \end{cases} \tag{3.9}$$

The python code to generate signum function is shown in Fig. 3.18, and the corresponding output is shown in Fig. 3.19.

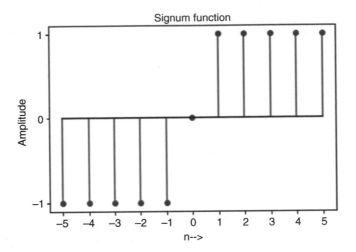

Fig. 3.19 Result of python code shown in Fig. 3.18

Inference

From Fig. 3.19, it is possible to observe that the signum function takes only three values, which are -1, 0 and 1; whenever $n < 0$, the signum function takes the value of -1. At $n = 0$, the signum function takes a value of '0'. For the positive values of 'n', the signum function takes the value of $+1$.

Task

1. Is it possible to obtain signum function from unit step function? If yes, write a python code to generate discrete signum signal from unit step signal.

3.2   Mathematical Operation on Discrete-Time Signals

This section discusses various mathematical operations that are performed on discrete-time signals. The basic mathematical operations that could be performed on the discrete-time signals are given in Fig. 3.20.

3.2.1   Amplitude Modification on DT Signal

The different signal operations that come under amplitude modification are discussed in this section.

(a) **Amplitude scaling**

   If $x[n]$ is the input signal, the scaling of the signal $x[n]$ by a factor of 'A' is represented as

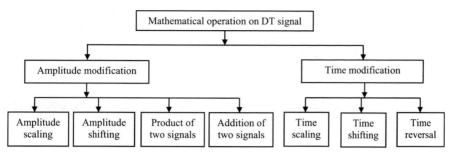

Fig. 3.20 Different mathematical operations on DT signal

```
import numpy as np
import matplotlib.pyplot as plt
#Step 1: Generating the input signal
n=np.arange(-10,11,1)
x=(n==0)
#Obtaining the output signals for different values of 'A'
A=[2,0.5,1] #Three different values of factor 'A'
y1=A[0]*x
y2=A[1]*x
y3=A[2]*x
#Step 2: Plotting the result
plt.subplot(2,2,1),plt.stem(n,x),plt.yticks([0,2]),plt.xlabel('n-->'),
plt.ylabel('Amplitude'),plt.title('x[n]'),plt.subplot(2,2,2),plt.stem(n,y1)
plt.yticks([0,2]),plt.xlabel('n-->'),plt.ylabel('Amplitude'),
plt.title('$y_1[n]$'),plt.subplot(2,2,3),plt.stem(n,y2)
plt.yticks([0,2]),plt.xlabel('n-->'),plt.ylabel('Amplitude'),
plt.title('$y_2[n]$'),plt.subplot(2,2,4),plt.stem(n,y3),plt.yticks([0,2]),
plt.xlabel('n-->'),plt.ylabel('Amplitude'),plt.title('$y_3[n]$')
plt.tight_layout()
```

Fig. 3.21 Python code to perform amplitude scaling

$$y[n] = Ax[n] \tag{3.10}$$

If $A > 1$, the operation is called as amplification, $A < 1$ represents attenuation. If $A = 1$, the output follows the input, it is called as input follower or buffer.

Experiment 3.10 Amplitude Scaling

Generate unit sample signal and perform the amplitude scaling for three different values of A, namely: $A = 2$, $A = 0.5$ and $A = 1$. Plot the input and output signal and comment on the observed output.

The python code, which performs the above-mentioned task, is shown in Fig. 3.21, and the corresponding output is shown in Fig. 3.22.

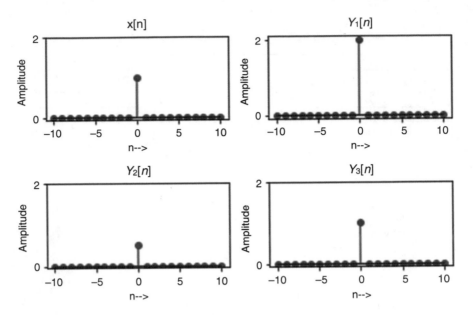

Fig. 3.22 Result of python code shown in Fig. 3.21

Inference

The following inferences can be made from this experiment:

1. From Fig. 3.22, it is possible to observe that $y_1[n]$ is the amplified version of $x[n]$, and $y_2[n]$ is the attenuated version of $x[n]$. If the gain is unity, the output follows the input, which is evident from the output $y_3[n]$.
2. This example illustrates the scaling of the amplitude axis for different values of the factor 'A'.

Task

1. Write a python code to illustrate the fact that amplitude scaling changes the energy of the signal.

(b) Amplitude Shifting

If $x[n]$ is the input signal, the amplitude shifting of the signal $x[n]$ by a factor of 'C' is represented as

$$y[n] = x[n] \pm C \tag{3.11}$$

Experiment 3.11 Amplitude Shifting (DC Offset)

Let $x[n]$ represent the discrete-time sinusoidal signal, and perform the DC offset of this signal $x[n]$ to obtain the signals $y_1[n] = x[n] + C$ and $y_2[n] = x[n] - C$. The value

```
#DC offset
import numpy as np
import matplotlib.pyplot as plt
#Step 1: Generation of input sinusoidal sequence
t=np.linspace(0,1,100)
x=np.sin(2*np.pi*5*t)
#Step 2: Perform DC offset
offset=[5,-5]
y1=x+offset[0]
y2=x+offset[1]
#Step 3: Ploting the input and output signals
plt.subplot(3,1,1),plt.stem(t,x),
plt.xlabel('n-->'),plt.ylabel('Amplitude'),plt.title('x[n]')
plt.subplot(3,1,2),plt.stem(t,y1)
plt.xlabel('n-->'),plt.ylabel('Amplitude'),plt.title('$y_1$[n]')
plt.subplot(3,1,3),plt.stem(t,y2)
plt.xlabel('n-->'),plt.ylabel('Amplitude'),plt.title('$y_2$[n]')
plt.tight_layout()
```

Fig. 3.23 Python code which performs DC offset

of 'C' for this experiment is to be chosen as 5. Write a python code to perform this task and comment on the observed output.

The python code, which performs the above-mentioned task, is shown in Fig. 3.23, and the corresponding output is shown in Fig. 3.24.

Inference

By observing Fig. 3.24, it is possible to infer that the reference for signal $y_1[n]$ is +5 V, whereas the reference for signal $y_2[n]$ is −5 V. This is termed as DC offset.

Task

1. Does amplitude shifting affect the energy of the signal? Write a python code to answer this question.

(c) **Product of Two Signals**

The product of two signals $x_1[n]$ and $x_2[n]$ is represented by

$$y[n] = x_1[n] \times x_2[n] \tag{3.12}$$

The amplitude of the resultant signal $y[n]$ gets modified. For example, consider

$$x_1[n] = \sin(2\pi f_1 n) \tag{3.13}$$

$$x_2[n] = \cos(2\pi f_2 n) \tag{3.14}$$

Substituting Eqs. (3.13) and (3.14) in Eq. (3.12), we get

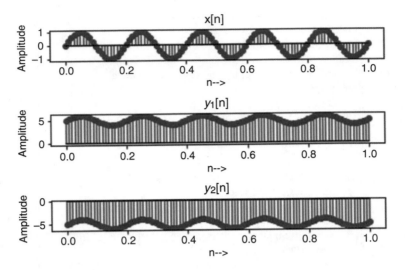

Fig. 3.24 Result of python code shown in Fig. 3.23

$$y[n] = \sin(2\pi f_1 n) \times \cos(2\pi f_2 n) \tag{3.15}$$

Using the formula

$$\sin A \cos B = \frac{1}{2}\{\sin(A + B) + \sin(A - B)\} \tag{3.16}$$

Equation (3.15) can be written as

$$y[n] = \frac{1}{2}\{\sin 2\pi(f_1 + f_2)n + \sin 2\pi(f_1 - f_2)n)\} \tag{3.17}$$

The amplitude of the output signal is different from the input signal $x[n]$.

Experiment 3.12 Product of Two Signals

Obtain the product of the two signals given by $x_1[n] = \sin(2\pi f_1 n)$ and $x_2[n] = \sin(2\pi f_2 n)$. In this example, consider $f_1 = f_2 = 5\,\text{Hz}$. Using the relation (3.17), the expression for the output signal is given by $y[n] = \frac{1}{2}\{\sin 2\pi(f_1 + f_2)n + \sin 2\pi(f_1 - f_2)n)\}$. In this case, $f_1 = f_2 = 5\,\text{Hz}$; hence, the expression for the output signal is given by $y[n] = \frac{1}{2}\{\sin 2\pi(10)n)\}$. The frequency of the resultant signal should be 10 Hz, whereas its amplitude is reduced by half. The python code, which performs this task, is shown in Fig. 3.25, and the corresponding output is shown in Fig. 3.26.

Inference

The following inferences can be drawn from this experiment:

```
#Product of two signals
import numpy as np
import matplotlib.pyplot as plt
#Step 1: Generation of input signals
t=np.arange(0,100,1)
Fs=100
x=np.sin(2*np.pi*(5/Fs)*t)
y=np.cos(2*np.pi*(5/Fs)*t)
#Step 2: Product of the two signals
z=np.multiply(x,y)
#Step 3: Plotting the result
plt.subplot(3,1,1),plt.stem(t,x),plt.xlabel('n-->'),plt.ylabel('Ampltitude'),plt.title('$x_1[n]$')
plt.subplot(3,1,2),plt.stem(t,y),plt.xlabel('n-->'),plt.ylabel('Ampltitude'),plt.title('$x_2[n]$'),
plt.subplot(3,1,3),plt.stem(t,z),plt.yticks([-1,1]),plt.xlabel('n-->'),plt.ylabel('Ampltitude'),
plt.title('$y[n]$')
plt.tight_layout()
```

Fig. 3.25 Python code to obtain the product of the two signals

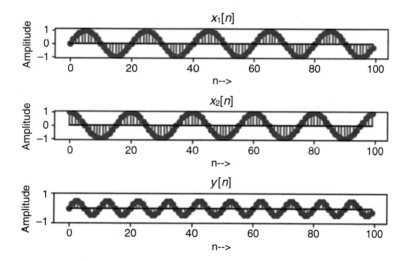

Fig. 3.26 Result of python code shown in Fig. 3.25

1. From Fig. 3.25, two signals of the same frequency are generated and multiplied.
2. From Fig. 3.26, it is possible to observe that $x_1[n]$ is a sine wave and $x_2[n]$ is a cosine wave. The resultant signal $y[n]$ is a sinusoidal signal with a frequency of 10 Hz, whereas the amplitude of the output waveform is reduced by a factor of half.

(d) Signal Addition

```
#Addition of two signals
import numpy as np
import matplotlib.pyplot as plt
#Step 1: Generation of input signals
n=np.arange(-10,11,1)
x=(n>=0)
y=(n>=0)
#Step 2: Addition of the two signals
z=np.add(x.astype('float32'),y.astype('float32'))
#Step 3: Plotting the result
plt.subplot(3,1,1),plt.stem(n,x),plt.xticks(n)
plt.yticks([0,2]),plt.xlabel('n-->'),plt.ylabel('Amplitude'),
plt.title('$x_1[n]$'),plt.subplot(3,1,2),plt.stem(n,y),plt.xticks(n)
plt.yticks([0,2]),plt.xlabel('n-->'),plt.ylabel('Amplitude'),
plt.title('$x_2[n]$'),plt.subplot(3,1,3),plt.stem(n,z),plt.xticks(n)
plt.yticks([0,2]),plt.xlabel('n-->'),plt.ylabel('Amplitude'),plt.title('$y[n]$')
plt.tight_layout()
```

Fig. 3.27 Python code to perform addition of two signals

The signal addition results in a change in the amplitude of the signal. Two signals $x_1[n]$ and $x_2[n]$ are added together to obtain the resultant output signal $y[n]$, which is given by

$$y[n] = x_1[n] + x_2[n] \tag{3.18}$$

Experiment 3.13 Signal Addition
In this example, let $x_1[n] = u[n]$ and $x_2[n] = u[n]$. The signal $y[n]$ is the addition of two unit step signals. The python code which performs this task is shown in Fig. 3.27, and the corresponding output is shown in Fig. 3.28.

Inferences
The following inferences are drawn from these Figs. 3.27 and 3.28:

1. By observing Fig. 3.27, it is possible to observe that the result of logical operation is converted to float using the command '.astype('float32')'.
2. By observing Fig. 3.28, the inputs $x_1[n]$ and $x_2[n]$ are unit step signal, whose amplitude takes value from 0 to 1, whereas the amplitude of the output signal $y[n]$ has variation from 0 to 2.
3. This experiment illustrates the fact that the amplitude of the signal can be changed by signal addition operation.

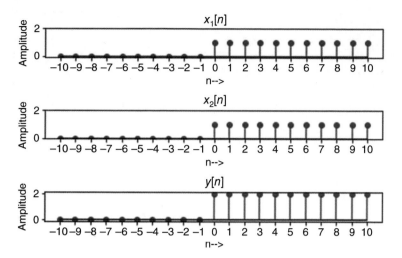

Fig. 3.28 Result of python code shown in Fig. 3.27

Task

1. Write a python code to illustrate the fact that '*signal addition is a commutative operation*'.

3.2.1.1   Time Scaling Operation

Time scaling operations can be classified into two types, namely, (1) downsampling and (2) upsampling.

(a) **Downsampling**

The downsampling of the signal $x[n]$ by a factor of 'M' is represented as

$$y[n] = x[Mn] \tag{3.19}$$

where 'M' is an integer. Here '$M - 1$' samples will be discarded between two consecutive samples. Downsampling by a factor of '2' is represented as

$$y[n] = x[2n] \tag{3.20}$$

Experiment 3.14 Downsampling

This experiment discusses the downsampling operation on the input signal. The python code to perform downsampling by a factor of '2' is shown in Fig. 3.29, and the corresponding output is shown in Fig. 3.30.

```
#Downsampling by a factor of M
import numpy as np
import matplotlib.pyplot as plt
#Step 1: Generating the input signal
n=np.arange(-10,11,1)
x=n
M=2  #Downsampling factor
m=np.arange(n[0]/2,(n[-1]/2)+1,1)
#Step 2: Performing downsampling operation
y=x[::M]
#Step 3: Plotting the input and downsampled signal
plt.subplot(2,1,1),plt.stem(n,x),plt.xlabel('n-->'),
plt.ylabel('Amplitude'),plt.title('x[n]'),plt.xticks(n)
plt.subplot(2,1,2),plt.stem(m,y),plt.xlabel('n-->'),
plt.ylabel('Amplitude'),plt.title('y[n]'),plt.xticks(n)
plt.tight_layout()
```

Fig. 3.29 Python code to perform downsampling operation

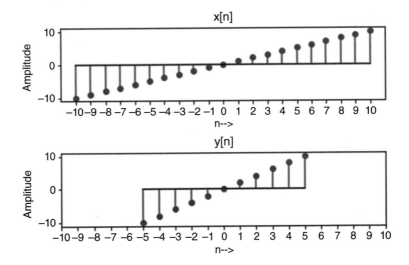

Fig. 3.30 Result of downsampling operation

Inferences

The following inferences can be drawn from this experiment:

1. By observing Fig. 3.30, the number of samples in the input signal $x[n]$ is 21, whereas the number of samples in the output signal $y[n]$ is 11.
2. Downsampling leads to a reduction in the number of samples.

Task

1. Write a python code to prove the fact that downsampling is an irreversible operation. That is, it is not possible to obtain the original signal from the downsampled signal because downsampling results in loss of signal samples.

(b) **Upsampling**

The upsampling of the signal $x[n]$ by a factor of 'L' is represented by

$$y[n] = x\left[\frac{n}{L}\right] \tag{3.21}$$

The upsampling operation is basically inserting '$L - 1$' zeros between two consecutive samples. For $L = 2$, the above expression can be written as

$$y[n] = x\left[\frac{n}{2}\right]$$

Experiment 3.15 Upsampling
This experiment deals with the upsampling process of discrete-time signal. The python code, which performs the upsampling operation by a factor of 2, is shown in Fig. 3.31, and the corresponding output is shown in Fig. 3.32.

Inference
The following observations can be made from this experiment:

By observing Fig. 3.32, it is possible to observe that in the case of upsampling by a factor of 2, one zero is inserted between successive samples. Generally, when upsampling by a factor of 'L', '$L - 1$' zeros will be inserted between successive samples. Also, it shows that the number of samples in the output increases to almost L times than the number of samples in the input signal.

Task

1. Write a python code to illustrate the fact that 'Upsampling is a reversible operation'. It is possible to obtain the original signal from the upsampled signal.

3.2.1.2 Time Shifting Operation

The time shifting operation can be broadly classified into two types: (1) delay operation and (2) advance operation.

(a) **Delay operation**

The delaying of the input signal by a factor of 'k' units is expressed as

```
#Upsampling by a factor of 2
import numpy as np
import matplotlib.pyplot as plt
#Step 1: Generating the input signal
L=2   #Upsampling factor
n=np.arange(-5,6,1)
N=len(n)
m=np.arange(-N+1,N+1,1)
x=np.ones(N)
#Step 2: Upsampling the input signal
y=np.zeros(L*N)
y[::2]=x
#Step 3: Plotting the input and output signal
plt.subplot(2,1,1),plt.stem(n,x),plt.xlabel('n-->'),plt.ylabel('Amplitude'),plt.title('x[n]'),
plt.xticks(m),plt.subplot(2,1,2),plt.stem(m,y),plt.xlabel('n-->'),
plt.ylabel('Amplitude'),plt.title('y[n]'),plt.xticks(m)
plt.tight_layout()
```

Fig. 3.31 Python code performs upsampling by a factor of 2

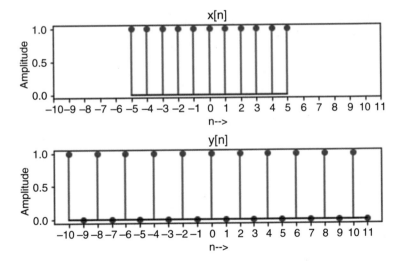

Fig. 3.32 Result of upsampling by a factor of 2

$$y[n] = x[n - k] \tag{3.22}$$

where 'k' must be a positive integer.

(b) **Advance operation**

The advance of the input signal $x[n]$ by a factor of 'k' units is expressed as

Fig. 3.33 Delay and advance of unit step sequence

```
#Delay and advance operation
import numpy as np
import matplotlib.pyplot as plt
n=np.arange(-10,11)
x1=(n>=0) #u[n]
k=5
x2=(n>=k) #Delay of u[n]
x3=(n>=-k)#Advance of u[n]
plt.subplot(3,1,1),plt.stem(n,x1),plt.xticks(n)
plt.xlabel('n-->'),plt.ylabel('Amplitude'),plt.title('u[n]')
plt.subplot(3,1,2),plt.stem(n,x2),plt.xticks(n)
plt.xlabel('n-->'),plt.ylabel('Amplitude'),plt.title('u[n-5]')
plt.subplot(3,1,3),plt.stem(n,x3),plt.xticks(n)
plt.xlabel('n-->'),plt.ylabel('Amplitude'),plt.title('u[n+5]')
plt.tight_layout()
```

$$y[n] = x[n + k] \tag{3.23}$$

where 'k' must be a positive integer.

Experiment 3.16 Time Shifting Operation

This experiment performs both delay and advance operations by a factor of 'k' units on the unit step signal. First, the unit step signal is generated; then, it is delayed by a factor of 5 units. The unit step signal is advanced by the factor of 5 units. The python code, which performs this task, is shown in Fig. 3.33, and the corresponding output is shown in Fig. 3.34.

Inference

This experiment illustrates the concept of shifting operation on the signal. Delay of the signal $u[n]$ by a factor of '5' units results in $u[n - 5]$, whereas advance of the signal $u[n]$ by a factor of 5 units results in $u[n + 5]$. It is to be observed that shifting operation on the signal will not alter the energy of the signal.

Task

1. Write a python code to illustrate the fact that the signal energy is unaltered due to signal shifting.

3.2.1.3   Time Reversal Operation

The time reversal of the signal $x[n]$ is denoted as $x[-n]$. This refers to flipping the signal $x[n]$ from left to right and right to left. It can be considered as a signal reflection about the origin. A discrete-time signal can be reversed in time by

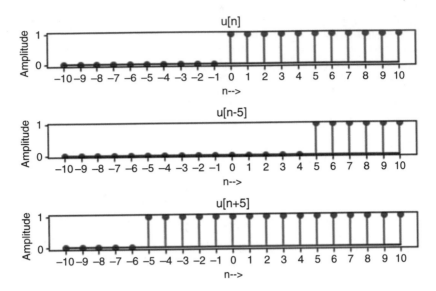

Fig. 3.34 Delay and advance of unit step sequence

changing the sign of the independent variable for all instances. Two different ways to perform time reversal operation in python are given below.

Experiment 3.17 Time Reversal Without Built-In Function
This experiment deals with the time reversal operation using python without built-in function. In this method, the signal $x[n]$ is flipped from left to right using the command" $x[::-1]$", the python code which performs the task of time reversal is shown in Fig. 3.35, and the corresponding output is shown in Fig. 3.36.

Inference
Figure 3.36 clearly indicates that the left side of the input signal is moved into the right side of the output signal and the right side of the input signal is moved into the left side of the output signal.

Experiment 3.18 Time Reversal Using Built-In Function
This experiment tries to obtain the time reversal using a python built-in function. In this method, the built-in function '$np.fliplr()$' is used to perform a time reversal operation. The python code, which performs this task, is shown in Fig. 3.37, and the corresponding output is shown in Fig. 3.38.

Inference
This experiment confirms that the time reversal can be done using '$np.fliplr$' built-in function.

Task
1. Write a python code to illustrate that flipping operation does not alter the signal's energy.

```
#Time reversal operation
import numpy as np
import matplotlib.pyplot as plt
n=np.arange(-10,11,1)
x=(n)
y=x[::-1]
plt.subplot(2,1,1),plt.stem(n,x)
plt.xticks(n),plt.xlabel('n-->'),plt.ylabel('Amplitude'),plt.title('x[n]')
plt.subplot(2,1,2),plt.stem(n,y)
plt.xticks(n),plt.xlabel('n-->'),plt.ylabel('Amplitude'),plt.title('y[n]')
plt.tight_layout()
```

Fig. 3.35 Method-1 to perform time reversal operation

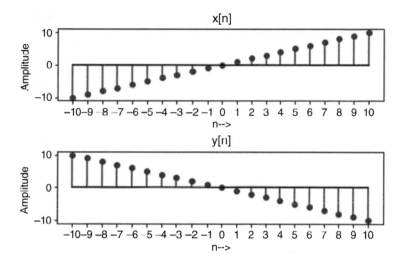

Fig. 3.36 Result of python code shown in Fig. 3.35

3.3 Convolution

Convolution is an important operation in digital signal processing, because many DSP algorithms use convolution operations in one form or other. The most common application of convolution operation is filtering. It can be used for signal enhancement. The relationship between the input and output of a linear time-invariant system shown in Fig. 3.39.

The relationship between the input and output of the system is given by

```
#Time reversal operation
import numpy as np
import matplotlib.pyplot as plt
n=np.arange(-10,11,1)
x=(n)
y=np.fliplr([x])[0]
plt.subplot(2,1,1),plt.stem(n,x)
plt.xticks(n),plt.xlabel('n-->'),plt.ylabel('Amplitude'),plt.title('x[n]')
plt.subplot(2,1,2),plt.stem(n,y)
plt.xticks(n),plt.xlabel('n-->'),plt.ylabel('Amplitude'),plt.title('y[n]')
plt.tight_layout()
```

Fig. 3.37 Method-2 to perform time reversal operation

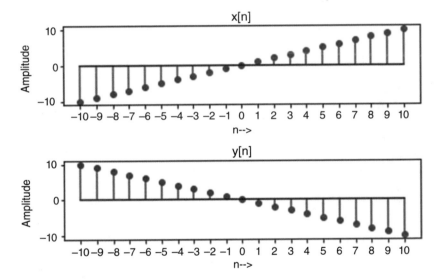

Fig. 3.38 Result of python code shown in Fig. 3.37

$$y[n] = x[n] * h[n] \tag{3.24}$$

In the above expression, '*' denotes the convolution operation. The above expression can be written as

$$y[n] = \sum_{k=-\infty}^{\infty} x[k]h[n-k] \tag{3.25}$$

Convolution obeys commutative property; hence, the above equation can be expressed as

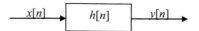

Fig. 3.39 Representation of the LTI system

$$y[n] = \sum_{k=-\infty}^{\infty} h[k]x[n-k] \qquad (3.26)$$

Experiment 3.19 Convolution of Given Signal with Unit Sample Signal
This experiment illustrates the fact that the convolution of any signal ($x[n]$) with unit sample signal ($\delta[n]$) will result in the same signal $x[n]$. This is expressed as

$$x[n] * \delta[n] = x[n] \qquad (3.27)$$

The python code, which illustrates the above concept, is shown in Fig. 3.40, and the corresponding output is shown in Fig. 3.41.

Inferences
The following inferences can be drawn from this experiment:

1. From Fig. 3.41, the input signal ($x[n]$) generated is a triangular signal.
2. The impulse response ($h[n]$) is unit sample signal ($\delta[n]$).
3. The signal $x[n]$ is convolved with unit sample signal to obtain the output signal y [n]. It can be observed that the output signal $y[n]$ resembles the input signal $x[n]$.

Experiment 3.20 Convolution of the Signal $x[n]$ with Shifted Unit Sample Signal
This experiment illustrates the fact that the signal $x[n]$ can be shifted by convolving it with $\delta[n \pm k]$. Convolving the signal $x[n]$ with $\delta[n-k]$ results in delaying the signal $x[n]$ by a factor of 'k'. Convolving the signal $x[n]$ with $\delta[n+k]$ results in advancing the signal $x[n]$ by a factor of 'k'. This is expressed as

$$x[n] * \delta[n \pm k] = x[n \pm k] \qquad (3.28)$$

The python code, which performs this task, is shown in Fig. 3.42, and the corresponding output is shown in Fig. 3.43.

Inferences
The task performed by the python program is summarized in Fig. 3.44.

1. The input signal $x[n]$ is applied to two systems with impulse responses $h_1[n] = \delta[n-k]$ and $h_2[n] = \delta[n+k]$ to obtain the output signals $y_1[n]$ and $y_2[n]$ respectively.
2. By comparing the input signal $x[n]$ with the output signal $y_1[n]$, it is possible to observe that the output signal $y_1[n]$ is a shifted version (delayed version) of the input signal $x[n]$.

```
#Convolution with unit sample sequence
import numpy as np
import matplotlib.pyplot as plt
n=np.arange(-5,6)
N=len(n)
n1=np.arange(-N+1,N)
#Step 1: Generation of triangular signal
x=5-np.abs(n)
#Step 2: Generation of unit sample signal
h=(n==0)
#Step 3: Perform the convolution
y=np.convolve(x,h,mode='full')
#Step 4: Displaying the result
plt.subplot(3,1,1),plt.stem(n,x),plt.xticks(n),plt.xlabel('n-->'),
plt.ylabel('Amplitude'),plt.title('x[n]'),plt.subplot(3,1,2),plt.stem(n,h),
plt.xticks(n),plt.xlabel('n-->'),plt.ylabel('Amplitude'),plt.title('h[n]')
plt.subplot(3,1,3),plt.stem(n1,y),plt.xticks(n1),plt.xlabel('n-->'),
plt.ylabel('Amplitude'),plt.title('y[n]')
plt.tight_layout()
```

Fig. 3.40 Convolution of the signal $x[n]$ with unit sample signal $\delta[n]$

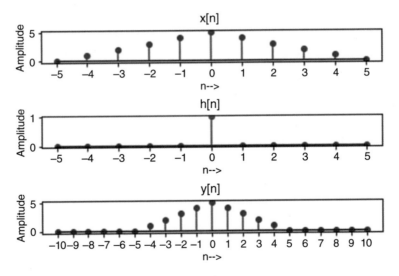

Fig. 3.41 Result of python code shown in Fig. 3.40

3. By comparing the input signal $x[n]$ with the output signal $y_2[n]$, it is possible to observe that the output signal $y_2[n]$ is a shifted version (advanced version) of the input signal $x[n]$.

4. This experiment illustrates the fact that signal shifting can be accomplished using convolution operation.

```
#Convolution with shifted unit sample sequence
import numpy as np
import matplotlib.pyplot as plt
n=np.arange(-5,6)
N=len(n)
n1=np.arange(-N+1,N)
#Step 1: Generation of triangular signal
x=5-np.abs(n)
#Step 2: Generation of shifted unit sample signals
k=5
h1=(n==k)
h2=(n==-k)
#Step 3: Perform the convolution
y1=np.convolve(x,h1,mode='full')
y2=np.convolve(x,h2,mode='full')
#Step 4: Displaying the result
plt.subplot(3,2,1),plt.stem(n,x),plt.xticks(n)
plt.xlabel('n-->'),plt.ylabel('Amplitude'),plt.title('x[n]')
plt.subplot(3,2,2),plt.stem(n,x),plt.xticks(n)
plt.xlabel('n-->'),plt.ylabel('Amplitude'),plt.title('x[n]')
plt.subplot(3,2,3),plt.stem(n,h1),plt.xticks(n)
plt.xlabel('n-->'),plt.ylabel('Amplitude'),plt.title('$h_1[n]$')
plt.subplot(3,2,4),plt.stem(n,h2),plt.xticks(n)
plt.xlabel('n-->'),plt.ylabel('Amplitude'),plt.title('$h_2[n]$')
plt.subplot(3,2,5),plt.stem(n1,y1),plt.xlabel('n-->'),plt.ylabel('Amplitude'),
plt.title('$y_1[n]$'),plt.subplot(3,2,6),plt.stem(n1,y2),plt.xlabel('n-->'),
plt.ylabel('Amplitude'),plt.title('$y_2[n]$')
plt.tight_layout()
```

Fig. 3.42 Python code to perform convolution of signal $x[n]$ with shifted unit sample signal

Task

1. Repeat the above experiment with a rectangular pulse signal instead of a triangular one.

Experiment 3.21 Commutative Property of Convolution

The motive of this experiment is to prove the commutative property of convolution. The commutative property of convolution is expressed as

$$x[n] * h[n] = h[n] * x[n] \tag{3.29}$$

The python code to illustrate the commutative property of convolution is given in Fig. 3.45, and the corresponding output is shown in Fig. 3.46.

Inferences

The following inferences can be drawn from Fig. 3.46:

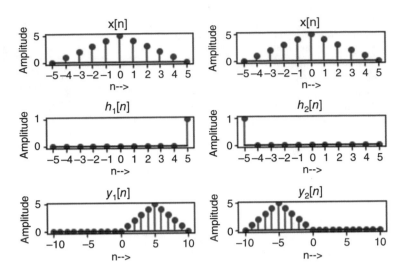

Fig. 3.43 Result of python code shown in Fig. 3.42

Fig. 3.44 Task performed by the python example

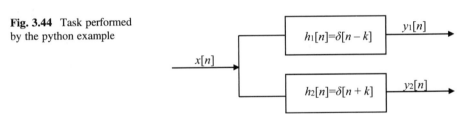

1. The input signal $x[n]$ is a pulse signal. Similarly, the signal $h[n]$ is a pulse signal. The signals $x[n]$ and $h[n]$ are the same.
2. The signal $y_1[n]$ is obtained by convolving $x[n]$ with $h[n]$, whereas the signal $y_2[n]$ is obtained by convolving $h[n]$ with $x[n]$. From Fig. 3.46, the signals $y_1[n]$ and $y_2[n]$ are the same.
3. This experiment illustrates that *convolution is commutative*. Also, the convolution of two pulse signals results in a triangular signal.

Task
1. In the above experiment, let L_1 and L_2 be the length of the signals $x[n]$ and $h[n]$. Then, the length of the convolved signal is $L_1 + L_2 - 1$. Write a python code to illustrate that linear convolution results in stretching the length of the signal.

Experiment 3.22 Associative Property of Convolution
The associative property of convolution is expressed as

$$(x[n] * h_1[n]) * h_2[n] = x[n] * (h_1[n] * h_2[n]) \tag{3.30}$$

To illustrate this property, the input signal $x[n]$ chosen is $x[n] = e^{j\pi n}$, which toggles between $+1$ and -1. The impulse response $h_1[n] = \delta[n - k]$ and the impulse

```
#Commutative property of convolution
import numpy as np
import matplotlib.pyplot as plt
n=np.arange(-5,6)
N=len(n)
n1=np.arange(-N+1,N)
#Step 1: Generating x[n]
x=np.array([0,0,0,0,1,1,1,0,0,0,0])
#Step 2: Generating h[n]
h=x
#Step 3: Obtaining the outputs
y1=np.convolve(x,h,mode='full')
y2=np.convolve(h,x,mode='full')
#Sep 4: Plotting the results
plt.subplot(2,2,1),plt.stem(n,x),plt.xticks(n)
plt.xlabel('n-->'),plt.ylabel('Amplitude'),plt.title('x[n]')
plt.subplot(2,2,2),plt.stem(n,h),plt.xlabel('n-->'),plt.ylabel('Amplitude'),
plt.title('h[n]'),plt.subplot(2,2,3),plt.stem(n1,y1)
plt.xlabel('n-->'),plt.ylabel('Amplitude'),plt.title('$y_1[n]$= x[n]*h[n]')
plt.subplot(2,2,4),plt.stem(n1,y2)
plt.xlabel('n-->'),plt.ylabel('Amplitude'),plt.title('$y_2[n]$=h[n]*x[n]')
plt.tight_layout()
```

Fig. 3.45 Python code to illustrate the commutative property of convolution

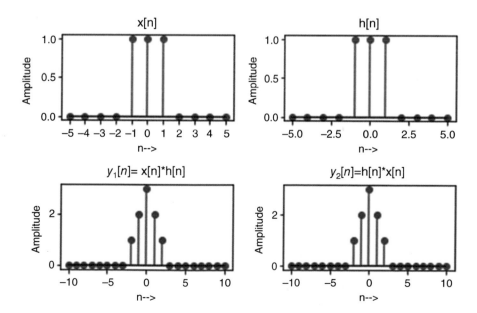

Fig. 3.46 Result of python code shown in Fig. 3.45

```
#Associative property of convolution
import numpy as np
import matplotlib.pyplot as plt
n=np.arange(-5,6)
#Step 1: Generation of triangular signal
x=np.exp(1j*np.pi*n)
#Step 2: Generation of shifted unit sample signals
k=5
h1=(n==k)
h2=(n==-k)
#Step 3: Perform the convolution
u1=np.convolve(x,h1,mode='full')
y1=np.convolve(u1,h2,mode='full')
v1=np.convolve(h1,h2,mode='full')
y2=np.convolve(x,v1,mode='full')
N=len(y1)
n1=np.arange(-N/2,N/2)
#Step 4: Displaying the result
plt.figure(1),plt.subplot(3,1,1),plt.stem(n,x),plt.xticks(n),plt.xlabel('n'),
plt.ylabel('Amplitude'),plt.title('x[n]'),plt.subplot(3,1,2),plt.stem(n,h1),plt.xticks(n)
plt.xlabel('n-->'),plt.ylabel('Amplitude'),plt.title('$h_1$[n]'),plt.subplot(3,1,3),
plt.stem(n,h2),plt.xticks(n),plt.xlabel('n-->'),plt.ylabel('Amplitude'),plt.title('$h_2$[n]')
plt.tight_layout()
plt.figure(2),plt.subplot(2,1,1),plt.stem(n1,y1),
plt.title('$y_1[n]$=(x[n]*$h_1$[n])*$h_2$[n]'), plt.xlabel('n-->'),
plt.ylabel('Amplitude'),plt.subplot(2,1,2),plt.stem(n1,y1)
plt.title('$y_2[n]$=x[n]*($h_1$[n])*$h_2$[n])'),plt.xlabel('n-->'),plt.ylabel('Amplitude')
plt.tight_layout()
```

Fig. 3.47 Python code to illustrate associative property of convolution

response $h_2[n] = \delta[n + k]$. The python code, which illustrates the associative property of the convolution operation, is given in Fig. 3.47, and the corresponding outputs are shown in Figs. 3.48 and 3.49, respectively.

Inferences

The following are the inferences from this experiment:

1. The input signal $x[n] = (-1)^n$, $-5 \leq n \leq 5$. The impulse response $h_1[n] = \delta[n - 5]$ and $h_2[n] = \delta[n + 5]$, which is shown in Fig. 3.48.
2. The output $y_1[n] = (x[n]*h_1[n])*h_2[n]$, whereas the output $y_2[n] = x[n]*(h_1[n] *h_2[n])$. From Fig. 3.49, it is possible to observe that the output $y_1[n] = y_2[n]$, which shows that associative property of convolution is verified.

Experiment 3.23 Distributive Property of Convolution

The distributive property of convolution is expressed as

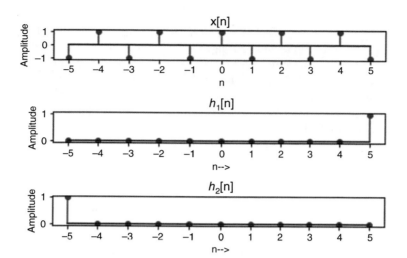

Fig. 3.48 Input signal and impulse response

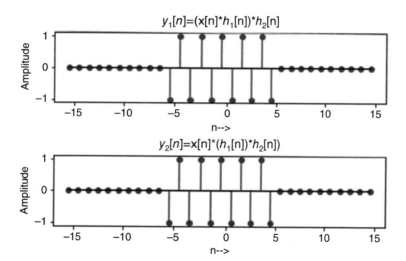

Fig. 3.49 Output signal

$$x[n] * (h_1[n] + h_2[n]) = x[n] * h_1[n] + x[n] * h_2[n] \qquad (3.31)$$

For illustration, the signal $x[n]$ is chosen as $x[n] = \delta[n + 1] + 2\delta[n] + \delta[n - 1]$, $h_1[n] = \delta[n + 1] + \delta[n] + \delta[n - 1]$ and $h_2[n] = -\delta[n + 1] - \delta[n - 1]$ such that $h_1[n] + h_2[n]$ results in unit sample signal. The python code, which illustrates the distributive of convolution, is shown in Fig. 3.50, and the corresponding outputs are shown in Figs. 3.51 and 3.52, respectively.

Inferences

1. From Fig. 3.51, it is possible to observe that the input signal and the impulse responses are all finite-duration signals. The input signal is expressed as $x[n] = \delta[n + 1] + 2\delta[n] + \delta[n - 1]$. The impulse responses are given by $h_1[n] = \delta[n + 1] + \delta[n] + \delta[n - 1]$ and $h_2[n] = -\delta[n + 1] - \delta[n - 1]$.

2. The sum of the impulse responses results in a unit sample signal, which is expressed as $h_1[n] + h_2[n] = \delta[n]$. Also, convolution of any input signal $x[n]$ with unit sample signal results in the same signal, which is expressed as $x[n] * \delta[n] = x[n]$. Because of this property, the output signal $y_1[n]$ is same as the input signal $x[n]$.

3. By observing the output signals $y_1[n]$ and $y_2[n]$, it is possible to infer $y_1[n] = y_2[n]$, which implies that the distributive property of convolution is illustrated through this experiment.

Experiment 3.24 Convolution of a Square Wave with Lowpass Filter Coefficient

In this experiment, a square wave of fundamental frequency 5 Hz is generated. It is then passed through moving average filter with $M = 5, 7, 9$, and 11. The block diagram of the experiment performed is shown in Fig. 3.53.

The impulse response of lowpass filter (moving average filter) is given by

$$h[n] = \frac{1}{M} \sum_{k=0}^{M-1} \delta[n - k] \tag{3.32}$$

In this experiment, the value of 'M' is chosen as 5, 7, 9 and 11.

The expression for the output signal is given by

$$y[n] = x[n] * h[n]$$

The python code which accomplishes this task is shown in Fig. 3.54, and the corresponding output is shown in Figs. 3.55 and 3.56.

Inferences

The following inferences can be drawn from Figs. 3.55 and 3.56:

1. The input to the system is a square wave of a fundamental frequency 5 Hz.
2. The system is passed through lowpass filter to obtain a triangular waveform.
3. By observing the input and output waveform, it is possible to observe that the system converts drastic change (square waveform) to a gradual change (sawtooth waveform). The system basically performs lowpass filtering of the input signal.
4. The extent of smoothing is governed by the value of 'M'. Increasing the value of 'M' increases the extent of smoothing the input signal.

```
#Distributive property of convolution
import numpy as np
import matplotlib.pyplot as plt
n=np.arange(-5,6)
#Step 1: Generation of input signal
x=np.array([0,0,0,0,1,2,1,0,0,0,0])
#Step 2: Generation of h1 and h2
k=5
h1=np.array([0,0,0,1,1,1,0,0,0,0])
h2=np.array([0,0,0,-1,0,-1,0,0,0,0])
#Step 3: Perform the convolution
h=h1+h2
y1=np.convolve(x,h,mode='full')
y2=np.convolve(x,h1,mode='full')+np.convolve(x,h2,mode='full')
N=len(y1)
n1=np.arange(-N/2,N/2)
#Step 4: Displaying the result
plt.figure(1),plt.subplot(3,1,1),plt.stem(n,x),plt.xticks(n),plt.xlabel('n-->'),
plt.ylabel('Amplitude'),plt.title('x[n]'),plt.subplot(3,1,2),plt.stem(n,h1),plt.xticks(n)
plt.xlabel('n-->'),plt.ylabel('Amplitude'),plt.title('$h_1$[n]'),plt.subplot(3,1,3),
plt.stem(n,h2),plt.xticks(n),plt.xlabel('n-->'),plt.ylabel('Amplitude'),plt.title('$h_2$[n]')
plt.tight_layout()
plt.figure(2),plt.subplot(2,1,1),plt.stem(n1,y1),plt.xlabel('n-->'),
plt.ylabel('Amplitude'),plt.title('$y_1[n]$=(x[n]*$h_1$[n]*$h_2$[n]')
plt.subplot(2,1,2),plt.stem(n1,y1),plt.xlabel('n-->'),plt.ylabel('Amplitude'),
plt.title('$y_2[n]$=x[n]*($h_1$[n]*$h_2$[n])')
plt.tight_layout()
```

Fig. 3.50 Python code to illustrate distributive property of convolution

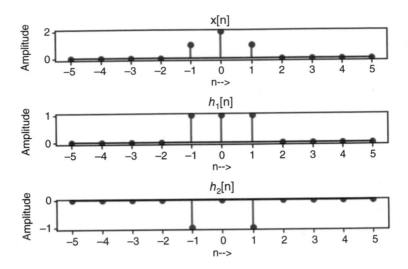

Fig. 3.51 Plot of input signal and the impulse responses

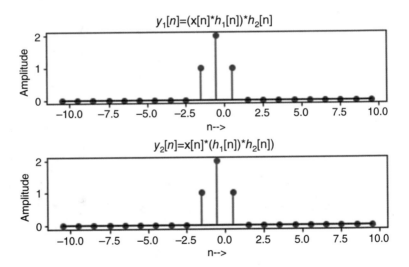

Fig. 3.52 Plot of the output signals

Fig. 3.53 Block diagram of problem statement

Input signal → Low pass filter → Output signal
Square wave   ?

Task

1. In the above experiment, replace the square wave input with sine wave with a spike signal. That is a sine wave with an abrupt change in amplitude in a few time instants. Now pass this sine wave through the moving average filter and comment on the observed signal.

```
#Low pass filtering of square wave
import numpy as np
import matplotlib.pyplot as plt
from scipy import signal
#Step 1: Generation of input signal
t=np.linspace(0,1,100)
x=signal.square(2*np.pi*5*t)
#Step 2: Generation of low pass filter coefficient
M=[5,7,9,11]
fig1=plt.figure(1)
plt.plot(t,x),plt.xlabel('Time'),plt.ylabel('Amplitude'),plt.title('Input signal')
for i in range(len(M)):
    h=1/M[i]*np.ones(M[i])
#Step 3: Obtaining the output signal
    y=np.convolve(x,h,mode='full')
    fig2=plt.figure(2)
#Step 4: Plotting the results
    plt.subplot(2,2,i+1),plt.plot(t,y[0:len(t)]),   plt.xlabel('Time'),
    plt.ylabel('Amplitude'),plt.title('Output signal for M={}'.format(M[i]))
    plt.tight_layout()
```

Fig. 3.54 Python code to perform lowpass filtering of square wave

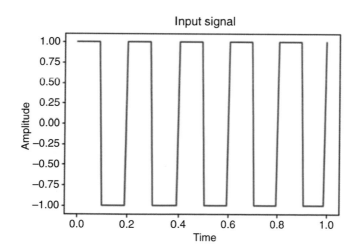

Fig. 3.55 Input square waveform

Experiment 3.25 Convolution of a Square Wave with Highpass Filter Coefficient

In this experiment, the square wave is passed through highpass filter whose impulse response is $h[n] = \{1/2, -1/2\}$. The highpass filter is basically a change detector. When a square wave is fed to highpass filter, the resultant waveform is a spike

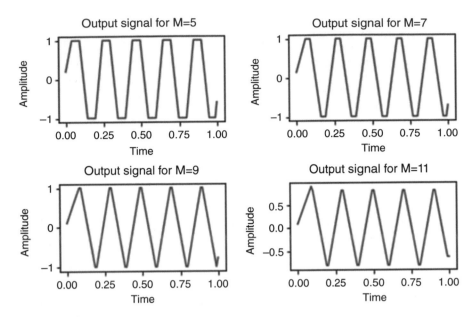

Fig. 3.56 Lowpass filtered square waveform

waveform. The python code, which performs this task, is shown in Fig. 3.57, and the corresponding output is shown in Fig. 3.58.

Inferences

The following inferences can be made from this experiment:

1. From Fig. 3.57, it is possible to infer that the input signal is a square wave, the impulse response of highpass filter is $h[n] = \{1/2, -1/2\}$.
2. From Fig. 3.58, it is possible to observe that the output waveform is a spike waveform. It is due to the fact that differentiation of a constant is zero. In a square wave, major portion is constant in magnitude; hence, differentiation of a constant is zero. Highpass filter is a change detector; hence, it gives spike waveform as the output for the input square waveform.

Task

1. Generate sine wave of 5 Hz frequency. Add white noise, which follows normal distribution to this sine wave. Now pass this noisy sine wave through highpass filter. Plot the clean sine wave, noisy sine wave and highpass filtered signal. Write a python code to answer the query 'Does highpass filter tend to amplify the noise?'

```
#Square wave through high pass filter
import numpy as np
import matplotlib.pyplot as plt
from scipy import signal
#Step 1: Generation of input signal
t=np.linspace(0,1,100)
x=signal.square(2*np.pi*5*t)
#Step 2: Generation of high pass filter coefficient
h=np.array([0.5,-0.5])
#Step 3: To obtain the output signal
y=np.convolve(x,h,mode='full')
#Step 4: Plotting the input and output signal
plt.subplot(3,1,1),plt.plot(t,x),plt.xlabel('time'),plt.ylabel('Amplitude')
plt.title('Input signal'),plt.subplot(3,1,2),plt.stem(h),plt.xlabel('n-->'),
plt.ylabel('Amplitude'),plt.title('h[n]'),plt.subplot(3,1,3),
plt.plot(t,y[0:len(t)]),plt.xlabel('time'),plt.ylabel('Amplitude'),plt.title('Output signal')
plt.tight_layout()
```

Fig. 3.57 Python code to perform highpass filtering of square wave

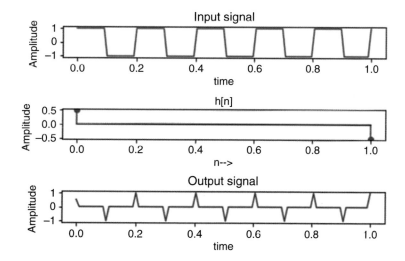

Fig. 3.58 Spike waveform obtained by differentiating input square wave

3.4   Correlation

Correlation is a tool to find the relative similarity between two signals. Correlation has two variants, namely: autocorrelation and cross-correlation. Autocorrelation involves the correlation of a signal with itself. Cross-correlation is performed when two different signals are correlated with one another.

The expression for autocorrelation of the sequence $x[n]$ is given by

$$r_{xx}(l) = x[l] * x[-l] \tag{3.33}$$

Equation (3.33) gives the relationship between correlation and convolution. Convolving the folded version of the sequence $x[n]$ with the signal $x[n]$ results in autocorrelation. Equation (3.33) can be expressed as

$$r_{xx}(l) = \sum_{n=-\infty}^{\infty} x[n]x[n+l] \tag{3.34}$$

Some of the properties of the autocorrelation function are summarized below:

1. Autocorrelation function is an even function. It is expressed as $r_{xx}(-l) = r_{xx}(l)$.
2. Autocorrelation attains its maximum value at zero lag. It is expressed as $r_{xx}(0) \geq |r_{xx}(k)|$ for all 'k'.

The cross-correlation between two signals $x[n]$ and $y[n]$ is expressed as

$$r_{xy}(l) = x[l] * y[-l] \tag{3.35}$$

The above equation can be expressed as

$$r_{xy}(l) = \sum_{k=-\infty}^{\infty} x[k]y[l+k] \tag{3.36}$$

Experiment 3.26 Autocorrelation and Cross-correlation of Sine and Cosine Waves

In this experiment, two signals, namely, sine wave and cosine wave of frequency 5 Hz, are generated. Then, the autocorrelation between the sinewave and cosine wave and the cross-correlation between sine and cosine wave is computed. The results of autocorrelated and cross-correlated signals are plotted. The python code, which performs the above-mentioned task, is shown in Fig. 3.59, and the corresponding output is shown in Fig. 3.60.

Inferences
The following observation can be made from this experiment:

1. The autocorrelation between the sine waves is represented by $r_{xx}(l)$. The autocorrelation result is observed to be even symmetric. The maximum value is obtained at zero lag.
2. The autocorrelation between the cosine waves is represented by $r_{yy}(l)$. The autocorrelation is an even symmetric function with the maximum value obtained at zero lag.
3. The cross correlation between sine and cosine waves is not even symmetric. Also, it is possible to observe that $r_{xy}(l)$ is not equal to $r_{yx}(l)$.

```
#Autcorrelation and cross-correlation
import numpy as np
import matplotlib.pyplot as plt
#Step 1: Generation of sine and cosine wave
t=np.linspace(0,1,100)
f=5
x=np.sin(2*np.pi*f*t)
y=np.cos(2*np.pi*f*t)
N=len(x)
#Step 2: Perform autocorreation and cross-correlation
rxx=np.correlate(x,x,mode='full')
ryy=np.correlate(y,y,mode='full')
rxy=np.correlate(x,y,mode='full')
ryx=np.correlate(y,x,mode='full')
lag = np.arange(-N+1,N)
#Step 3: Plot the results
plt.subplot(2,2,1),plt.plot(lag,rxx),plt.xlabel('Lag'),plt.ylabel('Autocorrelation')
plt.title('$r_{xx}(l)$'),plt.subplot(2,2,2),plt.plot(lag,ryy)
plt.xlabel('Lag'),plt.ylabel('Autocorrelation'),plt.title('$r_{yy}(l)$')
plt.subplot(2,2,3),plt.plot(lag,rxy),plt.xlabel('Lag'),plt.ylabel('Cross correlation')
plt.title('$r_{xy}(l)$'),plt.subplot(2,2,4),plt.plot(lag,ryx),plt.xlabel('Lag'),
plt.ylabel('Cross correlation'),plt.title('$r_{yx}(l)$')
plt.tight_layout()
```

Fig. 3.59 Autocorrelation and cross-correlation between signals

4. The autocorrelation and cross-correlation are used to find the relative similarity between the two signals.

Tasks

1. Write a python code to illustrate the fact that maximum value of autocorrelation occurs at zero lag.
2. Write a python code to illustrate the fact that correlation can be performed in terms of convolution. That is convolution of a signal with its folded version results in autocorrelation.

Experiment 3.27 Autocorrelation of Sine Wave to Itself and Noisy Signal

In this experiment, sine wave of 5 Hz is generated. It is stored as the variable 'x'. The sine wave is then corrupted by random noise, which follows normal distribution to obtain the signal 'y'. The autocorrelation of clean sine wave is obtained as $r_{xx}(l)$, and the cross-correlation between the clean and noisy sine wave is obtained as $r_{xy}(l)$. The python code, which performs this task, is shown in Fig. 3.61, and the corresponding output is shown in Fig. 3.62.

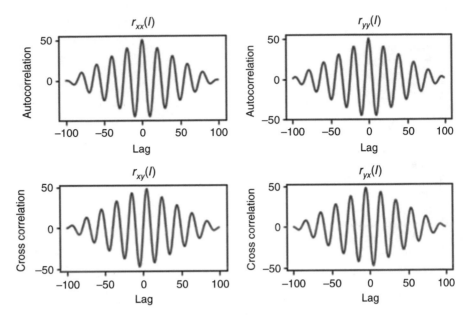

Fig. 3.60 Autocorrelation and cross-correlation results

Inferences

1. In Fig. 3.62, $x(t)$ represents clean sine wave of 5 Hz frequency, and $y(t)$ represents noisy sine wave. The noisy sine wave is obtained by adding random noise to the clean sine wave.
2. In Fig. 3.62, $r_{xx}(l)$ represents the autocorrelation of a clean sine wave. The autocorrelation function exhibits even symmetry, with the maximum value occurring at zero lag.
3. In Fig. 3.62, $r_{xy}(l)$ represents the cross-correlation between clean and noisy sine waves. The cross-correlation is not exhibiting even symmetry relation. Comparing $r_{xx}(l)$ and $r_{xy}(l)$, the maximum value is obtained in autocorrelation function. Thus, the autocorrelation reveals the relative similarity between the signals.

Experiment 3.28 Delay Estimation Using Autocorrelation

In this experiment, unit step sequence (signal ×1) is generated, it is then shifted by a factor of '5' units to the right to obtain the signal ×2. The autocorrelation of the signal ×1 to itself and the correlation between the signals ×1 and ×2 are used to estimate the delay. The python code, which performs this function, is shown in Fig. 3.63, and the corresponding output is shown in Fig. 3.64.

Inference

Upon displaying the result, the answer in the variable 'td' is '5', which is a measure of delay between the two signals $x_1[n]$ and $x_2[n]$. Thus, autocorrelation can be used to measure or estimate the delay between the two signals.

```
#Autocorrelation and cross-correlation
import numpy as np
import matplotlib.pyplot as plt
#Step 1: Generation of sine and cosine wave
t=np.linspace(0,1,100)
f=5
x=np.sin(2*np.pi*f*t)
#Step 2: Generation of noisy signal
w=2.5*np.random.randn(len(t))
y=x+w
N=len(x)
#Step 2: Perform autocorreation and cross-correlation
rxx=np.correlate(x,x,mode='full')
rxy=np.correlate(x,y,mode='full')
lag = np.arange(-N+1,N)
#Step 3: Plot the results
plt.subplot(2,2,1),plt.plot(t,x),plt.xlabel('Time'),plt.ylabel('Amplitude')
plt.title('Sine wave (x(t))'),plt.subplot(2,2,2),plt.plot(t,y)
plt.xlabel('Time'),plt.ylabel('Amplitude'),plt.title('Noisy sine wave(y(t))')
plt.subplot(2,2,3),plt.plot(lag,rxx),plt.xlabel('Lag'),plt.ylabel('Autocorrelation')
plt.title('$r_{xx}(l)$'),plt.subplot(2,2,4),plt.plot(lag,rxy)
plt.xlabel('Lag'),plt.ylabel('Cross correlation'),plt.title('$r_{xy}(l)$')
plt.tight_layout()
```

Fig. 3.61 Python code to perform autocorrelation of clean and noisy sine wave

Exercises

1. Generate the following sequences (a) $x_1[n] = \delta[n + 1] + \delta[n - 1]$ (b) $x_1[n] = \delta[n + 1] - \delta[n - 1]$ (c) $x_3[n] = \delta[n] + 2\delta[n - 1] + \delta[n - 2]$ and (d) $x_4[n] = \delta[n] - \delta[n - 1] + \delta[n - 2]$, and plot it using a subplot, which consists of two rows and two columns. The time index should vary from -5 to $+5$.

2. Write a python code to generate the finite length discrete-time signals (a) $x_1[n] = u[n] - u[n - 5]$, (b) $x_2[n] = \delta[n]$, (c) $x_3[n] = u[n + 5] - u[n - 5]$ and (d) $x_4[n] = \begin{cases} n, 0 \le n \le 5 \\ 0, \text{otherwise} \end{cases}$ in the interval $-10 \le n \le 10$. Use subplot to plot the generated signals.

3. Generate a complex exponential signal $x[n] = e^{j\frac{\pi}{4}n}$, $-10 \le n \le 10$. Perform the following: (a) Extract the real and imaginary part of this signal. (b) Reconstruct the signal $x[n]$ from the real and imaginary parts using the relation $x[n] = \text{Re}\{x[n]\} + j\,\text{Im}\{x[n]\}$.

4. Generate a complex exponential signal of the form $x[n] = e^{j\frac{\pi}{8}n}$, $-10 \le n \le 10$. Obtain the signal $y[n]$, which is expressed as $y[n] = x[n] \times x^*[n]$, and comment on the nature of the signal $y[n]$.

5. Write a python code to generate the following sequences:

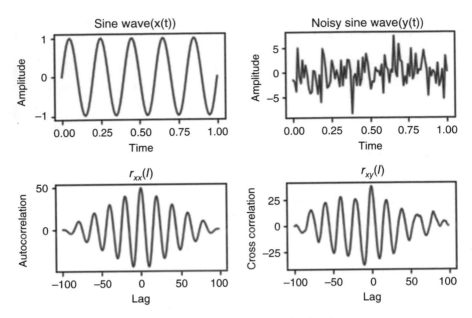

Fig. 3.62 Autocorrelation and cross-correlation of clean and noisy sine wave

```
#Delay estimation using correlation
import numpy as np
import matplotlib.pyplot as plt
n=np.arange(-20,21)
#Step 1: Generation of unit step signal
x1=(n>=0)
#Step 2: Delay signal by a factor of 5 units
x2=(n>=5)
N=len(x1)
lag=np.arange(-N+1,N)
#Step 3: Perform autocorrelation of signal x1
rxx=np.correlate(x1,x1,mode='full')
#Step 4: Perform the cross-correlation between x1 and x2
ryx=np.correlate(x2,x1,mode='full')
#Step 4: Estimate the delay
td=np.argmax(ryx)-np.argmax(rxx)
#Step 5: Plot the signal and its delayed version
print('Time delay={}'.format(td))
plt.subplot(2,1,1),plt.stem(n,x1),plt.xlabel('n-->'),plt.ylabel('Amplitude'),plt.title('$x_1[n]$')
plt.subplot(2,1,2),plt.stem(n,x2),plt.xlabel('n-->'),plt.ylabel('Amplitude'),plt.title('$x_2[n]$')
plt.tight_layout()
```

Fig. 3.63 Python code to perform delay estimation

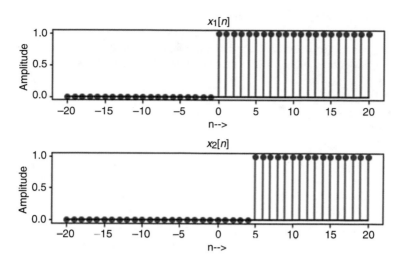

Fig. 3.64 Signal and its delayed version

(a) $x_1[n] = \delta[n] + \delta[n-1] + \delta[n-2] + \delta[n-3] + \delta[n-4] + \delta[n-5] + \delta[n-6] + \delta[n-7]$

(b) $x_2[n] = \delta[n] - \delta[n-1] + \delta[n-2] - \delta[n-3] + \delta[n-4] - \delta[n-5] + \delta[n-6] - \delta[n-7]$

(c) $x_3[n] = \delta[n] + \delta[n-1] - \delta[n-2] - \delta[n-3] + \delta[n-4] + \delta[n-5] - \delta[n-6] - \delta[n-7]$

(d) $x_4[n] = \delta[n] + \delta[n-1] + \delta[n-2] + \delta[n-3] - \delta[n-4] - \delta[n-5] - \delta[n-6] - \delta[n-7]$

Compute the energy of these sequences and comment on the obtained result.

6. Sketch the following signals in the range $-5 \le n \le 5$ (a) $x_1[n] = 2^n\delta[n-2]$ (b) $x_2[n] = n[\delta[n+2] + \delta[n-2]]$.

7. Generate the signal $x[n] = \begin{cases} 5 - |n|, |n| \le 5 \\ 0, \text{otherwise} \end{cases}$ in the range $-10 \le n \le 10$. Extract the even and odd part of the signal. Try to reconstruct the signal from the even and odd part and comment on the observed output.

8. Write a python code to demonstrate the following facts:

   (a) Product to two even signals is an even signal.
   (b) Product of two odd signals is an even signal.
   (c) Product of an even and odd signal is odd signal.

9. Read a speech signal and perform the autocorrelation of the speech signal, and observe whether the autocorrelation function is an even function.

10. Read a 'male' and 'female' voice. Perform the following

(a) Autocorrelation of the male voice (x)
(b) Autocorrelation of the female voice (y)
(c) Cross-correlation between male and female voice
(d) Cross-correlation between female and male voice

Comment on the observed output.

Objective Questions

1. The python code segment shown below generates

```
n=np.arange(-10,11,1)
x=(n>=0)
```

A. Unit sample signal
B. Unit step signal
C. Unit ramp signal
D. Real exponential signal

2. The value of the signal 'x' shown in the following python code is high at $n =$?

```
n=np.arange(-10,11,1)
x=(n==-2)
```

A. -1
B. -2
C. 0
D. 2

3. If the variable 'x' contains the signal of interest, then the variable 'y' in the following python code returns

```
y=np.sum(np.abs(x**2))
```

A. Maximum value of the signal
B. Minimum value of the signal
C. Energy of the signal
D. Power of the signal

4. The signal generated in the variable 'x' after executing the following segment of code is

```
n=np.arange(-10,11)
x=(n==-1)|(n==0)|(n==1)
```

A. $x[n] = \delta[n] + \delta[n-1] - \delta[n+1]$
B. $x[n] = \delta[n+1] + \delta[n] + \delta[n-1]$
C. $x[n] = \delta[n+1] + \delta[n] - \delta[n-1]$
D. $x[n] = \delta[n+1] + 2\delta[n] + \delta[n-1]$

5. The signal generated in the variable '*x*' after executing the following segment of code is

```
n=np.arange(-10,11,1)
x1=n*(n>=0)
x2=(n+1)*(n+1>=0)
x=x2-x1
```

   A. Unit sample sequence
   B. Unit step sequence
   C. Unit ramp sequence
   D. Real exponential sequence

6. What would be the energy of the signal '*x*' which is stored in variable '*E*' if the following code segment is executed?

```
n=np.arange(-10,11)
x=(n==-1)|(n==0)|(n==1)
E=np.sum(np.abs(x**2))
```

   A. 1J
   B. 2J
   C. 3J
   D. 4J

7. What operation is performed on the input signal '*x*' if the following segment of code is executed?

```
x=[1,2,3]
y=np.convolve(x,x[::-1],mode='full')
```

   A. Convolution of signal '*x*' with itself
   B. Correlation of the signal '*x*' with itself
   C. Power spectral estimation of the signal '*x*'
   D. Energy density estimation of signal '*x*'

8. A square wave is fed to a lowpass filter, the resulting signal is

   A. Sinc wave
   B. Cosine wave
   C. Triangular wave
   D. Inverted square wave

9. The energy of the signal is unaltered by the following mathematical operation

   A. Downsampling of the signal by a factor of '*M*'
   B. Upsampling the signal by a factor of '*L*'
   C. Amplitude scaling
   D. Folding of the signal

10. The energy of the signal is unaltered by the following mathematical operation:

    A. Downsampling of the signal by a factor of 'M'
    B. Upsampling the signal by a factor of 'L'
    C. Delaying or advancing the signal by a factor of 'k'
    D. Amplitude scaling of the signal

11. Upsampling by a factor of 'L' inserts

    A. 'L' zeros between successive samples
    B. '$L-1$' zeros between successive samples
    C. '$L+1$' zeros between successive samples
    D. '$L+2$' zeros between successive samples

12. If a discrete-time signal $x[n]$ obeys the relation $x[-n] = x[n]$, then the signal is

    A. Odd signal
    B. Even signal
    C. Either even or odd signal
    D. Neither even nor odd signal

13. Sum of elements of finite duration discrete-time odd signal is

    A. Infinite
    B. One
    C. Zero
    D. Always negative

14. The python code shown below generates the following signal in the variable 'x'

```
n=np.arange(-20,21)
k=5
x1=(n>=k)
```

    A. $u[n]$
    B. $u[-n]$
    C. $u[n+5]$
    D. $u[n-5]$

15. The product of two odd signal results in

    A. Odd signal
    B. Even signal
    C. Either even or odd signal depending on the length of the signals
    D. Neither even nor odd signal

16. Identify the statement which is FALSE

    A. Autocorrelation is finding the relative similarity of the signal to itself.
    B. Autocorrelation is an even function.

C. Autocorrelation attains its maximum value at zero lag.

D. Auto correlation is an odd function.

17. What will be the fundamental period of the signal 'x' if the following python code is executed?

```
n=np.arange(-10,11,dtype='float')
x=(-1)**n
```

A. 1
B. 2
C. 3
D. 4

18. Assertion: Highpass filter act as change detector

    Reason: Highpass filter has the ability to detect the change in the input signal

A. Both assertion and reason are true.
B. Assertion is true, reason is false.
C. Assertion is false, reason may be true.
D. Both assertion and reason are false.

19. What will be the length of the signal 'y' if the following code segment is executed?

```
n=np.arange(-5,6)
x=np.array([0,0,0,0,1,2,1,0,0,0,0])
h= np.array([0,0,0,0,1,1,1,0,0,0,0])
y=np.convolve(x,h,mode='full')
```

A. 11
B. 21
C. 31
D. 41

20. What will be the impulse response $(h[n])$ if the following code segment is executed?

```
n=np.arange(-5,6)
h1=np.array([0,0,0,0,1,1,1,0,0,0,0])
h2=np.array([0,0,0,0,-1,0,-1,0,0,0,0])
h= h1+h2
```

A. $h[n] = \delta[n]$
B. $h[n] = \delta[n-1]$
C. $h[n] = u[n]$
D. $h[n] = u[n-1]$

21. Identify the statement that is WRONG with respect to 'folding' or 'time reversal' operation

    A. Folding operation does not alter the energy of the signal.
    B. Folding increases the length of the signal.
    C. If the folded version of the signal is equal to the signal itself, then the signal is even signal.
    D. If the folded version of the signal is equal to the signal itself, then the signal is odd signal.

22. If $x[n]$ is a unit step signal, then the following signal ($y[n]$) generated from $x[n]$ is

```
n=np.arange(-5,6)
y=n*x
```

    A. Unit sample signal
    B. Unit step signal
    C. Unit ramp signal
    D. Real exponential signal

23. The fundamental frequency of the signal generated by executing the following code is

```
n=np.arange(-10,11,dtype='float')
x=(-1)**n
```

    A. $\omega = \pi/2$ rad/sample
    B. $\omega = \pi$ rad/sample
    C. $\omega = \pi/4$ rad/sample
    D. $\omega = \pi/8$ rad/sample

Bibliography

1. Lonnie C. Ludeman, "Fundamentals of Digital Signal Processing", John Wiley and Sons, 1986.
2. S. Esakkirajan, T. Veerakumar and Badri N Subudhi, "Digital Signal Processing", McGraw Hill, 2021.
3. Sophocles Orfanidis, "Introduction to Signal Processing", Pearson, 1995.
4. Hwei P. Hsu, "Signals and Systems", Schaum's outline series, McGraw Hill Education, 2017.
5. Maurice Charbit, "Digital Signal Processing with Python Programming", Wiley-ISTE, 2017.

Chapter 4
Discrete-Time Systems

Learning Objectives

After reading this chapter, the reader is expected to

- Obtain the impulse response and step response of the discrete-time system.
- Plot the magnitude and phase response of the discrete-time system.
- Plot the pole-zero plot of the discrete-time system.
- Verifying the linearity, time-invariance, causal and stable properties of the discrete-time system.

Roadmap of the Chapter

This chapter begins with different types of representations of discrete-time system, including difference equation, block diagram and state-space. Properties of a discrete-time system which includes linearity, time-invariance, causal and stable are verified with python illustration. In this chapter, discrete-time system responses, including impulse response, step response and frequency response, are plotted, and the obtained results are interpreted.

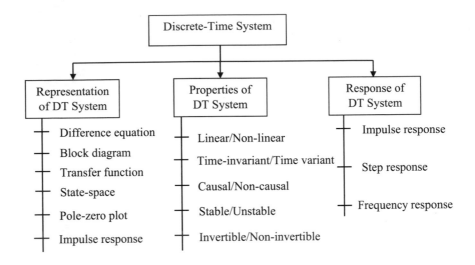

PreLab Questions

1. List different forms of representation of discrete-time system.
2. When a discrete-time system is said to be a *relaxed system*?
3. When a discrete-time system is said to be linear? Give a few examples of linear system.
4. Is it always true that the cascade of two non-linear systems will result in a non-linear system? Justify your answer.
5. Why real-time (real-world) systems are considered as causal systems?
6. 'All memoryless systems are causal, whereas all the causal systems are not memoryless'. Justify this statement.
7. Is it always true that a cascade of two time-variant systems will result in a time-variant system? Justify your answer.
8. When a discrete-time system is said to be invertible? Give an example of the invertible and non-invertible systems.
9. Is it possible to test the causality and stability of a linear time-invariant discrete-time system from its impulse response? If so, how?
10. Distinguish between static and dynamic discrete-time systems. Cite an example for static and dynamic discrete-time systems.
11. When is a discrete-time system said to be non-recursive, and when is it said to be recursive? Give examples for each class of the discrete-time system.
12. The pole-zero plot of a discrete-time system exhibits a zero at $z = 1$; what can you infer about this system?
13. When is a discrete-time system invertible? Give an example of a discrete-time system, which is invertible, and an example of a non-invertible discrete-time system.
14. Mention two advantages of state-space representation of the discrete-time system.

15. What do you understand by the statement 'Discrete-time system is characterised by its impulse response ($h[n]$)'?

4.1 Discrete-Time System

Discrete-time (DT) system accepts a discrete-time signal as input and generates a discrete-time signal as the output. The input to the discrete-time system is termed as 'excitation' and the output of the system is termed as 'response'. The block diagram of DT system is shown in Fig. 4.1.

4.2 Representation of DT Systems

Different forms of representation of DT systems include (1) block diagram, (2) difference equation, (3) transfer function, (4) pole-zero plot, (5) state-space, etc. Python illustration with respect to different forms of representation of discrete-time system and python examples to obtain different DT system responses, including impulse, step, magnitude and phase responses, are discussed in this chapter.

4.2.1 Difference Equation Representation of Discrete-Time Linear Time-Invariant System

The relationship between the input and output of a discrete-time linear time-invariant (LTI) system is expressed in terms of linear constant coefficient difference equation (LCCDE) as

$$\sum_{k=0}^{N-1} a_k y[n-k] = \sum_{k=0}^{M-1} b_k x[n-k] \tag{4.1}$$

where $\{a_k\}$ and $\{b_k\}$ are the output and input coefficients respectively. The above equation represents the fact that weighted sum of input is equal to the weighted sum of output. Equation (4.1) can be expanded as

Fig. 4.1 Block diagram of discrete-time system

$$a_0y[n] + a_1y[n-1] + \cdots + a_Ny[n-N] = b_0x[n] + b_1x[n-1] + \cdots \\ + b_Mx[n-M]$$ (4.2)

If $a_0 = 1$, the above expression can be written as

$$y[n] + a_1y[n-1] + \cdots + a_Ny[n-N] = b_0x[n] + b_1x[n-1] + \cdots \\ + b_Mx[n-M]$$ (4.3)

Equation (4.3) can be expressed as

$$y[n] = b_0x[n] + b_1x[n-1] + \cdots + b_Mx[n-M] \\ - \{a_1y[n-1] + \cdots + a_Ny[n-N]\}$$ (4.4)

If the current output is not a function of the previous output of the system, the system is said to be a non-recursive system. The input-output relationship of a non-recursive system is given by

$$y[n] = b_0x[n] + b_1x[n-1] + \cdots + b_Mx[n-M]$$ (4.5)

An example of non-recursive system is a finite impulse response (FIR) filter. In the case of infinite impulse response (IIR) filter, the current output is a function of the current input, previous input and previous output.

Experiment 4.1 Solution of Difference Equation with Zero Initial Condition
This experiment discusses solving the difference equation with zero initial conditions. The relationship between the input and output of the discrete-time system is given by

$$y[n] = \frac{1}{2}y[n-1] + x[n]$$ (4.6)

Let us consider the input to the system as a unit step signal; hence, the above equation can be written as

$$y[n] = \frac{1}{2}y[n-1] + u[n]$$ (4.7)

Substituting $n = 0$ in the expression (4.7), we get

$$y[0] = \frac{1}{2}y[-1] + u[0]$$

If the initial condition is zero, $y[-1] = 0$ and $u[0] = 1$; hence, the above equation can be written as

$$y[0] = 1$$

Substituting $n = 1$ in Eq. (4.7), we get

$$y[1] = \frac{1}{2}y[0] + u[1]$$

Substituting the value of $y[0]$ as 1 and $u[1] = 1$ in the above equation, we get

$$y[1] = \frac{1}{2} \times 1 + 1 = \frac{3}{2}$$

Substituting $n = 2$ in Eq. (4.7), we get

$$y[2] = \frac{1}{2} \times \frac{3}{2} + 1$$

Simplifying the above equation, we get

$$y[2] = \frac{5}{4}$$

The output of the system is given by

$$y[n] = \left\{ 1, \frac{3}{2}, \frac{5}{4}, \cdots \right\}$$

The python code, which obtains the output of the system, is given in Fig. 4.2, and the corresponding output is shown in Fig. 4.3.

```
import numpy as np
from scipy import signal
import matplotlib.pyplot as plt
n=np.arange(0,4)
x=np.ones(len(n)) #Input
#Defining the system
num=[1]
den=[1,-1/2]
#Obtaining the output
y=signal.lfilter(num,den,x)
#Displaying the result
plt.stem(y),plt.xlabel('n-->'),plt.ylabel('Amplitude'),plt.title('y[n]')
plt.tight_layout()
```

Fig. 4.2 Python code to obtain the output of discrete-time system

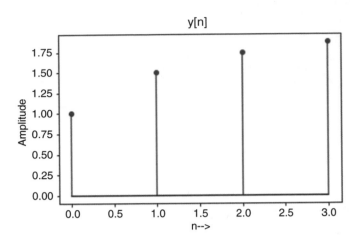

Fig. 4.3 Simulation result

Inference
The output of the system obtained using python example is in agreement with the theoretical result.

Task
1. Write a python code to obtain the impulse response of a discrete-time system, whose input and output are related as $y[n] = x[n] + 2y[n-1]$. Also, try to answer the question "whether the impulse response is absolutely summable or not".

Experiment 4.2 Solution of Difference Equation with Initial Condition
This experiment deals with solving the difference equation with the initial condition. The relationship between the input and output of a discrete-time system is given by

$$y[n] = \frac{1}{2}y[n-1] + x[n] \tag{4.8}$$

If $x[n] = \left(\frac{1}{4}\right)^n u[n]$ and $y[-1] = 1$.
Substituting $x[n] = \left(\frac{1}{4}\right)^n u[n]$ in Eq. (4.8), we get

$$y[n] = \frac{1}{2}y[n-1] + \left(\frac{1}{4}\right)^n u[n] \tag{4.9}$$

Substituting $n = 0$ in Eq. (4.9), we get

$$y[0] = \frac{1}{2}y[-1] + \left(\frac{1}{4}\right)^0 u[0]$$

Substituting $y[-1] = 1$ and $\left(\frac{1}{4}\right)^0 u[0] = 1$ in the above equation, we get

```
import numpy as np
from scipy import signal
n=np.arange(0,4)
x=(1/4)**n
num=[1]
den=[1,-1/2]
y1=signal.lfiltic(num,den,y=[1.])
y=signal.lfilter(num,den,x,zi=y1)
print(y[0])
```

[1.5   1.   0.5625 0.296875]

(a) Python Code   **(b) Result**

Fig. 4.4 Python code to obtain the solution of difference equation. (**a**) Python Code (**b**) Result

$$y[0] = \frac{1}{2} \times 1 + 1 = \frac{3}{2}$$

Substituting $n = 1$ in Eq. (4.9), we get

$$y[1] = \frac{1}{2}y[0] + \left(\frac{1}{4}\right)^1 u[1]$$

Substituting $y[0] = 3/2$ and $u[1] = 1$ in the above equation, we get

$$y[1] = \frac{1}{2} \times \frac{3}{2} + \frac{1}{4} = 1$$

Substituting $n = 2$ in Eq. (4.9), we get

$$y[2] = \frac{1}{2}y[1] + \left(\frac{1}{4}\right)^2 u[2]$$

Substituting $y[1] = 1$ in the above expression, we get

$$y[2] = \frac{1}{2} \times 1 + \frac{1}{16} = \frac{9}{16} = 0.5625.$$

The python code, which obtains the solution of the difference equation with a non-zero initial condition, is given in Fig. 4.4a, and the corresponding output is shown in Fig. 4.4b.

Inference
The built-in function '*lfiltic*' and '*lfilter*' is used to obtain the solution of difference equation. The experimental result is in agreement with the theoretical result.

4.2.2   State-Space Model of a Discrete-Time System

The state-space model describes the system's dynamics through two equations, namely: 'state equation' and 'output equation'. The state equation describes how the input influences the state, and the output equation describes how the state and the input directly influence the output. It is to be noted that the state-space representations of a particular system's dynamics are not unique. The two equations are given below

$$x[k+1] = Ax[k] + Bu[k] \tag{4.10}$$

$$y[k] = Cx[k] + Du[k] \tag{4.11}$$

where $u[k] \in \mathfrak{R}^m$ is the input, $y[k] \in \mathfrak{R}^p$ is the output and $x[k] \in \mathfrak{R}^n$ is the state vector. $'A'$ is the system matrix, $'B'$ and $'C'$ are the input and output matrices, and $'D'$ is the feed forward matrix.

4.2.2.1   State-Space to Transfer Function

Taking Z-transform on both sides of Eq. (4.10), we get

$$zX(z) - zx_0 = AX(z) + BU(z)$$

Assuming an zero initial condition (x_0), the above equation can be expressed as

$$zX(z) = AX(z) + BU(z)$$

The above equation can be expressed as

$$(zI - A)X(z) = BU(z)$$

The expression for $X(z)$ is given by

$$X(z) = B(zI - A)^{-1}U(z) \tag{4.12}$$

Taking Z-transform on both sides of the Eq. (4.11), we get

$$Y(z) = CX(z) + DU(z) \tag{4.13}$$

Substituting the expression for $X(z)$ from Eq. (4.12) in Eq. (4.13), we get

$$Y(z) = CB(zI - A)^{-1}U(z) + DU(z)$$

The expression for the transfer function is given by

$$H(z) = \frac{Y(z)}{U(z)} = C(zI - A)^{-1}B + D \tag{4.14}$$

Thus, the transfer function of the system is represented in terms of state-space model.

Experiment 4.3 State-Space to Transfer Function
The state-space representation of discrete-time system is given by

$$x[k + 1] = Ax[k] + Bu[k]$$

and

$$y[k] = Cx[k] + Du[k],$$

where $A = \begin{bmatrix} -1 & 2 \\ 1 & 0 \end{bmatrix}$, $B = \begin{bmatrix} 1 \\ 0 \end{bmatrix}$, $C = [-1 \ 2]$ and $D = [1]$. Obtain the transfer function of the system using python.

The relationship between the transfer function and state-space representation is given by

$$H(z) = C(zI - A)^{-1}B + D$$

Step 1: To determine $(zI - A)^{-1}$

$$zI - A = \begin{bmatrix} z & 0 \\ 0 & z \end{bmatrix} - \begin{bmatrix} -1 & 2 \\ 1 & 0 \end{bmatrix}$$

Upon simplifying the above equation, we get

$$zI - A = \begin{bmatrix} z+1 & -2 \\ -1 & z \end{bmatrix}$$

$$(zI - A)^{-1} = \begin{bmatrix} \dfrac{z}{z^2 + z - 2} & \dfrac{2}{z^2 + z - 2} \\ \dfrac{1}{z^2 + z - 2} & \dfrac{z+1}{z^2 + z - 2} \end{bmatrix}$$

Step 2: To determine $H(z)$
The expression for the transfer function $H(z)$ is given by

Fig. 4.5 Python code to
obtain the transfer function
from state-space model

```
#State-space to transfer function
from scipy import signal
import numpy as np
#Step 1: Defining the state-space model
A=[[-1,2],[1,0]]
B=[[1], [0]]
C = [[-1, 2]]
D = 1
#Step 2: Obtaining the transfer function
[num,den]=signal.ss2tf(A,B,C,D)
print('numerator=',num)
print('denominaor=',den)
```

Fig. 4.6 Transfer function
of the discrete-time system

```
numerator= [[1 0 0]]
denominaor= [ 1.  1. -2.]
```

$$H(z) = C(zI - A)^{-1}B + D$$

$$H(z) = [-1 \quad 2] \begin{bmatrix} \dfrac{z}{z^2+z-2} & \dfrac{2}{z^2+z-2} \\ \dfrac{1}{z^2+z-2} & \dfrac{z+1}{z^2+z-2} \end{bmatrix} \begin{bmatrix} 1 \\ 0 \end{bmatrix} + [1]$$

Simplifying the above expression, we get

$$H(z) = \begin{bmatrix} \dfrac{-z+2}{z^2+z-2} & \dfrac{2z}{z^2+z-2} \end{bmatrix} \begin{bmatrix} 1 \\ 0 \end{bmatrix} + [1]$$

Upon simplifying the above equation, the transfer function of the system is
given by

$$H(z) = \frac{-z+2}{z^2+z-2} + 1$$

The transfer function of the system is given by

$$H(z) = \frac{z^2}{z^2+z-2}$$

The built-in function 'ss2tf' available in *scipy* library can be used to obtain the
transfer function of the system from the state-space representation. The python
code, which performs this task, is shown in Fig. 4.5, and the corresponding output
is shown in Fig. 4.6.

Fig. 4.7 State-space model from the transfer function

```
#Transfer function to state-space
import numpy as np
from scipy import signal
#Step 1: Defining the transfer function
num=[1, 0, 0]
den=[1,1,-2]
#Step 2: Obtaining the state-space model
A,B,C,D=signal.tf2ss(num,den)
print("A=",A,"\n","B=",B,"\n","C=",C,"\n","D=",D)
```

Fig. 4.8 Result of python code shown in Fig. 4.7

```
A= [[-1. 2.]
 [ 1. 0.]]
B= [[1.]
 [0.]]
C= [[-1. 2.]]
D= [[1.]]
```

Upon executing the code shown in Fig. 4.5, the transfer function of the system obtained is given in Fig. 4.6.

Inference

The transfer function obtained using the built-in function '*ss2tf*' is in agreement with the theoretical result.

Experiment 4.4 Transfer Function to State-Space

The objective of this experiment is to obtain the state-space representation of the discrete-time system, whose transfer function is given by $H(z) = \frac{z^2}{z^2+z-2}$. As per the previous experiment, the value of the state-space model parameters should be $A = \begin{bmatrix} -1 & 2 \\ 1 & 0 \end{bmatrix}$, $B = \begin{bmatrix} 1 \\ 0 \end{bmatrix}$, $C = \begin{bmatrix} -1 & 2 \end{bmatrix}$ and $D = [1]$. The built-in function '*tf2ss*' in *scipy* library can be used to obtain the state-space representation of discrete-time system from the transfer function. The python code, which performs this task, is shown in Fig. 4.7, and the corresponding output is shown in Fig. 4.8.

Inference

From Fig. 4.8, it is possible to observe that the state-space model parameters are $A = \begin{bmatrix} -1 & 2 \\ 1 & 0 \end{bmatrix}$, $B = \begin{bmatrix} 1 \\ 0 \end{bmatrix}$, $C = \begin{bmatrix} -1 & 2 \end{bmatrix}$ and $D = [1]$.

Task

1. What will be the value of the state-space parameter '*c*' if the numerator and the denominator polynomial of the transfer function are same? For example, $H(z) = \frac{z^2+z+1}{z^2+z+1}$.

4.2.3 Impulse Response and Step Response of Discrete-Time System

Impulse response is the reaction of the discrete-time system to unit sample input signal, whereas step response is the reaction of the system to unit step input signal. The discrete-time system is completely characterised by its impulse response. The meaning is, if one knows the impulse response of the system, it is possible to infer whether the system is causal and stable from it.

Figure 4.9 depicts the input-output relationship of a discrete-time system in which $x[n]$ represents the input signal, $h[n]$ represents the impulse response and $y[n]$ represents the output of the system. If the system is a LTI system, then the output of the system is expressed as

$$y[n] = x[n]^* h[n] \tag{4.15}$$

In the above expression, '*' indicates the convolution operation. If the input to the system is unit sample signal, then the output of the system is given by

$$y[n] = \delta[n]^* h[n] \tag{4.16}$$

Convolution of any signal with unit sample signal results in the same signal; hence, the above equation can be expressed as

$$y[n] = h[n] \tag{4.17}$$

Thus, the impulse response of the system is the reaction of an LTI system to unit sample input signal.

The reaction of LTI system to unit step input signal is termed as step response of the system. It is denoted as $s[n]$. The relationship between step response ($s[n]$) and impulse response ($h[n]$) is given by

$$s[n] = \sum_{k=-\infty}^{n} h[k] \tag{4.18}$$

Experiment 4.5 Impulse and Step Responses of the System
This experiment deals with the computation of impulse and step responses from the LCCDE. An LTI discrete-time system is defined by the difference equation $y[n] = x[n] + y[n-1]$. Plot the impulse response and step response of the system.

Fig. 4.9 Input-output of a discrete-time system

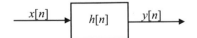

```
#Impulse and step response of LTI DT system
import numpy as np
import matplotlib.pyplot as plt
from scipy import signal
#Step 1: Generation of unit sample and unit step signal
x1=np.zeros(100)
x1[0]=1
x2=np.ones(100)
#Step 2: Define the system
num=[1]
den=[1,-1]
#Step 3: Obtaining the impulse and step response
h=signal.lfilter(num,den,x1)
s=signal.lfilter(num,den,x2)
#Plotting the result
plt.subplot(2,1,1),plt.stem(h),plt.xlabel('n-->'),plt.ylabel('h[n]')
plt.title('Impulse response (h[n])')
plt.subplot(2,1,2),plt.stem(s),plt.xlabel('n-->'),plt.ylabel('s[n]')
plt.title('Step response (s[n])')
plt.tight_layout()
```

Fig. 4.10 Python code to obtain the impulse and step response of the system

Impulse response is the response of the system to unit sample input signal, and step response is the response of the system to unit step input signal. The python code, which computes the impulse and step response of the system, is shown in Fig. 4.10, and the corresponding output is shown in Fig. 4.11.

Inferences

The following inferences can be made from this experiment:

1. From Fig. 4.11, it is possible to observe that the impulse response of the system is unit step signal. This implies $h[n] = u[n]$. The impulse response of the system is not absolutely summable; hence, this system is not stable system.
2. The step response of the system is a ramp signal. This implies $s[n] = nu[n]$.

Task

1. In the above experiment if $h[n] = \delta[n]$, what will be the step response of the system?

Experiment 4.6 Computation of Impulse and Step Responses of the System from the Difference Equation

This experiment also discusses the computation of the impulse and step response from the difference equation. Let us consider the discrete-time LTI system, whose difference equation is given by $y[n] - 0.5y[n-1] = x[n] + x[n-1]$. The python code, which obtains the impulse and step responses of the given discrete-time system, is given in Fig. 4.12, and the corresponding output is shown in Fig. 4.13.

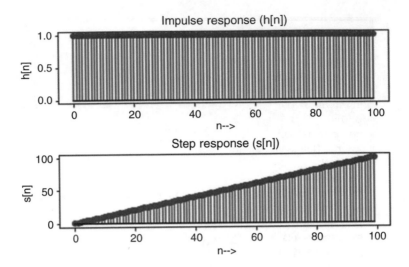

Fig. 4.11 Impulse response and step response of the given system

Inferences

The following inferences can be drawn from this experiment:

1. The impulse response decays to zero. The impulse response is absolutely summable; hence, the given system is a stable system.
2. The step response settles to a finite value after a short span.

Task

1. In the above experiment, from the impulse response, will it be possible to comment on the stability of the system? Write a python code to find whether the impulse response is absolutely summable or not?

Experiment 4.7 Computation of Step Response from the Impulse Response

This experiment discusses the computation of step response from the impulse response. The impulse response of the system is unit step signal. The step response is obtained by repeatedly adding the impulse response. The python command '*np. cumsum*' can be used to obtain the step response of the system from impulse response of the system. The relationship between input and output of a linear time-invariant system is given by

$$y[n] = x[n]{}^*h[n] \tag{4.19}$$

It is given that the impulse response is unit step signal; hence, $h[n] = u[n]$. Also, unit step response of the system implies that the input to the system is unit step signal; hence, $x[n] = u[n]$. The output of the system is given by

```
#Impulse and step responses of LTI DT system
import numpy as np
import matplotlib.pyplot as plt
from scipy import signal
#Step 1: Generation of unit sample and unit step signal
x1=np.zeros(100)
x1[0]=1
x2=np.ones(100)
#Step 2: Define the system
num=[1,1]
den=[1,-0.5]
#Step 3: Obtaining the impulse and step response
h=signal.lfilter(num,den,x1)
s=signal.lfilter(num,den,x2)
#Plotting the result
plt.subplot(2,1,1),plt.stem(h),plt.xlabel('n-->'),plt.ylabel('h[n]')
plt.title('Impulse response (h[n])')
plt.subplot(2,1,2),plt.stem(s),plt.xlabel('n-->'),plt.ylabel('s[n]')
plt.title('Step response (s[n])')
plt.tight_layout()
```

Fig. 4.12 Python code to obtain the impulse and step response of the system

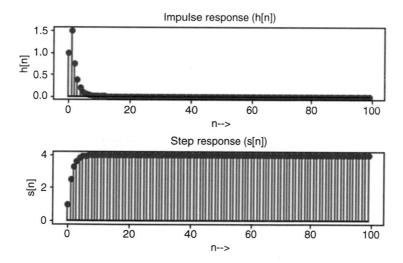

Fig. 4.13 Result of python code shown in Fig. 4.12

$$y[n] = u[n]^* u[n] \qquad (4.20)$$

The above equation can be simplified as

```
#Step response from the impulse response
import numpy as np
import matplotlib.pyplot as plt
#Step 1: Impulse response of the system
h=np.ones(10)
#Step 2: Step response of the system
s=np.cumsum(h,axis=0)
#Step 3: Plotting the result
plt.subplot(2,1,1),plt.stem(h),plt.xlabel('n-->'),plt.ylabel('h[n]')
plt.title('Impulse response (h[n])')
plt.subplot(2,1,2),plt.stem(s),plt.xlabel('n-->'),plt.ylabel('s[n]')
plt.title('Step response (s[n])')
plt.tight_layout()
```

Fig. 4.14 Step response from impulse response

$$y[n] = (n + 1)u[n]$$

The python code, which obtains the step response from the impulse response, is given in Fig. 4.14, and the corresponding output is shown in Fig. 4.15.

Inferences
The following inferences can be made from this experiment:

1. The impulse response of the system is unit step signal.
2. The step response of the system is $(n + 1)u[n]$, which is similar to that of a ramp signal.
3. The step response obtained using '$np.cumsum$' command is in agreement with the theoretical result.

Experiment 4.8 Impulse Response from Step Response
This experiment tries to obtain the impulse response from the step response. If $s[n]$ represents the step response of discrete-time system, then the impulse response of the system is given by

$$h[n] = s[n] - s[n - 1] \qquad (4.21)$$

In this experiment, the step response of the discrete-time system is chosen as unit step signal. This implies $s[n] = u[n]$. Upon taking the impulse response as per Eq. (4.21), one should obtain

$$h[n] = u[n] - u[n - 1] \qquad (4.22)$$

The above equation can be simplified as

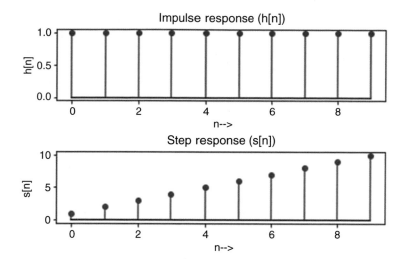

Fig. 4.15 Impulse response and step response of the system

$$h[n] = \delta[n] \tag{4.23}$$

The impulse response of the system is obtained as unit sample signal. The python code, which obtains the impulse response from the step response, is shown in Fig. 4.16, and the corresponding output is shown in Fig. 4.17.

Inferences

From this experiment the following inferences can be drawn:

1. From Fig. 4.17, the step response of the system is unit step sequence.
2. From the step response, the impulse response is derived which is unit sample signal, which is in agreement with the theoretical result.

4.2.4   Pole-Zero Plot of Discrete-Time System

The pole-zero plot of a discrete-time system is plotted in the Z-plane. The position on the complex plane is represented by $re^{j\omega}$. The transfer function of the discrete-time system is given by

$$H(z) = \frac{B(z)}{A(z)} \tag{4.24}$$

The zeros are value of 'z' for which $B(z) = 0$. In other words, zeros are the complex frequencies that make the overall gain of the transfer function is zero. The

```
#Impulse response from step response
import numpy as np
import matplotlib.pyplot as plt
#Step 1: Step response of the system
s=np.ones(100)
s1=np.zeros(len(s)+1)
s1[1:]=s
h=s-s1[0:len(s)] #s[n]-s[n-1]
plt.subplot(2,1,1),plt.stem(s),plt.xlabel('n-->'),plt.ylabel('s[n]'),
plt.title('Step response (s[n])')
plt.subplot(2,1,2),plt.stem(h),plt.xlabel('n-->'),plt.ylabel('h[n]'),
plt.title('Impulse response (h[n])')
plt.tight_layout()
```

Fig. 4.16 Impulse response from the step response

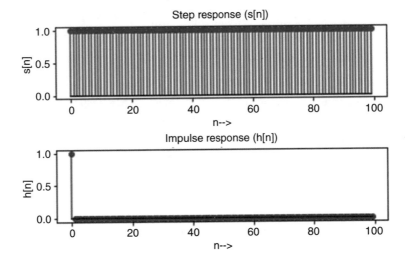

Fig. 4.17 Result of python code shown in Fig. 4.16

poles are the values of 'z' for which $A(z) = 0$. The poles are the complex frequencies that make the overall gain of the transfer function is infinite. The poles and zeros in the Z-plane are indicated by the symbol 'x' and 'o', respectively.

Experiment 4.9 Plotting the Zeros of Non-recursive System

The objective of this experiment is to plot the zeros of the non-recursive system. Consider two discrete-time system with the transfer function $H_1(z) = 1 - z^{-1}$ and $H_2(z) = 1 + z^{-1}$. The built-in function 'tf2zpk' in *scipy* library is utilized to plot the Z-plane of the two systems. The python code does this task is shown in Fig. 4.18, and the corresponding output is shown in Fig. 4.19.

```
#Pole-zero plot
import numpy as np
import matplotlib.pyplot as plt
from scipy import signal
#To plot the unit circle
theta=np.linspace(0,2*np.pi,100)
#Defining system-1
num1, den1=[1,-1],[1]
z1,p1,k1=signal.tf2zpk(num1,den1)
#Defining system-2
num2,den2=[1,1],[1]
z2,p2,k2=signal.tf2zpk(num2,den2)
#To plot unit circle
plt.subplot(1,2,1),plt.plot(np.real(z1),np.imag(z1),'ko')
plt.plot(np.real(p1),np.imag(p1),'rx'),plt.plot(np.cos(theta),np.sin(theta))
plt.title('Z-plane of system-1'),plt.xlabel('$\sigma$'),plt.ylabel('$j\omega$')
plt.subplot(1,2,2), plt.plot(np.real(z2),np.imag(z2),'ko'),
plt.plot(np.real(p2),np.imag(p2),'rx'),plt.plot(np.cos(theta),np.sin(theta)),
plt.title('Z-plane of system-2'),plt.xlabel('$\sigma$'),plt.ylabel('$j\omega$')
plt.tight_layout()
```

Fig. 4.18 Python code to plot the Z-plane of the given system

Inferences

The following inferences can be made for discrete-time systems 1 and 2:

1. System-1 has zero at $z = 1$. This implies that the zero occurs at $\omega = 0$. This zero will block all low frequency components; hence, the system will act like a high pass filter.
2. System-2 has zero at $z = -1$. This implies that the zero occurs at $\omega = \pi$. The system-2 will block all high frequency components. Thus, the system-2 will act like a low pass filter.

Task

1. From the pole-zero plot, will it be possible to find whether systems-1 and -2 are minimum phase system or not?

Experiment 4.10 Plot the Magnitude and Phase Responses of Non-recursive System

The objective of this experiment is to plot the magnitude and phase responses of the given non-recursive systems. Consider two discrete-time system with the transfer function $H_1(z) = 1 - z^{-1}$ and $H_2(z) = 1 + z^{-1}$. The built-in function '*freqz*' in *scipy* library can be used to obtain the magnitude and phase response of the system. The python code, which does this task, is shown in Fig. 4.20, and the corresponding output is shown in Fig. 4.21.

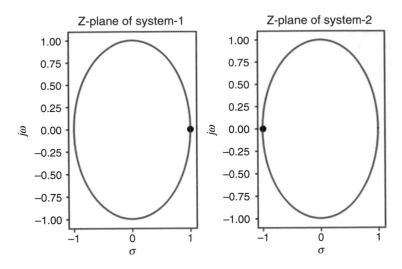

Fig. 4.19 Result of python code shown in Fig. 4.18

Inferences

The following inferences are made from this experiment:

1. From the magnitude responses of system-1 and system-2, it is possible to infer that system-1 acts like a high pass filter and system-2 acts like low pass filter.
2. From the phase responses of the two systems, it is possible to infer that both the systems exhibit linear phase characteristics in the pass band. This means that the phase angle varies linearly with respect to frequency. The linear phase response implies that the system will not exhibit phase distortion. The linear phase characteristics is an important attribute of finite impulse response filter.

Task

1. In the above experiment, what is the purpose of the command '*np.unwrap*'? What will happen if this command is not included in the program?

4.3   Responses of Discrete-Time System

Response of discrete-time system refer to how the discrete-time system react to different types of test signals. The response of discrete-time system to unit sample input signal is referred to impulse response of the system. The response of discrete-time system unit step input signal is referred as step response of the system. The response of discrete-time system to complex exponential signal is referred as frequency response of the system. The frequency response of the system comprises of magnitude response and phase response.

```
#Magnitude and phase response of discrete-time systems
import numpy as np
import matplotlib.pyplot as plt
from scipy import signal
#To plot the unit circle
theta=np.linspace(0,2*np.pi,100)
#Defining system-1
num1,den1=[1,-1],[1]
#Defining system-2
num2,den2=[1,1],[1]
#Magnitude and phase response of systems
w1,H1=signal.freqz(num1,den1)
w2,H2=signal.freqz(num2,den2)
#Plotting the magnitude and phase response of the systems
plt.subplot(2,2,1),plt.plot(w1,20*np.log10(np.abs(H1)))
plt.xlabel('$\omega$'),plt.ylabel('Magnitude'),plt.title('Magnitude response-System1')
plt.subplot(2,2,2),plt.plot(w2,20*np.log10(np.abs(H2)))
plt.xlabel('$\omega$'),plt.ylabel('Magnitude'),plt.title('Magnitude response-System2')
plt.subplot(2,2,3),plt.plot(w1,np.unwrap(np.angle(H1)))
plt.xlabel('$\omega$'),plt.ylabel('Phase'),plt.title('Phase response-System1')
plt.subplot(2,2,4),plt.plot(w2,np.unwrap(np.angle(H2)))
plt.xlabel('$\omega$'),plt.ylabel('Phase'),plt.title('Phase response-System2')
plt.tight_layout()
```

Fig. 4.20 Python code to plot the magnitude and phase response of the system

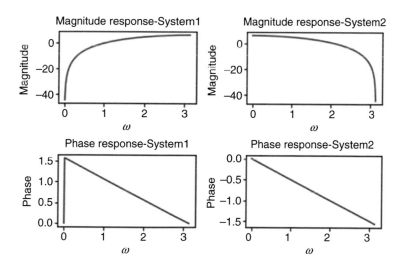

Fig. 4.21 Result of python code shown in Fig. 4.20

Experiment 4.11 Responses of Discrete-Time System
This experiment deals the computation of impulse response, pole-zero plot and
frequency response of the discrete-time system which is given by $h[n] = a^n u$
$[n]$. Let us choose the value of a is 1. The python code, which obtains the impulse
response, pole-zero plot, magnitude and phase response of the discrete-time system,
is shown in Fig. 4.22, and the corresponding output is shown in Fig. 4.23.

Inferences
The following inferences can be drawn from this experiment:

1. The impulse response of the signal is obtained as $h[n] = u[n]$. It is possible to
   observe that the impulse response of the system takes a value of '1'. The impulse
   response is not absolutely summable. Hence, the system is not stable.
2. The pole-zero plot indicates a pole at $z = 1$. This means that the pole is placed at
   $\omega = 0$. For the discrete-time system to be stable, the pole should lie inside the unit
   circle. Here the pole lies on the unit circle. Hence, the system is not BIBO stable.
   The order of the system is one.
3. The magnitude response indicates that the gain of the system at low frequency is
   high and it decreases with increase in frequency.
4. The phase angle varies linearly with respect to frequency.

```python
#Responses of discrete-time system
import numpy as np
import matplotlib.pyplot as plt
from scipy import signal
h=np.zeros(100)
h[0]=1
#Defining system 1
num,den=[1],[1,-1]
#Obtaining the impulse responses of the system
h1=signal.lfilter(num,den,h)
#Obtaining the pole-zero plot of the system
z,p,k=signal.tf2zpk(num,den)
#To obtain the frequency responses of the three systems
w,H=signal.freqz(num,den)
theta=np.linspace(0,2*np.pi,100)
plt.subplot(2,2,1),plt.stem(h1),plt.xlabel('n-->'),plt.ylabel('h[n]'),
plt.title('Impulse response (h[n])')
plt.subplot(2,2,2),plt.plot(w,20*np.log10(np.abs(H)))
plt.xlabel('$\omega$-->'),plt.ylabel('|H(j$\omega$)|'),plt.title('Magnitude response')
plt.subplot(2,2,3),plt.plot(np.real(z),np.imag(z),'ko'),
plt.plot(np.real(p),np.imag(p),'rx'),plt.plot(np.cos(theta),np.sin(theta))
plt.title('Z-plane'),plt.xlabel('$\sigma$'),plt.ylabel('$j\omega$')
plt.subplot(2,2,4),plt.plot(w,np.angle(H))
plt.xlabel('$\omega$-->'),plt.ylabel('$\u2220$H(j$\omega$)'),plt.title('Phase response')
plt.tight_layout()
```

Fig. 4.22 Python code to obtain the responses of the discrete-time system

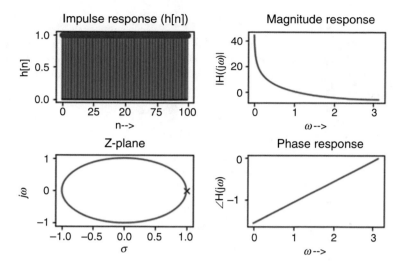

Fig. 4.23 Result of python code shown in Fig. 4.22

Task

1. From the impulse response and the pole-zero plot, will it be possible to comment on the stability of the system?

4.4   Different Representations and Response of Unit Delay Discrete-Time System

Consider a discrete-time system, which introduces unit delay between the input and output signal. The different representations of unit delay system are discussed first.

(a) **Block diagram representation**

    The block diagram, which represents unit delay system, is given in Fig. 4.24.

(b) **Transfer function representation**

    The transfer function representation of unit delay system is given by

$$H(z) = z^{-1} \tag{4.25}$$

(c) **Impulse response of unit delay system**

    The impulse response of discrete-time system is obtained by taking inverse Z-transform of the transfer function. It is given by

Fig. 4.24 Block diagram representation of unit delay system

$$h[n] = Z^{-1}\{H(z)\} \tag{4.26}$$

Substituting the expression of $H(z)$ from Eq. (4.25) in Eq. (4.26), we get

$$h[n] = Z^{-1}\{z^{-1}\}$$

Upon taking inverse Z-transform, the expression for impulse response is obtained as

$$h[n] = \delta[n-1] \tag{4.27}$$

(d) **Step response of the system**

The step response of the system is the response of the system to unit step input signal. For a linear time-invariant system, the relationship between the input and output signal is given by

$$y[n] = x[n] * h[n] \tag{4.28}$$

In the above equation, '*' represents the convolution. To obtain the step response of the system, $x[n] = u[n]$ and $h[n] = \delta[n-1]$, which is obtained from the previous result.

$$s[n] = u[n] * \delta[n-1]$$

The above equation can be simplified by using the property of delta function as

$$s[n] = u[n-1]$$

Experiment 4.12 Unit Delay DT System Analysis

The objective of this experiment is to realize unit delay system using python. The program consists of two sections. First section obtains the response of the unit delay system like, impulse response, step response, magnitude and phase response. The next section deals with exciting the unit delay system with a sinusoidal input signal and obtaining the output signal. The output signal should be a delayed (one unit

```
#Unit delay system
import numpy as np
import matplotlib.pyplot as plt
from scipy import signal
#Step1: Defining the system
num,den=[0,1],[1]
#Part1: Obtaining the responses of discrete-time system
#Impulse response of the system
h1=np.zeros(10)
h1[0]=1
h=signal.lfilter(num,den,h1)
#Step response of the system
s1=np.ones(10)
s=signal.lfilter(num,den,s1)
#Magnitude and phase response of the system
w,H=signal.freqz(num,den)
#Plotting different responses of the system
plt.figure(1),plt.subplot(2,2,1),plt.stem(h),plt.xlabel('n-->'),plt.ylabel('h[n]'),
plt.title('Unit delay system (h[n])'),plt.subplot(2,2,2),plt.plot(w,np.abs(H))
plt.xlabel('$\omega$-->'),plt.ylabel('|H(j$\omega$)|'),plt.title('Magnitude response')
plt.subplot(2,2,3),plt.stem(s),plt.xlabel('n-->'),plt.ylabel('s[n]'),
plt.title('Step response'),plt.subplot(2,2,4),plt.plot(w,np.angle(H))
plt.xlabel('$\omega$-->'),plt.ylabel('$\u2220$H(j$\omega$)'),plt.title('Phase response')
plt.tight_layout()
#Part2: Input and output of the system
t=np.linspace(0,1,100)
x=np.sin(2*np.pi*5*t)
y=signal.lfilter(num,den,x)
plt.figure(2),plt.plot(t,x,t,y),plt.xlabel('Time'),plt.ylabel('Amplitude')
plt.legend(['Input','Output'],loc=1),plt.title('Input and output of Unit delay system')
plt.tight_layout()
```

Fig. 4.25 Python code to obtain the response of unit delay system

delay) version of the input signal. The python code, which accomplishes this task, is shown in Fig. 4.25, and the corresponding outputs are shown in Figs. 4.26 and 4.27, respectively.

Inferences

From this experiment following inferences can be made:

1. From Fig. 4.25, it is possible to observe that the python code consists of two sections. Section 1 obtains the responses of the unit delay system, which include (a) impulse response, (b) step response, (c) magnitude response and (d) phase response.

2. The section of python code simulates sinusoidal signal of 5 Hz frequency as the input signal, and it is fed to the unit delay system to obtain the output signal. The input is represented by the variable 'x', and the variable represents the output 'y'.

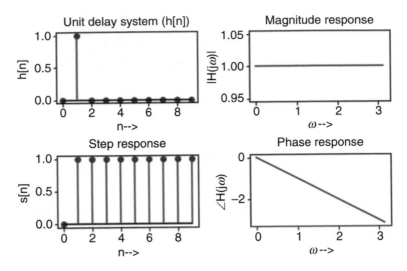

Fig. 4.26 Responses of unit delay system

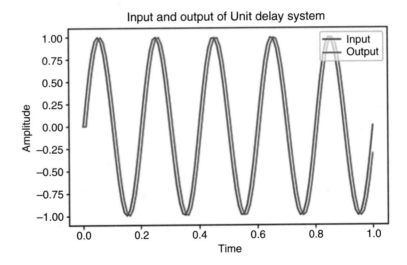

Fig. 4.27 Input and output of unit delay system

3. From Fig. 4.26, the impulse response of the filter is obtained as $h[n] = \delta[n - 1]$, which is in agreement with the theoretical result. The step response of the system is given by $s[n] = u[n - 1]$, which is in agreement with the theoretical result. The magnitude response of the unit delay system is unity, which implies that the system will not affect the magnitude of the input signal. The phase response varies linearly with respect to frequency.

4. Figure 4.27 depicts the input and output signal. The input signal is a sinusoidal signal of 5 Hz frequency. From this figure, it is possible to observe that the output signal is a delayed version of the input signal. The delay between the input and output signal is one unit, which justifies the term the system is a unit delay system.

Task

1. From the impulse response, will it be possible to comment on the stability of the system? If so, state whether the system is stable or unstable.

4.5 Properties of Discrete-Time System

The properties of discrete-time system discussed in this section include (1) linearity property, (2) time shift property, (3) causality and (4) stability.

4.5.1 Linearity Property

A discrete-time system is linear if it obeys superposition principle. According to the superposition principle, the system should obey both the homogeneity and additivity properties. According to the homogeneity property, scaling of the input should result in scaling of the output. Both these properties are expressed in Table 4.1.

For a discrete-time system to be linear, the system should be a relaxed system, and it should obey superposition principle. For a relaxed system, zero input should result in zero output.

Experiment 4.13 Testing the Linearity Property of Discrete-Time System
This experiment tries to check whether the given discrete-time system is a linear or nonlinear. The relationship between the input and output of the system is given by $y[n] = nx[n]$. For the system to be linear, the response of the system to the weighted sum of input is equal to the sum of the weighted responses. The python code, which examines whether the given system is linear or not, is given in Fig. 4.28, and the corresponding output is shown in Fig. 4.29.

Inferences
The following inferences can be made from this experiment:

From Fig. 4.29, it is possible to observe that the response of the system to the weighted sum of input is equal to the sum of the weighted responses. The system obeys superposition principle; hence, it is a linear system.

Task

1. The relationship between the input and output of the system is given by $y[n] = nx[n] + 5$. An offset being added. Will it affect the linearity of the system? Write a python code to illustrate that modified system is a non-linear system.

Table 4.1 Superposition principle

S. No	Property	Representation	Meaning
1	Homogeneity property		Scaling of the input $x[n]$ by a factor 'α' results in scaling of the output $y[n]$ by the same factor 'α'
2	Additivity property		Response of the system to sum of inputs is equal to sum of individual responses

```
#Test for linearity property
import numpy as np
import matplotlib.pyplot as plt
n=np.arange(-10,11,1)
#Step1: Defining the two inputs
x1=(n==1)
x2=(n==2)
#Step2: Defining the scaling factors
alpha,beta=2,4
y1=n*x1
y2=n*x2
#Step3: Response due to weighted sum of input
y_1=n*(alpha*x1+beta*x2)
#Step3:Sum of weighted response
y_2=alpha*y1+beta*y2
#Step4:Plotting the results
plt.subplot(2,1,1),plt.stem(n,y_1),plt.xticks(n),plt.xlabel('n-->'),
plt.ylabel('Amplitude'),plt.title('Response due to weighted sum of inputs')
plt.subplot(2,1,2),plt.stem(n,y_2),plt.xticks(n),plt.xlabel('n-->'),
plt.ylabel('Amplitude'),plt.title('Sum of weighted responses')
plt.tight_layout()
```

Fig. 4.28 Python code to test the linearity of the given discrete-time system

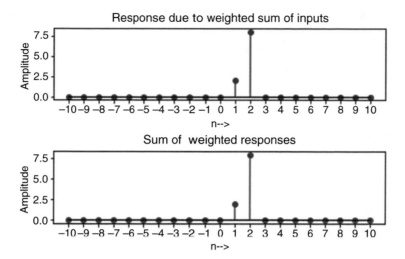

Fig. 4.29 Result of python code shown in Fig. 4.28

4.5.2 Time-Invariant and Time-Variant System

A DT system is said to be time-invariant or shift-invariant if its input-output characteristics do not change with time. This implies time shift of the input causes a corresponding shift in the output. This implies that the system response is independent of time.

Property	Representation		Meaning
Time shift property	$x[n]$ → Time-invariant system → $y[n]$ ⟹ $x[n-k]$ → Time-invariant system → $y[n-k]$		Shift in the input signal $x[n]$ by a factor of 'k' should result in shift in the output signal $y[n]$ by the same factor 'k'

Experiment 4.14 Testing the Time-Invariant Property of Discrete-Time System
The aim of this experiment is to check the given DT system function is time varying or time invarying system by python code. The relationship between the input and output of the discrete-time system is given by $y[n] = ne^{x[n]}$. A discrete-time system is time-invariant, if a time shift in the input signal should result in a time shift in the output signal. The python code, which tests the time-invariance property of the given discrete-time system, is given in Fig. 4.30, and the corresponding output is shown in Fig. 4.31.

Inference
From Fig. 4.31, the output due to time shift in the input is not equal to the time shift in the output; hence, the system is time-variant.

```
#Test for time-invariance property
import numpy as np
import matplotlib.pyplot as plt
n=np.arange(-10,11,1)
#Step1: Defining the input
x=(n==0)
k=5  #Shift parameter
#Step2: Time shift in the input
x1=(n==k)
y1=n*np.exp(x1)
#Step 3: Time shift in the output
y2=(n-k)*np.exp(x1)
#Step4:Plotting the results
plt.subplot(2,1,1),plt.stem(n,y1),plt.xticks(n),plt.xlabel('n-->'),
plt.ylabel('Amplitude'),plt.title('Output due to time shift in the input')
plt.subplot(2,1,2),plt.stem(n,y2),plt.xticks(n),plt.xlabel('n-->'),
plt.ylabel('Amplitude'),plt.title('Time shift in the output')
plt.tight_layout()
```

Fig. 4.30   Python code to test the time-invariance property

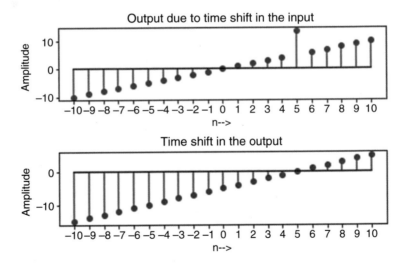

Fig. 4.31   Result of python code shown in Fig. 4.30

Task

1. The relationship between the input and output of the system is given by $y[n] = x[-n]$. Examine whether the system is time-variant or not.

4.5.3 Causal and Non-causal System

A discrete-time system is causal if it is non-anticipatory. The output of the system should not be dependent on the future value of the input. A DT-LTI system is causal if the impulse response of the system is zero for $n < 0$.

Experiment 4.15 Check the DT System Is Causal or Non-causal
This experiment uses python to discuss whether the given DT system is causal or not. Let us consider four discrete-time systems with the impulse responses: $h_1[n] = \delta[n + 1] + \delta[n] + \delta[n - 1]$, $h_2[n] = \delta[n] + \delta[n - 1] + \delta[n - 2]$, $h_3[n] = u[n] - u[n - 1]$ and $h_4[n] = \delta[n + 2] + \delta[n + 1] + \delta[n]$ for $-10 \leq n \leq 10$. The python code, which plots the impulse responses of the above-mentioned discrete-time systems, is shown in Fig. 4.32, and the corresponding output is shown in Fig. 4.33.

Inferences
For the discrete-time system to be causal, the impulse response should be equal to zero for $n < 0$.

1. The impulse response of system-1 is non-zero for $n < 0$; hence, discrete-time system-1 is a non-causal system.
2. The impulse response of system-2 is zero for $n < 0$; hence, discrete-time system-2 is a causal system.

```
#Impulse response of DT systems
import numpy as np
import matplotlib.pyplot as plt
N = 10
n= np.arange(-N, N + 1,1, dtype = float)
#Defining the impulse response of the four systems
h1=np.zeros(2*N+1,dtype='float')
h2=np.zeros(2*N+1,dtype='float')
h3=np.zeros(2*N+1,dtype='float')
h4=np.zeros(2*N+1,dtype='float')
h1=[1 if (i==-1) | (i==0)|(i==1) else 0 for i in n]
h2=[1 if (i==0) | (i==1)|(i==2) else 0 for i in n]
h3=[1 if (i==0)  else 0 for i in n]
h4=[1 if (i==-2) | (i==-1)|(i==0) else 0 for i in n]
#Plotting the impulse response
plt.subplot(2,2,1),plt.stem(n,h1),plt.xlabel('n-->'),plt.ylabel('Amplitude'),
plt.title('$h_1[n]$'),plt.subplot(2,2,2),plt.stem(n,h2),plt.xlabel('n-->'),
plt.ylabel('Amplitude'),plt.title('$h_2[n]$'),plt.subplot(2,2,3),plt.stem(n,h3),
plt.xlabel('n-->'),plt.ylabel('Amplitude'),plt.title('$h_3[n]$'),plt.subplot(2,2,4),
plt.stem(n,h4),plt.xlabel('n-->'),plt.ylabel('Amplitude'),plt.title('$h_4[n]$')
plt.tight_layout()
```

Fig. 4.32 Python code to plot the impulse response of the system

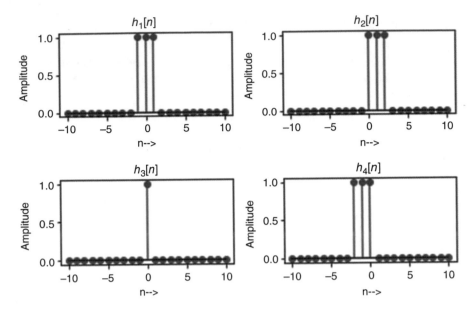

Fig. 4.33 Impulse responses of the discrete-time systems

3. The impulse response of system-3 is zero for $n < 0$; hence, discrete-time system-3 is a causal system.
4. The impulse response of system-4 is non-zero for $n < 0$; hence, discrete-time system-4 is a non-causal system.
5. From the impulse response, it is possible to infer whether the discrete-time system is causal or not.

Task
1. The relationship between the input and output of the system is expressed as $y[n] = x[-n]$. Examine whether the system is causal or not?

4.5.4   Stability of Discrete-Time System

A discrete-time system is stable if the following criterion are met.

(a) **BIBO stability criterion**: A discrete-time system is stable if bounded input results in bounded output.
(b) **Stability criterion with respect to impulse response**: A discrete-time system is stable if the impulse response of the system is absolutely summable.
(c) **Stability with respect to position of pole**: A discrete-time system is stable if the pole of the discrete-time system lies within the unit circle.

It is to be noted that all the above-mentioned criteria are not independent criteria. It means that one criterion implies the other.

Experiment 4.16 BIBO Stability Criterion

The aim of this experiment is to obtain the given DT system is stable or not. The relationship between the input and output of a linear time-invariant discrete-time system is given by $y[n] = x[n] + y[n-1]$. The relationship between the input and output of the system is given by

$$y[n] = x[n] + y[n-1]$$

Taking Z-transform on both sides of the above equation, we get

$$Y(z) = X(z) + z^{-1}Y(z)$$

The transfer function of the system is obtained as

$$H(z) = \frac{Y(z)}{X(z)} = \frac{1}{1 - z^{-1}}$$

The python code, which applies the unit step input signal to the above-mentioned system, is given in Fig. 4.34, and the corresponding output is shown in Fig. 4.35.

Inference

The following inferences can be made from this experiment:

Figure 4.36 shows that the input signal ($x[n]$) is a unit step signal, a bounded input signal. By observing the output signal ($y[n]$) is not a bounded signal. The output signal is a ramp signal, which is not bounded. This shows that bounded input signal to the system does not result in bounded output signal. Hence, the system is not BIBO stable.

Experiment 4.17 Stability Criterion Based on the Impulse Response

This experiment discusses the stability of the DT system to be checked from the impulse response of it. Let us consider two discrete-time LTI systems with impulse responses $h_1[n] = \left(\frac{1}{2}\right)^n u[n]$ and $h_2[n] = (2)^n u[n]$.

For the system to be stable, the impulse response should be absolutely summable. The impulse response of system-1 is absolutely summable; hence, it is stable. On the other hand, the impulse response of the system-2 is not absolutely summable; hence, it is unstable. The python code, which plots the impulse response and obtains the absolute sum of the impulse response of the above-mentioned discrete-time LTI systems, is given in Fig. 4.36, and the corresponding output is shown in Figs. 4.37 and 4.38.

```
#BIBO stability criterion
import numpy as np
import matplotlib.pyplot as plt
from scipy import signal
#Step1: Defining the system
num,den=[1],[1,-1]
#Step 2: Generation of unit step input signal
N=50
n=np.arange(N)
x=np.ones(N)
#Step 3: Obtain the output of the system
y=signal.lfilter(num,den,x)
#Step 4: Plot the input and output of the system
plt.subplot(2,1,1),plt.stem(n,x),plt.xlabel('n-->'),
plt.ylabel('Amplitude'),plt.title('x[n]'),plt.subplot(2,1,2),
plt.stem(n,y),plt.xlabel('n-->'),plt.ylabel('Amplitude'),plt.title('y[n]')
plt.tight_layout()
```

Fig. 4.34 Python code to check the BIBO criterion of discrete-time LTI system

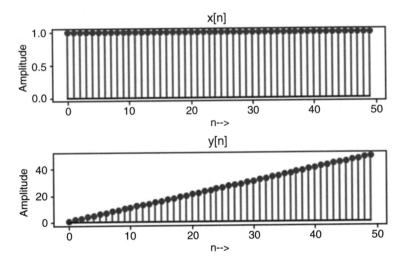

Fig. 4.35 Input-output of the discrete-time LTI system

Inferences

The following inferences can be made from this experiment:

1. From Fig. 4.37, the impulse response of system-1 is absolutely summable, whereas the impulse response of system-2 is not absolutely summable. Hence, system-1 is a stable system, whereas system-2 is not a stable system.

```
#Stability based on impulse response
import numpy as np
import matplotlib.pyplot as plt
N=50
n=np.arange(N)
#Step 1: Defining the impulse response of the two systems
h1=0.5**n
h2=2.0**n
#Step 2: Obtaining the absolute sum of the impulse response
print('The absolute sum of impulse response of system 1 is:',np.sum(abs(h1)))
print('The absolute sum of impulse response of system 2 is:',np.sum(abs(h2)))
#Step 3: Plotting the impulse response of the two systems
plt.subplot(1,2,1),plt.stem(n,h1)
plt.xlabel('n-->'),plt.ylabel('Amplitude'),plt.title('$h_1[n]$')
plt.subplot(1,2,2),plt.stem(n,h2)
plt.xlabel('n-->'),plt.ylabel('Amplitude'),plt.title('$h_2[n]$')
plt.tight_layout()
```

Fig. 4.36 Python code to test the stability of discrete-time system based on the impulse response

The absolute sum of impulse response of system 1 is: **1.999999999999982**

The absolute sum of impulse response of system 2 is: **1125899906842623.0**

Fig. 4.37 The absolute sum of impulse response of the two systems

2. From Fig. 4.38, the impulse response of system-1 ($h_1[n]$) is converging, whereas the impulse response of system-2 ($h_2[n]$) is diverging. Therefore, system-1 is stable, and system-2 is unstable.

Experiment 4.18 Stability Based on the Location of Poles of the Discrete-Time System

This experiment discusses the verification of the stability of the DT system based on the location of the poles of the DT system. Let us consider the transfer function of two discrete-time LTI systems given by $H_1(z) = \frac{2}{(1-0.2z^{-1})(1-0.4z^{-1})}$ and $H_2(z) = \frac{1}{(1-2z^{-1})(1-4z^{-1})}$. For the discrete-time system to be stable, the poles should lie within the unit circle. The poles of system-1 defined the transfer function $H_1(z)$ lies within the unit circle; hence, the system is stable, whereas the poles of the system-2 defined by the transfer function $H_2(z)$ lies outside the unit circle; hence, the system is unstable. The python code to obtain the pole-zero plot of the above-mentioned discrete-time systems is given in Fig. 4.39, and the corresponding output is shown in Fig. 4.40.

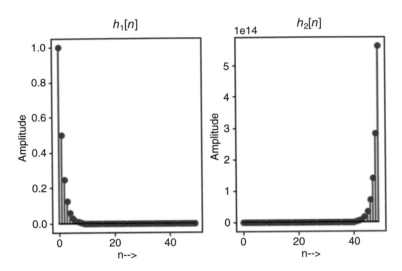

Fig. 4.38 Plot of impulse responses of the discrete-time systems

Inference
From Fig. 4.40, the poles of discrete-time system-1 lies within the unit circle; hence, the system is stable. The poles of discrete-time system-2 lies outside the unit circle; hence, the system is unstable.

Task
1. Comment on the stability of a discrete-time system, whose current output depends on current and past input signal values. Write a python code to validate your answer.

4.5.5   Invertibility of Discrete-Time System

A discrete-time system is invertible if distinct input results in distinct output.

Experiment 4.19 Examining the Invertibility of Discrete-Time System
This experiment tries to examine the invertibility of two discrete-time systems, whose input-output relationship is given by (1) system 1: $y[n] = x[2n]$ and (2) system 2: $y[n] = x[n/2]$. The python code, which examines the invertibility of discrete-time system, is shown in Fig. 4.41, and the corresponding output is shown in Figs. 4.42 and 4.43.

```
#Pole-zero plot of discrete-time systems
import numpy as np
import matplotlib.pyplot as plt
from scipy import signal
#Defining system 1
num1,den1=[2],[1,-0.6,0.032]
#Defining system 2
num2,den2=[1],[1,-6,8]
#Obtaining the pole-zero plot of the system
z1,p1,k1=signal.tf2zpk(num1,den1)
z2,p2,k2=signal.tf2zpk(num2,den2)
theta=np.linspace(0,2*np.pi,100)
re=0;
for i in range(len(p1)):
    if (p1[i].real>1) or (p1[i].imag>1):re=re+1;
    else:re=0;
if re==0:print('System is Stable')
else:print('System is Unstable')
re1=0;
for i in range(len(p2)):
    if (p2[i].real>1) or (p2[i].imag>1):re1=re1+1;
    else:re1=0;
if re1==0:print('System is Stable')
else:print('System is Unstable')
#Plotting the pole-zero plot
plt.subplot(2,1,1),plt.plot(np.real(z1),np.imag(z1),'ko'),plt.plot(np.real(p1),
np.imag(p1),'rx'),plt.plot(np.cos(theta),np.sin(theta)),plt.title('Z-plane of system-1')
plt.xlabel('$\sigma$'),plt.ylabel('$j\omega$'),plt.subplot(2,1,2),
plt.plot(np.real(z2),np.imag(z2),'ko'),plt.plot(np.real(p2),np.imag(p2),'rx')
plt.plot(np.cos(theta),np.sin(theta)),plt.title('Z-plane of system-2'),
plt.xlabel('$\sigma$'),plt.ylabel('$j\omega$')
plt.tight_layout()
```

Fig. 4.39 Python code to obtain the pole-zero plot of the given discrete-time systems

Inferences

The following are the inferences:

1. Two discrete-time systems considered in this example are the following: (1) System-1, downsampling by a factor of 2, and (2) system-2, upsampling by a factor of 2.
2. The input signals considered to excite the discrete-time signals are denoted as $x_1[n]$ and $x_2[n]$. $x_1[n]$ is a DC signal, whereas $x_2[n]$ is the highest frequency digital signal. The output signals of system-1 for the inputs $x_1[n]$ and $x_2[n]$ are denoted as $s_{11}[n]$ and $s_{22}[n]$, respectively.

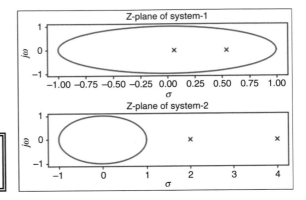

System is Stable

System is Unstable

Fig. 4.40 Result and pole-zero plot of the discrete-time systems

```
#Invertibility of DT system
import numpy as np
import matplotlib.pyplot as plt
N=10
n=np.arange(N)
x1=np.ones(N)
x2=np.exp(1j*np.pi*n)
#System 1 output
s11=x1[::2]
s12=x2[::2]
#System 2 output
s21=np.zeros(2*len(x1))
s21[::2]=x1
s22=np.zeros(2*len(x2))
s22[::2]=x2
plt.figure(1),plt.subplot(2,2,1),plt.stem(n,x1),plt.xlabel('n-->'),plt.ylabel('Amplitude'),
plt.title('$x_1[n]$'),plt.subplot(2,2,2),plt.stem(s11),plt.xlabel('n-->'),
plt.ylabel('Amplitude'),plt.title('$s_{11}[n]$'),plt.subplot(2,2,3),plt.stem(n,x2)
plt.xlabel('n-->'),plt.ylabel('Amplitude'),plt.title('$x_2[n]$'),plt.subplot(2,2,4),
plt.stem(s12),plt.xlabel('n-->'),plt.ylabel('Amplitude'),plt.title('$s_{12}[n]$')
plt.tight_layout()
plt.figure(2),plt.subplot(2,2,1),plt.stem(n,x1),plt.xlabel('n-->'),plt.ylabel('Amplitude'),
plt.title('$x_1[n]$'),plt.subplot(2,2,2),plt.stem(s21),plt.xlabel('n-->'),
plt.ylabel('Amplitude'),plt.title('$s_{21}[n]$'),plt.subplot(2,2,3),plt.stem(n,x2)
plt.xlabel('n-->'),plt.ylabel('Amplitude'),plt.title('$x_2[n]$'),plt.subplot(2,2,4),
plt.stem(s22),plt.xlabel('n-->'),plt.ylabel('Amplitude'),plt.title('$s_{22}[n]$')
plt.tight_layout()
```

Fig. 4.41 Python code to examine the invertibility of the given discrete-time systems

3. The output of discrete-time system-2 for the input signals $x_1[n]$ and $x_2[n]$ is denoted as $s_{21}[n]$ and $s_{22}[n]$, respectively.

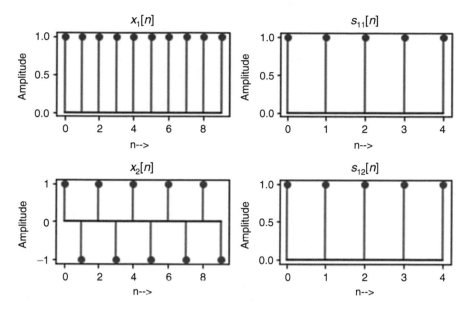

Fig. 4.42 Input-output signals of discrete-time system-I

4. Figure 4.42 shows the input and output signals corresponding to system-1 (downsample by a factor of 2). From Fig. 4.42, it is possible to observe that the input signals $x_1[n]$ and $x_2[n]$ are different but the output $s_{11}[n]$ and $s_{12}[n]$ are the same. This shows that the system produces same output for distinct inputs; hence, the system-1 is a non-invertible system.

5. Figure 4.43 shows the input and output signals corresponding to system-2 (upsample by a factor of 2). From Fig. 4.43, it is possible to observe that the system produces different output for distinct inputs. Hence, the system-2 is an invertible system.

Task

1. Write a python code to prove that cascade connection of *accumulator* and *backward difference* system results in an invertible system. System-1 is an accumulator, whose input-output relation is given by $y[n] = x[n] + y[n-1]$; system-2 is a backward system, whose difference equation is given by $y[n] = x[n] - x[n-1]$.

Exercises

1. Write a python code to plot the impulse response of the discrete-time system whose input-output relationship is given by $y[n] = x[n] + \frac{1}{2}y[n-1]$. From the impulse response plot, will it be possible to comment on the stability of the system?

2. Write a python code to obtain the state-space representation of the discrete-time system whose transfer function is given by $H(z) = \frac{z^2}{z^2 + 4z - 2}$.

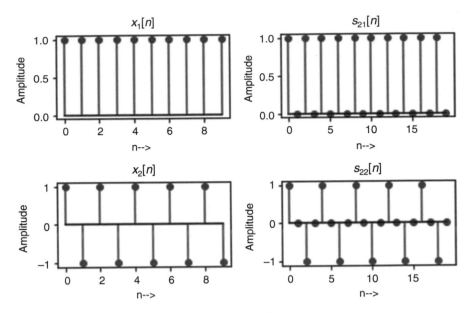

Fig. 4.43 Input-output signals of discrete-time system-II

3. The state-space representation of discrete-time system is given by $x[k+1] = Ax[k] + Bu[k]$ and $y[k] = Cx[k] + Du[k]$, where $A = \begin{bmatrix} 0 & 1 \\ -\frac{1}{6} & -\frac{5}{6} \end{bmatrix}$, $B = \begin{bmatrix} 0 \\ 1 \end{bmatrix}$, $C = [1 \quad 0]$ and $D = [0]$. Write a python code to obtain the transfer function of the system.

4. The impulse response of a discrete-time system $h[n] = (\alpha)^n, -10 \le n \le 10$. Plot the impulse response for $\alpha = \frac{1}{4}, \frac{3}{4}, 1, 2$. Use subplot to plot the impulse responses and comment on the observed result.

5. The relationship between the input and output of a discrete-time system is given by $y[n] = x\left[\frac{n}{2}\right]$. The system is a linear, time-variant system. Write a python code to validate the property.

6. Plot the pole-zero plot of the discrete-time systems whose transfer functions are given by $H_1(z) = 1 + \frac{1}{2}z^{-1}$, $H_2(z) = \frac{1}{1-\frac{3}{4}z^{-1}}$ and $H_3(z) = \frac{1+\frac{1}{2}z^{-1}}{1-\frac{3}{4}z^{-1}}$. Comment on the observed output.

7. Write a python code to obtain the magnitude and phase response of discrete-time system whose input-output relationship is given by $y[n] = \frac{1}{3}\{x[n] + x[n-1] + x[n-2]\}$. Examine whether the system exhibits linear phase characteristics from the phase response.

8. The impulse response of a discrete-time system is given by $h[n] = \delta[n] - \delta[n-1]$. Use subplot to plot the impulse and step responses of the system. Comment on the observed result.

9. Two discrete-time systems with impulse responses $h_1[n] = u[n]$ and $h_2[n] = \delta[n] - \delta[n-1]$ are connected in cascade. Write a python code to plot the magnitude and phase responses of the cascaded system and comment on the observed result.

10. Two discrete-time systems with transfer functions $H_1(z) = X(z^2)$ and $H_2(z) = \frac{1}{2}\left\{X\left(z^{\frac{1}{2}}\right) + X\left(-z^{\frac{1}{2}}\right)\right\}$ are connected in cascade. A sine wave of 5 Hz frequency is fed to the cascaded system. What will be the output of the system? Write a python code to plot the input and output signals and comment on the observed result.

Objective Questions

1. If '$h[n]$' represents the impulse response of the system, then $y[n] = \sum\limits_{k=-\infty}^{n} h[k]$ represents

   A. Magnitude response of the system
   B. Phase response of the system
   C. Shifted impulse response of the system
   D. Step response of the system

2. If the variable 'h' represents the impulse response of the system, then the variable 'y' in the following code results in

   ```
   h=np.ones(10)
   y=np.cumsum(h,axis=0)
   ```

   A. Magnitude response of the system
   B. Phase response of the system
   C. Shifted impulse response of the system
   D. Step response of the system

3. A discrete-time system is linear if it obeys

   A. Superposition theorem
   B. Thevenin's theorem
   C. Tellegen's theorem
   D. Norton's theorem

4. Assertion: Causal systems are non-anticipatory system.
   Reason: In causal system, the current output will not depend on the future value of the input.

   A. Both assertion and reason are true.
   B. Assertion is true; reason is false.
   C. Assertion is false; reason may be true.
   D. Both assertion and reason are false.

5. Identify the system which is NOT a relaxed system:

   A. $y[n] = nx[n]$
   B. $y[n] = x[-n]$
   C. $y[n] = Ax[n]$
   D. $y[n] = Ax[n] + B$

6. The relationship between the input and output of a discrete-time is given by $y[n] = \alpha x^2[n] + \beta x[n] + \gamma$. For the system to be linear

   A. $\alpha = 0$
   B. $\beta = 0$
   C. $\gamma = 0$
   D. $\alpha = 0$ and $\gamma = 0$

7. The impulse response of discrete-time linear, time-invariant system is given by $h[n] = \left[1 - \left(\frac{1}{2}\right)^n\right]u[n]$. The system is

   A. Causal and stable system
   B. Non-causal and stable system
   C. Causal and unstable system
   D. Non-causal and unstable system

8. A linear time-invariant discrete-time system is given by $y[n] = Ax^2[n] + Bx[n - 1] + Cy[n - 1]$. For the system to be static system.

   A. $A = 0$
   B. $B = 0$
   C. $C = 0$
   D. $B = C = 0$

9. The transfer function ($H(z)$) of the system derived from the state-space model is expressed as

   A. $C(zI - A)^{-1}B + D$
   B. $D(zI - A)^{-1}B + C$
   C. $D(zI - B)^{-1}A + C$
   D. $C(zI - B)^{-1}A + D$

10. The transfer function of a discrete-time system is represented as $H(z) = 1 - z^{-1}$. The system has zero at

   A. $\omega = 0$
   B. $\omega = \pi/4$
   C. $\omega = \pi/2$
   D. $\omega = \pi$

11. The transfer function of a discrete-time system is given by $H(z) = \frac{z-1}{\left(z-\frac{1}{2}\right)\left(z+\frac{3}{4}\right)}$.
The poles of the system are at

A. $-\frac{1}{2}, +\frac{3}{4}$
B. $\frac{1}{2}, -\frac{3}{4}$
C. $\frac{1}{2}, \frac{3}{4}$
D. $-\frac{1}{2}, -\frac{3}{4}$

12. The transfer function of a discrete-time system is represented as $H(z) = z^{-1}$. The impulse response of the system is given by

A. $h[n] = \delta[n]$
B. $h[n] = \delta[n-1]$
C. $h[n] = \delta[n-2]$
D. $h[n] = \delta[n-3]$

13. Among the transfer function of the discrete-time systems, identify the system which is NOT BIBO stable system:

A. $H(z) = \frac{1}{1-\frac{1}{4}z^{-1}}$
B. $H(z) = \frac{1}{1-\frac{1}{2}z^{-1}}$
C. $H(z) = \frac{1}{1-\frac{3}{4}z^{-1}}$
D. $H(z) = \frac{1}{1-z^{-1}}$

14. Among the input-output relationship of a given discrete-time systems, identify the system which is NOT a relaxed system:

A. $y[n] = nx[n]$
B. $y[n] = x[-n]$
C. $y[n] = e^{x[n]}$
D. $y[n] = x^2[n]$

Bibliography

1. Charles L. Phillips, John M. Parr, and Eve A. Riskin, "Signals, Systems, and Transforms", Pearson, 2013.
2. David J. Defatta, Joseph G. Lucas, and Villiam S. Hodgkiss, "Digital Signal Processing: A System Design Approach", Wiley India Pvt Ltd, 2009.
3. Simon Haykin, and Bary Van Veen, "Signals and Systems", Wiley, 2005.
4. S. Esakkirajan, T. Veerakumar and Badri N Subudhi, "Digital Signal Processing", McGraw Hill, 2021.
5. Jose Unpingco, "Python for Signal Processing", Springer, 2013.

Chapter 5
Transforms

Learning Objectives
After completing this chapter, the reader is expected to

- Compute the forward and inverse Z-transform.
- Analyse discrete-time system using Z-transform and discrete-time Fourier transform.
- Compute the spectrum of continuous-time and discrete-time signals.
- Plot and infer the spectrogram of stationary and non-stationary signals.
- Plot and interpret the scalogram of non-stationary signal.

Roadmap of the Chapter
Different transforms discussed in this chapter are given below as a flow diagram. Transforms are widely used for signal as well as system study. To analyse discrete-time system, Z-transform is widely used. DTFT can be considered as an evaluation of Z-transform on a unit circle. DTFT is used to obtain the frequency response of the system.

© The Author(s), under exclusive license to Springer Nature Singapore Pte Ltd. 2024   167
S. Esakkirajan et al., *Digital Signal Processing*,
https://doi.org/10.1007/978-981-99-6752-0_5

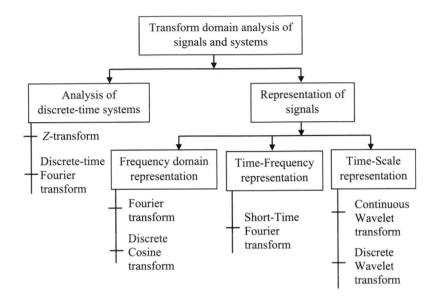

Signals can be analysed entirely in the frequency domain. Example of transform, which gives frequency domain representation, includes Fourier transform and discrete cosine transform. Short-time Fourier transform gives a joint time-frequency representation of the signal. STFT is an effective tool to analyse the non-stationary signal. Example of timescale representation includes continuous and discrete wavelet transform, which are effective in providing multi-resolution (MRA) analysis of the signals.

PreLab Questions
1. What do you understand by the term 'spectrum'?
2. What do you mean by unilateral and bilateral Z-transform?
3. What is the region of convergence (ROC) in the context of Z-transform?
4. Mention the essential condition for the function to be called as basis function? What is the basis of Fourier transform?
5. When applying Fourier analysis to a signal, under which circumstances should Fourier series analysis be employed, and under which circumstances Fourier transform be employed?
6. What is the relationship between discrete-time Fourier transform (DTFT) and Z-transform?
7. What do you mean by a double-sided spectrum of a signal?
8. What is the need of transform in signal analysis?
9. If the DFT of the signal $x[n]$ has to be real, what property should the signal $x[n]$ satisfy?
10. Compare Fourier transform and short-time Fourier transform.
11. Mention the significant features of discrete cosine transform?
12. Mention one significant advantage of wavelet transform over short-time Fourier transform.

5.1 Introduction to Transform

Transform is a tool to analyse signals and systems. Signals are converted from time or spatial domain to frequency domain using transform. Frequency domain is used to describe the signal in terms of frequency components. Each frequency has its own amplitude and phase. From the spectrum, it is possible to interpret the frequencies present in the signal. Thus, the time domain and the frequency domain representation of the signal are equivalent. It is possible to transform the signal from time domain to frequency domain and vice versa without any loss of information. Mathematically, transform takes the inner product of the signal with the basis function. The inner product is one way of quantifying the similarity or the dissimilarity of two signals.

5.2 Z-Transform

The Z-transform is a powerful tool to analyse linear, time-invariant discrete-time systems. The Z-transform for discrete-time signals is the counterpart of the Laplace transform for continuous-time signals. It simplifies the solution of discrete-time problems by converting LTI difference equations to algebraic equations and convolution to multiplication. The Z-transform of a discrete-time signal $x[n]$ is defined as

$$X(z) = \sum_{n=-\infty}^{\infty} x[n]z^{-n} \tag{5.1}$$

The above expression is often termed as two-sided Z-transform. Here, z is a complex variable. The Z-transform of right-sided sequence is expressed as

$$X(z) = \sum_{n=0}^{\infty} x[n]z^{-n} \tag{5.2}$$

The Z-transform of left-sided sequence is expressed as

$$X(z) = \sum_{n=-\infty}^{-1} x[n]z^{-n} \tag{5.3}$$

Region of Convergence The region of convergence of the Z-transform is the value of z for which $X(z)$ is finite. The region of convergence allows the unique inversion of the Z-transform. The ROC depends on the signal $x[n]$ being transformed. The ROC helps to characterize the system as causal or stable.

5.2.1   Z-Transform of Standard Test Sequences

The Z-transform of the standard test sequences is tabulated in Table 5.1.

Experiment 5.1 Z-Transform of the Unit Sample and Unit Step Sequences Using *sympy* Package

This experiment computes the Z-transform of the test sequences like unit sample and unit step sequences. The python code, which computes the Z-transform of test sequences, is shown in Fig. 5.1.

Inferences

The following inferences can be made from this experiment:

1. In this experiment *sympy* library package is utilized to compute the Z-transform of the test sequences.
2. The python command *sym.summation* is used for the summation computation, and *sym.KroneckerDelta* command is used to define the unit impulse sequence.
3. After executing the python code given in Fig. 5.1, the user has to enter '1' for the computation of Z-transform of unit sample sequence and '2' for the computation of Z-transform of unit step sequence.
4. The simulation result of this experiment is shown in Fig. 5.2. From this figure, it is evident that the simulation result is on par with the theoretical result.

Task

1. Write a python code to obtain the Z-transform of $x[n] = nu[n]$.

Experiment 5.2 Z-Transform of Unit Sample and Unit Step Sequences Using *lcapy* Package

This experiment discusses the *lcapy* package, which can be used to compute the Z-transform of unit sample and unit step sequences. The python code is shown in Fig. 5.3.

Inferences

Upon executing the code shown in Fig. 5.3, the result obtained is 1 and $1/(1 - 1/z)$, which is in agreement with the theoretical result. The python command *delta* defines the unit impulse sequence, '*us*' gives the unit step sequence and '*ZT*' obtains the Z-transform.

Experiment 5.3 Z-Transform of $x[n] = e^{jn}u[n]$ and $x[n] = \cos(n)$

This experiment deals with the computation of Z-transform of the given input sequences $x[n] = e^{jn}u[n]$ and $x[n] = \cos(n)$. The python code, which computes the Z-transform of $x[n] = e^{jn}u[n]$ and $x[n] = \cos(n)$, is shown in Fig. 5.4a, and the corresponding output is shown in Fig. 5.4b.

Inferences

1. From Fig. 5.4a, it is possible to observe that the package *lcapy* is used to define the exponential function. *lcapy* is a Python package for linear circuit analysis.

Table 5.1 Z-Transform of the standard test sequences

Sequence	Transform	ROC
$\delta[n]$	1	$\forall z$
$\delta[n-m]$	z^{-m}	$\forall z$ except 0 if $m>0$ or infinity if $m<0$
$u[n]$	$\frac{z}{z-1}$ or $\frac{1}{1-z^{-1}}$	$|z|>1$
$-u[-n-1]$	$\frac{z}{z-1}$ or $\frac{1}{1-z^{-1}}$	$|z|<1$
$nu[n]$	$\frac{z}{(z-1)^2}$ or $\frac{z^{-1}}{(1-z^{-1})^2}$	$|z|>1$
$(n+1)u[n]$	$\frac{z^2}{(z-1)^2}$ or $\frac{1}{(1-z^{-1})^2}$	$|z|>1$
$n^2u[n]$	$\frac{z(z+1)}{(z-1)^3}$ or $\frac{z^{-1}(1+z^{-1})}{(1-z^{-1})^3}$	$|z|>1$
$n^3u[n]$	$\frac{z(z^2+4z+1)}{(z-1)^4}$ or $\frac{z^{-1}(1+4z^{-1}+z^{-2})}{(1-z^{-1})^4}$	$|z|>1$
$n^4u[n]$	$\frac{z(z^3+11z^2+11z+1)}{(z-1)^5}$ or $\frac{z^{-1}(1+11z^{-1}+11z^{-2}+z^{-3})}{(1-z^{-1})^5}$	$|z|>1$
$n^5u[n]$	$\frac{z(z^4+26z^3+66z^2+26z+1)}{(z-1)^6}$ or $\frac{z^{-1}(1+26z^{-1}+66z^{-2}+26z^{-3}+z^{-4})}{(1-z^{-1})^6}$	$|z|>1$
$(-1)^nu[n]$	$\frac{z}{z+1}$ or $\frac{1}{1+z^{-1}}$	$|z|>1$
$-(-1)^nu[-n-1]$	$\frac{z}{z+1}$ or $\frac{1}{1+z^{-1}}$	$|z|<1$
$a^nu[n]$	$\frac{z}{z-a}$ or $\frac{1}{1-az^{-1}}$	$|z|>a$
$a^{|n|}$	$\frac{1-a^2}{(1-az^{-1})(1-az)}$	$a<|z|<\frac{1}{a}$
$a^{n-1}u[n-1]$	$\frac{1}{z-a}$ or $\frac{z^{-1}}{(1-az^{-1})}$	$|z|>a$
$(-a)^nu[n]$	$\frac{z}{z+a}$ or $\frac{1}{1+az^{-1}}$	$|z|>a$
$-a^nu[-n-1]$	$\frac{z}{z-a}$ or $\frac{1}{1-az^{-1}}$	$|z|<a$
$-a^{(n-1)}u[-n]$	$\frac{1}{z-a}$ or $\frac{z^{-1}}{1-az^{-1}}$	$|z|<a$
$a^n[u[n]-u[n-N]]$	$\frac{1-a^Nz^{-N}}{1-az^{-1}}$	$|z|>0$
$-nb^nu[-n-1]$	$\frac{zb}{(z-b)^2}$ or $\frac{bz^{-1}}{\left(1-bz^{-1}\right)^2}$	$|z|<b$
$na^nu[n]$	$\frac{za}{(z-a)^2}$ or $\frac{az^{-1}}{(1-az^{-1})^2}$	$|z|>a$
$\cos(\omega_0 n)u[n]$	$\frac{z^2-z\cos\omega_0}{z^2-2z\cos\omega_0+1}$ or $\frac{1-z^{-1}\cos\omega_0}{1-2z^{-1}\cos\omega_0+z^{-2}}$	$|z|>1$
$\sin(\omega_0 n)u[n]$	$\frac{z\sin\omega_0}{z^2-2z\cos\omega_0+1}$ or $\frac{z^{-1}\sin\omega_0}{1-2z^{-1}\cos\omega_0+z^{-2}}$	$|z|>1$
$[a^n\sin\omega_0 n]u[n]$	$\frac{az\sin\omega_0}{z^2-2az\cos\omega_0+a^2}$ or $\frac{az^{-1}\sin\omega_0}{1-2az^{-1}\cos\omega_0+a^2z^{-2}}$	$|z|>|a|$
$[a^n\cos\omega_0 n]u[n]$	$\frac{z(z-a\cos\omega_0)}{z^2-2az\cos\omega_0+a^2}$ or $\frac{(1-az^{-1}\cos\omega_0)}{1-2az^{-1}\cos\omega_0+a^2z^{-2}}$	$|z|>|a|$

2. Upon executing the commands shown in Fig. 5.4a, the result obtained is shown in Fig. 5.4b. The result of python code is in agreement with the theoretical result.

```
#Z-transform of unit sample and step sequences
import sympy as sym
n = sym.symbols('n', integer=True)
z = sym.symbols('z', complex=True)
S=int(input("Enter : (1= Unit Impulse, 2=Unit Step) : "));
if (S==1):
    X = sym.summation(sym.KroneckerDelta(n, 0) * z**(-n), (n, -sym.oo, sym.oo));
    print('X(z) = ', X)
elif(S==2):
    X = sym.summation(1*z**-n,(n,0,sym.oo));
    print('X(z) = ', X)
else:
    print('Please enter the correct number')
```

Fig. 5.1 Python code for Z-transform of unit sample sequence

```
Enter : (1= Unit Impulse, 2=Unit Step) : 1
X(z) = 1
Enter : (1= Unit Impulse, 2=Unit Step) : 2
X(z) = Piecewise((1/(1 - 1/z), 1/Abs(z) < 1), (Sum(z**(-n), (n , 0, oo)), True))
Enter : (1= Unit Impulse, 2=Unit Step) : 3
Please enter the correct number
```

Fig. 5.2 Simulation result of the python code given in Fig. 5.1

Fig. 5.3 Python code for Z-transform of unit sample and step sequences

```
#Z-transform of unit sample and unit step signal
from lcapy import n,delta,us
x =delta(n)
Xz=x.ZT()
print(Xz)
x1 = us(n)
Yz=x1.ZT()
print(Yz)
```

Experiment 5.4 Z-Transform of $x[n] = \left(\frac{1}{2}\right)^n u[n]$

This experiment discusses the python code to obtain the Z-transform of $x[n] = \left(\frac{1}{2}\right)^n u[n]$ and the corresponding output, which is shown in Fig. 5.5.

Inferences

1. Figure 5.5 shows that the *us* variable is called from the *lcapy* package as a unit step sequence and multiplied by $(1/2)^n$ to get $x[n]$.
2. The Z-transform of $x[n]$ is obtained using 'ZT' python command, and the result is displayed in Fig. 5.5. This result confirms the theoretical result.

```
import lcapy
from lcapy import n
x=lcapy.exp(1j*n)
y=lcapy.cos(n)
Xz=x.ZT()
print(Xz)
Yz=y.ZT()
print(Yz)
```

```
z/(z - exp(j))
z*(z - cos(1))/(z**2 - 2*z*cos(1) + 1)
```

(a) Pyhton code   **(b) Simulation result**

Fig. 5.4 Python code to Experiment 5.3. (**a**) Pyhton code. (**b**) Simulation result

```
from lcapy import n,us
x = (1/2)**n*us(n)
Xz=x.ZT()
print(Xz)
```

```
2*z/(2*z - 1)
```

Fig. 5.5 Python code and its simulation result

Task

1. Write a python code to obtain the Z-transform of $x[n] - \left(\frac{3}{4}\right)^n u[n]$.

5.3 Inverse Z-Transform

This section discusses some of the experiments related to the inverse Z-transform.

Experiment 5.5 Inverse Z-Transform of $X(z) = z^{-1}$
The python code computes the inverse Z-transform of $X(z) = z^{-1}$, and the corresponding output is shown in Fig. 5.6.

Inferences

1. From Fig. 5.6a, it is possible to infer that 'IZT' python command is used to obtain the inverse Z-transform.
2. After executing the python code given in Fig. 5.6a, the result obtained is shown in Fig. 5.6b. This result is in agreement with the theoretical result.

Task

1. Write a python code to obtain the inverse Z-transform of $X(z) = z^{-4}$.

Experiment 5.6 Inverse Z-Transform of $X(z) = 1/1 - z^{-1}$
The python code, which computes the inverse Z-transform of $X(z) = 1/1 - z^{-1}$ and the corresponding output, is shown in Fig. 5.7. From Fig. 5.7b, it is possible to

```
#Inverse z-transform of z^(-1)
from lcapy import z
X=z**(-1)
x=X.IZT()
print(x)
```

Piecewise((UnitImpulse(n - 1), n >= 0))

(a) Python Code   **(b) Simulation result**

Fig. 5.6 Python code to obtain the inverse z-transform of $X(z) = z^{-1}$ and its result. (**a**) Python code. (**b**) Simulation result

```
#Inverse Z-transform
import sympy
import lcapy
from lcapy import z
X=1/(1-z**(-1))
x=X.IZT()
print(x)
```

Piecewise((1, n >= 0))

(a) Python Code   **(b) Simulation result**

Fig. 5.7 Python code to obtain the inverse Z-transform of $X(z) = 1/1 - z^{-1}$ and its result. (**a**) Python code (**b**) Simulation result

observe that the result obtained using python code is in agreement with the theoretical result.

Inferences
The inverse Z-transform of $X(z) = 1/1 - z^{-1}$ will be $u[n]$, and the simulation result of the python code given in Fig. 5.7a is shown in Fig. 5.7b. This result is in agreement with the theoretical result.

Task
1. Write a python code to compute the inverse Z-transform of $X(z) = 1/1 - z^{-2}$.

5.4   Family of Fourier Series and Transforms

Based on the nature of the signal, the Fourier family can be classified into Fourier series or Fourier transform. Fourier series is an effective tool to analyse the periodic signal. If the signal is aperiodic, Fourier transform can be used to analyse the signal. Fourier transform can be viewed as the Fourier series when the period 'T' tends to infinity. The Fourier transform is a generalization of the Fourier series representation of functions. The Fourier series is limited to periodic functions, while Fourier

Table 5.2 Family of Fourier series and transforms

S. No.	Nature of the signal		Fourier family
	Continuous/ discrete	Periodic/ aperiodic	
1	Continuous	Periodic	Fourier series
2	Continuous	Aperiodic	Continuous-time Fourier transform (CTFT)
3	Discrete	Periodic	Discrete-time Fourier series
4	Discrete	Aperiodic	Discrete-time Fourier transform (DTFT)

Fig. 5.8 Spectrum of continuous-time signal

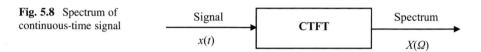

transform can be used for periodic and aperiodic functions. The family of Fourier series and transform is given in Table 5.2.

5.4.1 Continuous-Time Fourier Transform (CTFT)

The continuous-time Fourier transform (CTFT) of the signal $x(t)$ is represented as

$$X(\Omega) = \int_{-\infty}^{\infty} x(t) e^{-j\Omega t} dt \tag{5.4}$$

It can be interpreted as taking the inner product of the signal $x(t)$ with the basis function $e^{-j\Omega t}$. This is represented as

$$X(\Omega) = \langle x(t), e^{-j\Omega t} \rangle \tag{5.5}$$

Equations (5.4) and (5.5) are termed as 'analysis equation'. The result of continuous-time Fourier transform is termed as 'spectrum', which is illustrated in Fig. 5.8. The equations reveal that how an arbitrary signal $x(t)$ can be expanded as a sum of elementary harmonic functions. The elementary harmonic functions are termed as the basis function. The Fourier transform uses complex exponentials of various frequencies as its basis function.

CTFT is a complex function of 'Ω' in the range $-\infty < \Omega < \infty$. CTFT exists if the signal $x(t)$ satisfies Dirichlet conditions, which are given below:

1. The signal $x(t)$ has a finite number of discontinuities and a finite number of maxima and minima in any finite interval.
2. The signal $x(t)$ must be absolutely integrable, which is represented as

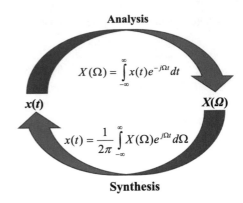

Fig. 5.9 Forward and inverse CTFT

Analysis

$$X(\Omega) = \int_{-\infty}^{\infty} x(t)e^{-j\Omega t}dt$$

$x(t)$ $X(\Omega)$

$$x(t) = \frac{1}{2\pi} \int_{-\infty}^{\infty} X(\Omega)e^{j\Omega t}d\Omega$$

Synthesis

$$\int_{-\infty}^{\infty} |x(t)|dt < \infty.$$

Inverse CTFT refers to obtaining the signal from the spectrum, which is also called as 'synthesis equation'. The inverse CTFT is given by

$$x(t) = \frac{1}{2\pi} \int_{-\infty}^{\infty} X(\Omega)e^{j\Omega t}d\Omega \tag{5.6}$$

From Eq. (5.6), it is possible to interpret that Fourier synthesis formula reconstructs a signal using a set of scaled complex exponentials.

Analysis refers to the decomposition of the signal into its constituent components specifying the weights of the basis functions in the expansion. Synthesis refers to the reconstruction of the signal from the basis functions chosen to represent the signal. The analysis and synthesis function of CTFT of the signal $x(t)$ is illustrated in Fig. 5.9.

(a) **Forward Fourier Transform**

The continuous-time Fourier transform of unit impulse signal is given by

$$\delta(\Omega) = \int_{-\infty}^{\infty} \delta(t)e^{-j\Omega t}dt \tag{5.7}$$

Upon simplifying the above equation, we get

$$\delta(\Omega) = 1 \tag{5.8}$$

From Eq. (5.8), it is possible to interpret that unit impulse contains a component at every frequency. Another way to interpret the result is to make up $\delta(t)$; one needs infinite number of equal frequency components.

(b) **Inverse Fourier transform of $\delta(\Omega)$**

The inverse CTFT of $\delta(\Omega)$ is given by

$$x(t) = \frac{1}{2\pi} \int_{-\infty}^{\infty} X(\Omega)e^{j\Omega t}d\Omega \tag{5.9}$$

Substituting $X(\Omega) = \delta(\Omega)$ in the above expression, we get

$$F^{-1}\{\delta(\Omega)\} = \frac{1}{2\pi} \int_{-\infty}^{\infty} \delta(\Omega)e^{j\Omega t}d\Omega \tag{5.10}$$

Using the sampling property of the impulse signal, the above expression can be simplified as

$$F^{-1}\{\delta(\Omega)\} = \frac{1}{2\pi} \tag{5.11}$$

From the above expression, it is possible to interpret that Fourier transform of a constant signal is

$$1 \leftrightarrow 2\pi\delta(\Omega)$$

Thus, Fourier transform of a DC signal results in an impulse signal in the frequency domain.

Experiment 5.7 Computation of Forward CTFT of the Impulse Signal and Inverse CTFT of the Resultant Forward CTFT

This experiment discusses the computation of forward CTFT of the impulse signal and inverse CTFT of the resultant forward CTFT. The python code that obtains the unit impulse signal spectrum is shown in Fig. 5.10, and the corresponding output is in Fig. 5.11.

Inferences

From Fig. 5.11, it is possible to observe that Fourier transform of an impulse function is a constant function in the frequency domain. The impulse function is a compact function in time domain, whereas its spectrum exists in all frequencies. Thus compression in time domain is equivalent to expansion in frequency domain and vice versa.

```
import numpy as np
import matplotlib.pyplot as plt
t=np.linspace(-5,6,100)
w=np.linspace(-50,60,1000)
x=(t==0)
plt.subplot(2,2,1),plt.plot(t,x),plt.xlabel('t-->'),plt.ylabel('$\u03B4[t]$'),plt.title('Input Signal')
y1=np.zeros(len(w))
for i in range(len(t)):
    y=x[i]*np.exp(-1j*w*t[i])
    y1=y1+y
y2=np.zeros(len(t))
for i in range(len(w)):
    y3=y1[i]*np.exp(1j*w[i]*t)
    y2=y2+y3
plt.subplot(2,2,2),plt.plot(w,np.abs(y1),linewidth=3),plt.title('Magnitude response')
plt.xlabel('$\Omega$-->'),plt.ylabel('|\u03B4($\{\Omega\}$|')
plt.subplot(2,2,3),plt.plot(w,np.angle(y1),linewidth=3),plt.title('Phase response')
plt.xlabel('$\Omega$-->'),plt.ylabel('$\phi(\{\Omega\})$')
plt.subplot(2,2,4),plt.plot(t,y2/np.max(y2)),plt.xlabel('t-->'),plt.ylabel('$\u03B4[t]$'),
plt.title('Reconstructed Signal')
plt.tight_layout()
```

Fig. 5.10 Python code to obtain the spectrum of unit impulse signal

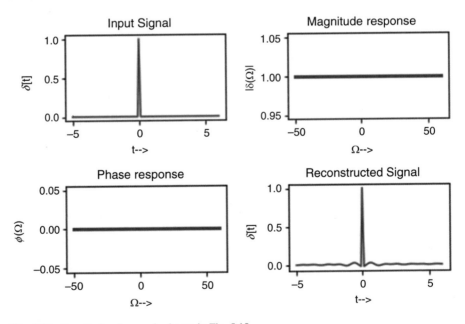

Fig. 5.11 Result of python code shown in Fig. 5.10

Task

1. Modify the above code to obtain the Fourier transform of $x(t) = \delta(t - 5)$ and comment on the observed result.

5.4.2 Fourier Transform of Standard Test Signals

This section focuses on obtaining the spectrum of standard test signals using CTFT. The standard test signals include sinusoidal signal, Gaussian function and pulse signal.

Experiment 5.8 CTFT of the Complex Exponential Signal ($e^{j\Omega_0 t}$ **and** $e^{-j\Omega_0 t}$)

In this experiment, the objective is to obtain the spectrum of the signal $e^{j\Omega_0 t}$ and $e^{-j\Omega_0 t}$. Both the spectrum should be given an impulse corresponding to the frequency 'Ω_0'. The python code which obtains the spectrum of the signals $e^{j\Omega_0 t}$ and $e^{-j\Omega_0 t}$ is shown in Fig. 5.12 and the corresponding output is shown in Fig. 5.13. To show the change in the spectrum between the signals $e^{j\Omega_0 t}$ and $e^{-j\Omega_0 t}$, double-sided spectrum is drawn instead of single-sided spectrum.

```python
import numpy as np
import matplotlib.pyplot as plt
t=np.linspace(-50,50,1000)
w=np.linspace(-5,5,100)
yy=np.exp(1j*(np.pi/4)*t)
xx=np.exp(-1j*(np.pi/4)*t)
plt.subplot(2,2,1),plt.plot(t,yy,linewidth=2),plt.xlabel('t-->'),plt.ylabel('x$_1$(t)')
plt.title('e$^{jΩot}$')
plt.subplot(2,2,2),plt.plot(t,xx,linewidth=2),plt.xlabel('t-->'),plt.ylabel('x$_2$(t)')
plt.title('e$^{-jΩot}$')
y1=np.zeros(len(w))
y2=np.zeros(len(w))
for i in range(len(t)):
    y=yy[i]*np.exp(-1j*w*t[i])
    y1=y1+y
    z=xx[i]*np.exp(-1j*w*t[i])
    y2=y2+z
plt.subplot(2,2,3),plt.plot(w,np.abs(y1)/len(t),linewidth=2)
plt.xlabel('$Ω$-->'),plt.ylabel('|X$_1$(${Ω}$|'),plt.title('Double sided Spectrum')
plt.subplot(2,2,4),plt.plot(w,np.abs(y2),linewidth=2)
plt.xlabel('$Ω$-->'),plt.ylabel('|X$_2$(${Ω}$)|'),plt.title('Double sided Spectrum')
plt.tight_layout()
```

Fig. 5.12 Python code to obtain the spectrum of complex exponential signal

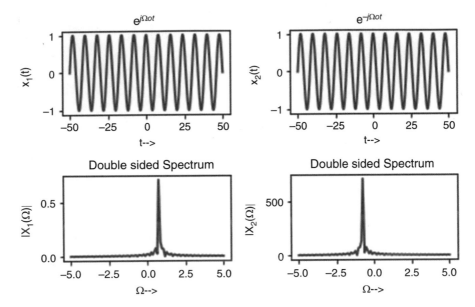

Fig. 5.13 Spectrum of the complex exponential signals

Inferences

From Fig. 5.13, it is possible to observe that both the signals $e^{j\Omega_0 t}$ and $e^{-j\Omega_0 t}$ produce single impulse at $\Omega = \Omega_0$ and at $\Omega = -\Omega_0$. In this case, the value of the frequency is 10 Hz; hence, it is possible to observe impulse at $\pi/4$ Hz and at $-\pi/4$ Hz, respectively, for the signal $e^{j\Omega_0 t}$ and $e^{-j\Omega_0 t}$.

Task

1. Obtain the CTFT of the signal $x(t) = e^{j\Omega_0 t} + e^{-j\Omega_0 t}$ and comment on the observed result.

Experiment 5.9 Fourier Transform of $x(t) = \cos(\Omega t)$

According to Euler's formula, the $\cos(\Omega t)$ can be expressed as

$$\cos(\Omega t) = \frac{e^{j\Omega t} + e^{-j\Omega t}}{2} \tag{5.12}$$

Hence, the signal $x(t)$ is expressed as

$$\cos(\Omega t) = \frac{1}{2}\left(e^{j\Omega t} + e^{-j\Omega t}\right) \tag{5.13}$$

Taking Fourier transform on both sides, we get

```
#Spectrum of Cosine wave
import numpy as np
import matplotlib.pyplot as plt
t=np.linspace(-50,50,1000)
w=np.linspace(-5,5,1000)
yy=np.cos((np.pi/4)*t)
plt.subplot(2,1,1),plt.plot(t,yy,linewidth=1.5),plt.xlabel('t-->'),plt.ylabel('x$_1$(t)')
plt.title('cos(Ω$_o$t)')
y1=np.zeros(len(w))
for i in range(len(t)):
   y=yy[i]*np.exp(-1j*w*t[i])
   y1=y1+y
plt.subplot(2,1,2),plt.plot(w,np.abs(y1)/len(t),linewidth=1.5)
plt.xlabel('$Ω$-->'),plt.ylabel('|X$_1$(${Ω}$)|'),plt.title('Double sided Spectrum')
plt.tight_layout()
```

Fig. 5.14 Python code to obtain the spectrum of cosine wave

$$\text{FT}\{\cos(\Omega t)\} = \frac{1}{2}\{\text{FT}(e^{j\Omega t}) + \text{FT}(e^{-j\Omega t})\} \tag{5.14}$$

From the previous example,

$$\text{FT}(e^{j\Omega t}) = 2\pi\delta(\Omega - \Omega_0) \tag{5.15}$$

$$\text{FT}(e^{-j\Omega t}) = 2\pi\delta(\Omega + \Omega_0) \tag{5.16}$$

Substituting Eqs. (5.15) and (5.16) in Eq. (5.14), we get

$$\text{FT}\{\cos(\Omega t)\} = \frac{1}{2}\{2\pi\delta(\Omega - \Omega_0) + 2\pi\delta(\Omega + \Omega_0)\} \tag{5.17}$$

Simplifying the above expression, we get

$$\text{FT}\{\cos(\Omega t)\} = \pi\{\delta(\Omega - \Omega_0) + \delta(\Omega + \Omega_0)\} \tag{5.18}$$

Thus, the spectrum of the cosine signal has two impulses placed symmetrically at the frequency of the cosine and its negative.

The python code, which obtains the spectrum of cosine wave, is shown in Fig. 5.14, and the corresponding output is shown in Fig. 5.15.

Inferences

From Fig. 5.15, it is possible to observe that the Fourier transform of the cosine signal has two impulses placed symmetrically at the frequency of the cosine and its negative which is in agreement with the theoretical result.

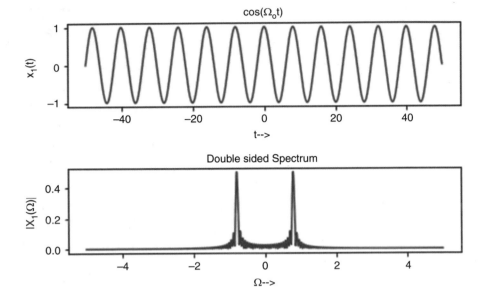

Fig. 5.15 Result of python code shown in Fig. 5.14

Fig. 5.16 Representation of
the signal $x(t)$

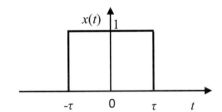

Task

1. Write a python code to illustrate the fact that the magnitude spectrum of sine wave and cosine wave of same amplitude, frequency and phase are alike.

Experiment 5.10 CTFT of the Signal $x(t) = \begin{cases} 1, & |t| < \tau \\ 0, \text{otherwise} \end{cases}$

The given signal is a rectangular pulse. It is shown in Fig. 5.16.

The expression for the CTFT of the signal $x(t)$ is given by

$$X(\Omega) = \int_{-\infty}^{\infty} x(t)e^{-j\Omega t}dt \tag{5.19}$$

In this case, the signal exists from $-\tau$ to τ; hence, the limit of integration is modified as

$$X(\Omega) = \int_{-\tau}^{\tau} x(t)e^{-j\Omega t}dt \qquad (5.20)$$

In the limit $-\tau$ to τ, the value taken by the signal $x(t)$ is one; hence, the above equation can be expressed as

$$X(\Omega) = \int_{-\tau}^{\tau} 1 \times e^{-j\Omega t}dt \qquad (5.21)$$

Upon performing the integration, we get

$$X(\Omega) = \left(\frac{e^{-j\Omega t}}{-j\Omega}\right)_{-\tau}^{\tau} \qquad (5.22)$$

Substituting the upper and lower limits in the above expression, we get

$$X(\Omega) = \frac{e^{-j\Omega\tau} - e^{j\Omega\tau}}{-j\Omega} \qquad (5.23)$$

The above equation can be written as

$$X(\Omega) = \frac{e^{j\Omega\tau} - e^{-j\Omega\tau}}{j\Omega}$$

The above equation can be simplified as

$$X(\Omega) = 2\frac{\sin(\Omega\tau)}{\Omega}$$

Multiplying and dividing the above equation by 'τ', we get

$$X(\Omega) = 2\tau\ \frac{\sin(\Omega\tau)}{\Omega\tau} = 2\tau \sin c(\Omega\tau) \qquad (5.24)$$

From the above expression, it is possible to conclude that Fourier transform of a rectangular function will result in a *sinc* function.

The objective is to write a python code to generate two rectangular functions with different width. Pass these two rectangular functions through Fourier transform to obtain their spectra. The python code, which generates two rectangular functions of different width and their corresponding spectra, is shown in Fig. 5.17, and the corresponding output is obtained in Fig. 5.18.

```
import numpy as np
import matplotlib.pyplot as plt
t=np.linspace(-50,50,1000)
w=np.linspace(-5,5,100)
yy=(abs(t)<15)
xx=(abs(t)<2)
plt.subplot(2,2,1),plt.plot(t,yy,linewidth=1.5),plt.xlabel('t-->'),plt.ylabel('x$_1$(t)')
plt.title('Rectangular Function-1')
plt.subplot(2,2,2),plt.plot(t,xx,linewidth=1.5),plt.xlabel('t-->'),plt.ylabel('x$_2$(t)')
plt.title('Rectangular Function-2')
y1=np.zeros(len(w))
y2=np.zeros(len(w))
for i in range(len(t)):
    y=yy[i]*np.exp(-1j*w*t[i])
    y1=y1+y
    z=xx[i]*np.exp(-1j*w*t[i])
    y2=y2+z
plt.subplot(2,2,3),plt.plot(w,np.abs(y1)/len(t),linewidth=1.5)
plt.xlabel('$\Omega$-->'),plt.ylabel('|X$_1$(${\Omega}$)|'),plt.title('Spectrum-1')
plt.subplot(2,2,4),plt.plot(w,np.abs(y2),linewidth=1.5)
plt.xlabel('$\Omega$-->'),plt.ylabel('|X$_2$(${\Omega}$)|'),plt.title('Spectrum-2')
plt.tight_layout()
```

Fig. 5.17 Python code to obtain the spectrum of rectangular function

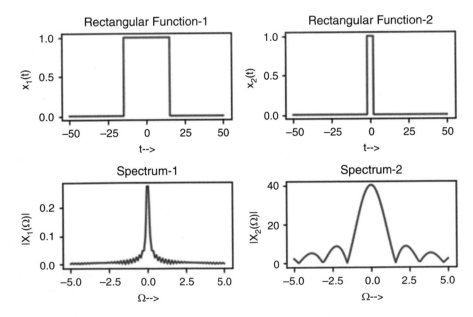

Fig. 5.18 Result of python code shown in Fig. 5.17

Inferences

The following inferences can be obtained by observing Fig. 5.18:

1. Two rectangular functions, rectangular function-1 and rectangular function-2, are generated.
2. The width of rectangular function-1 is larger than the width of rectangular function-2.
3. The spectrum of the rectangular function is observed to be a *sinc* function.
4. The main lobe width of spectrum-1 is narrower when compared to the main lobe width of spectrum-2.
5. This example illustrates the fact that compression in the time domain leads to expansion in the frequency domain and vice versa.

Task

1. Instead of rectangular pulse, obtain the magnitude spectrum of triangular pulse signal and comment on the observed result. Triangular pulse can be obtained by convolving two rectangular pulse signals.

Experiment 5.11 Inverse CTFT of $X(\Omega) = \begin{cases} 1, & |\Omega| < \Omega_0 \\ 0, & \text{otherwise} \end{cases}$

The expression for inverse continuous-time Fourier transform is given by

$$x(t) = \frac{1}{2\pi} \int_{-\infty}^{\infty} X(\Omega)e^{j\Omega t}d\Omega \tag{5.25}$$

The spectrum exists from $-\Omega_0$ to Ω_0; hence, the limit of integration has to be changed. This is represented as

$$x(t) = \frac{1}{2\pi} \int_{-\Omega_0}^{\Omega_0} X(\Omega)e^{j\Omega t}d\Omega \tag{5.26}$$

In the interval from $-\Omega_0$ to Ω_0, the value of the spectrum is unity. This is expressed as

$$x(t) = \frac{1}{2\pi} \int_{-\Omega_0}^{\Omega_0} 1 \times e^{j\Omega t}d\Omega \tag{5.27}$$

Upon performing the integration, we get

$$x(t) = \frac{1}{2\pi} \left(\frac{e^{j\Omega t}}{jt}\right)^{\Omega_0}_{-\Omega_0} \tag{5.28}$$

Substituting the upper and lower limits, we get

$$x(t) = \frac{1}{2\pi} \left(\frac{e^{j\Omega_0 t} - e^{-j\Omega_0 t}}{jt}\right) \tag{5.29}$$

The above equation can be written as

$$x(t) = \frac{1}{\pi t} \left(\frac{e^{j\Omega_0 t} - e^{-j\Omega_0 t}}{2j}\right) \tag{5.30}$$

The above equation can be expressed as

$$x(t) = \frac{\sin(\Omega_0 t)}{\pi t} \tag{5.31}$$

Thus, inverse Fourier transform of a rectangular function results in a sinc function. Comparing this example with the previous example, it is possible to write that rectangular function and sinc function are dual functions in Fourier domain.

The aim of this experiment is to prove that rectangular and sinc functions are dual functions in the Fourier domain. In the previous experiment, it is possible to prove that Fourier transform of rectangular function results in sinc function. In this experiment, the objective is to prove that Fourier transform of sinc function will result in a rectangular function. Execute the python code given in Fig. 5.19 and enter the number '1'. The simulation result of this python code is shown in Fig. 5.20.

Inferences
From Fig. 5.20, it is possible to observe that Fourier transform of sinc function results in a rectangular function. Also, it is possible to infer that compression in one domain (time) corresponds to expansion in another domain (frequency) and vice versa.

Task
1. What is the reason for ringing effect observed in the magnitude spectrum of sinc signal? Is there any way to minimize the ringing effect?

Experiment 5.12 CTFT of a Gaussian Function
The objective of this experiment is to prove that Fourier transform of a Gaussian function results in a Gaussian function. The expression for Gaussian function with mean μ and standard deviation σ is given by

```
#Fourier transform of Sinc and Gaussian functions
import numpy as np
import matplotlib.pyplot as plt
#Step 1: Generation of sinc function
t=np.linspace(-5,5,1000)
w=np.linspace(-60,60,1000)
S=int(input("Enter : (1 = Sinc, 2 = Gaussian) : "));
if (S==1):
  x1=np.sinc(t);
  x2=np.sinc(2*t)
  y1=np.zeros(len(w))
  y2=y1;
  for i in range(len(t)):
    yx1=x1[i]*np.exp(-1j*w*t[i]);
    yx2=x2[i]*np.exp(-1j*w*t[i]);
    y1=y1+yx1;#Step 2: Spectrum of sinc function
    y2=y2+yx2;#Step 2: Spectrum of sinc function
elif(S==2):
  mu,sigma1,sigma2=0,0.1,0.5; #Mean and sigma values
  x1=np.exp(-np.power(t - mu, 2.) / (2 * np.power(sigma1, 2.)));
  x2=np.exp(-np.power(t - mu, 2.) / (2 * np.power(sigma2, 2.)));
  y1=np.zeros(len(w))
  y2-y1;
  for i in range(len(t)):
    yx1=x1[i]*np.exp(-1j*w*t[i]);
    yx2=x2[i]*np.exp(-1j*w*t[i]);
    y1=y1+yx1;#Step 2: Spectrum of Gaussian function
    y2=y2+yx2;#Step 2: Spectrum of Gaussian function
else:
  print('Please enter the correct number')
  x1,x2,y1,y2=np.zeros(len(t)),np.zeros(len(t)),np.zeros(len(w)),np.zeros(len(w));
#Step 3: Plotting the results
plt.subplot(2,2,1),plt.plot(t,x1),plt.xlabel('t-->'),plt.ylabel('x$_1$(t)'),plt.title('Signal-1'),
plt.subplot(2,2,2),plt.plot(t,x2),plt.xlabel('t-->'),plt.ylabel('x$_2$(t)'), plt.title('Signal-2')
plt.subplot(2,2,3),plt.plot(w,np.abs(y1)),plt.xlabel('$\Omega$-->'),plt.ylabel('|X$_1$(${\Omega}$)|'),
plt.title('Spectrum of x$_1$(t)'),plt.subplot(2,2,4),plt.plot(w,np.abs(y2)), plt.xlabel('$\Omega$-->'),
plt.ylabel('|X$_2$(${\Omega}$)|'),plt.title('Spectrum of x$_2$(t)')
plt.tight_layout()
```

Fig. 5.19 Python code to obtain the spectrum of sinc and Gaussian function

$$x(t) = \frac{1}{\sigma\sqrt{2\pi}} e^{-\frac{(t-\mu)^2}{2\sigma^2}} \tag{5.32}$$

If the mean of the Gaussian function is zero, the above expression is given by

$$x(t) = \frac{1}{\sigma\sqrt{2\pi}} e^{-\frac{t^2}{2\sigma^2}} \tag{5.33}$$

Differentiating both sides with respect to t, we get

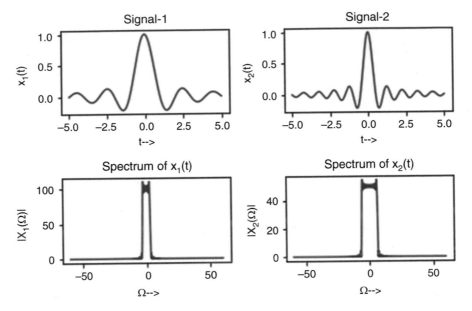

Fig. 5.20 Result of python code shown in Fig. 5.19

$$\frac{\mathrm{d}x(t)}{\mathrm{d}t} = \frac{1}{\sigma\sqrt{2\pi}} e^{-\frac{t^2}{2\sigma^2}} \times \frac{-2t}{2\sigma^2} \tag{5.34}$$

The above equation can be simplified as

$$\frac{\mathrm{d}x(t)}{\mathrm{d}t} = \frac{1}{\sigma\sqrt{2\pi}} e^{-\frac{t^2}{2\sigma^2}} \times \frac{-t}{\sigma^2} \tag{5.35}$$

Substituting Eq. (5.32) in Eq. (5.35), we get

$$\frac{\mathrm{d}x(t)}{\mathrm{d}t} = x(t) \times \frac{-t}{\sigma^2} \tag{5.36}$$

The above equation can be rearranged as

$$\frac{\mathrm{d}x(t)}{\mathrm{d}t} = -\frac{1}{\sigma^2} tx(t) \tag{5.37}$$

Taking Fourier transform on both sides, we get

$$\text{FT}\left\{\frac{dx(t)}{dt}\right\} = -\frac{1}{\sigma^2}\text{FT}\{tx(t)\} \tag{5.38}$$

Using the following fact

$$\text{FT}\left\{\frac{dx(t)}{dt}\right\} = j\Omega X(\Omega) \tag{5.39}$$

$$\text{FT}\{tx(t)\} = j\frac{dX(\Omega)}{d\Omega} \tag{5.40}$$

Substituting Eqs. (5.39) and (5.40) in Eq. (5.38), we get

$$j\Omega X(\Omega) = -\frac{1}{\sigma^2} \times j\frac{dX(\Omega)}{d\Omega} \tag{5.41}$$

Simplifying the above expression, we get

$$\Omega X(\Omega) = -\frac{1}{\sigma^2} \times \frac{dX(\Omega)}{d\Omega} \tag{5.42}$$

Upon rearranging the terms, we get

$$\frac{\frac{dX(\Omega)}{d\Omega}}{X(\Omega)} = -\sigma^2\Omega \tag{5.43}$$

Taking integral on both sides, we get

$$\int \frac{\frac{dX(\Omega)}{d\Omega}}{X(\Omega)} = -\sigma^2 \int \Omega d\Omega \tag{5.44}$$

Upon integration, we get

$$\ln[X(\Omega)] - \ln[X(0)] = -\sigma^2\frac{\Omega^2}{2} \tag{5.45}$$

Since the mean value of the Gaussian signal is assumed to be zero, the above equation can be written as

$$\ln[X(\Omega)] = -\sigma^2\frac{\Omega^2}{2} \tag{5.46}$$

Taking exponential on both sides, we get

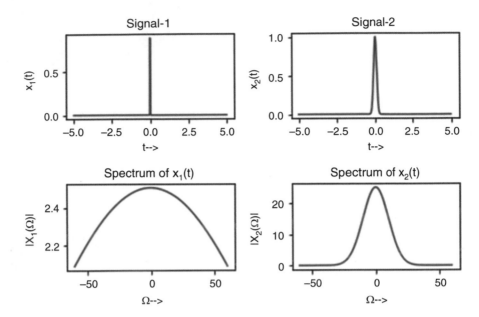

Fig. 5.21 Spectra of Gaussian functions

$$X(\Omega) = e^{-\frac{\sigma^2 \Omega^2}{2}} \tag{5.47}$$

From the above expression, it is possible to interpret that Fourier transform of a Gaussian function results in a Gaussian function.

In this experiment, two Gaussian signals are generated with zero mean and standard deviation as $\sigma_1 = 0.01$ and $\sigma_2 = 0.1$, respectively. After generating the two Gaussian signals, their spectra are obtained by taking the Fourier transform. The python code, which does this task, is shown in Fig. 5.19. After executing this code, enter the number '2'. The simulation result is shown in Fig. 5.21.

Inference
1. From Fig. 5.21, it is possible to observe that two Gaussian functions with zero mean and standard deviation $\sigma_1 = 0.01$ and $\sigma_2 = 0.1$ are generated.
2. Gaussian function-1 (x_1) has a narrow spread, whereas Gaussian function-2 (x_2) has a wider spread.
3. Upon obtaining the spectra, it is possible to infer the fact that if the signal spread is narrow in time domain (Gaussian function-1), the corresponding spectrum has wide spread (spectrum of Gaussian function-1).
4. On the other hand, if the Gaussian function has wide spread in time domain (Gaussian function-2), its spectrum is narrower (spectrum of Gaussian function-2).
5. This illustrates the fact that 'Compression in one domain leads to expansion in another domain and vice-versa'.

Task

1. In the above experiment, Signal-1 and Signal-2 are Gaussian functions. Now multiply Signal-1 and Signal-2 to obtain Signal-3. Obtain the spectrum of Signal-3 and comment on the observed result.

5.4.3   Discrete-Time Fourier Transform (DTFT)

Discrete-time Fourier transform is a transformation that maps the discrete-time signal into a complex valued function, which is given by

$$X\left(e^{j\omega}\right) = \sum_{n=-\infty}^{\infty} x[n]e^{-j\omega n} \tag{5.48}$$

DTFT is a way to represent the frequency content of discrete-time signal. The magnitude and phase form of DTFT representation is given by

$$X\left(e^{j\omega}\right) = \left|X\left(e^{j\omega}\right)\right|e^{j\Phi(e^{j\omega})} \tag{5.49}$$

In the above expression, $|X(e^{j\omega})|$ represents the magnitude of DTFT, and $\Phi(e^{j\omega})$ represents the phase of DTFT. The magnitude spectrum determines the relative presence of a sinusoid in the signal $x[n]$, whereas the phase spectrum determines how the sinusoids line up relative to one another to form the signal $x[n]$. The condition for the existence of DTFT is that the signal $x[n]$ should be absolutely summable. The signal $x[n]$ is absolutely summable if it obeys the following condition:

$$\sum_{n=-\infty}^{\infty} |x[n]| < \infty \tag{5.50}$$

The expression inverse discrete-time Fourier transform (IDTFT) expression is given by

$$x[n] = \frac{1}{2\pi} \int_{-\pi}^{\pi} X\left(e^{j\omega}\right)e^{j\omega n} d\omega \tag{5.51}$$

Experiment 5.13 DTFT of $x[n] = \begin{cases} 1, & |n| < N \\ 0, & \text{otherwise} \end{cases}$

The signal $x[n]$ represents a rectangular pulse. The DTFT of $x[n]$ is given by

```
# Python code for DTFT of rectangular pulse signal
import numpy as np
import matplotlib.pyplot as plt
n=np.arange(-5,6)
w=np.arange(-3*np.pi,3*np.pi,0.1)
x=(n>=0)
y=(n<=3)
z=x*y
plt.subplot(3,1,1),plt.stem(n,z),plt.xlabel('n-->'),plt.ylabel('x[n]')
y1=np.zeros(len(w))
for i in range(len(n)):
  y=z[i]*np.exp(-1j*w*n[i])
  y1=y1+y
print(y1)
plt.subplot(3,1,2),plt.plot(w,np.abs(y1),linewidth=3),plt.title('Magnitude response')
plt.xlabel('$\omega$-->'),plt.ylabel('|X(${j\omega)}$|')
plt.subplot(3,1,3),plt.plot(w,np.angle(y1),linewidth=3),plt.title('Phase response')
```

Fig. 5.22 Python code for Experiment 5.13

$$X\left(e^{j\omega}\right) = \sum_{n=-N}^{N} 1 \times e^{-j\omega n}$$

Using the summation formula $\displaystyle\sum_{n=-N}^{N} a^n = \begin{cases} \dfrac{a^{N+1} - a^{-N}}{a - 1}, & |a| < 1 \\ 2N + 1, & a = 1 \end{cases}$, the above

equation can be written as

$$X\left(e^{j\omega}\right) = \frac{e^{-j\omega(N+1)} - e^{j\omega N}}{e^{-j\omega} - 1}, \left|e^{-j\omega}\right| < 1$$

The above equation can be simplified as

$$X\left(e^{j\omega}\right) = \frac{e^{-j\omega N}e^{-j\omega} - e^{j\omega N}}{e^{-j\omega} - 1}$$

The above equation can be written as

$$X\left(e^{j\omega}\right) = \frac{\sin\left(N + \frac{1}{2}\right)\omega}{\sin\left(\frac{\omega}{2}\right)} \text{ if } \left|e^{-j\omega}\right| < 1$$

This shows that Fourier transform of a rectangular pulse signal will result in a sinc function.

Python code for the DTFT of rectangular pulse is given in Fig. 5.22, and its corresponding output is shown in Fig. 5.23.

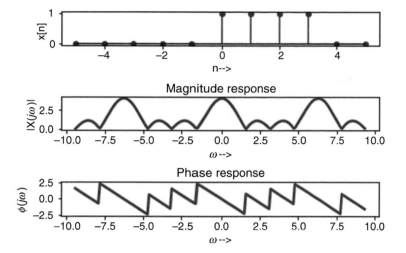

Fig. 5.23 Simulation result of python code given in Fig. 5.22

Fig. 5.24 Discrete-time
LTI system

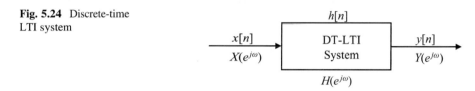

Inferences

From Fig. 5.23, it is possible to infer that the magnitude response of a rectangular function is a *sinc* function, which is in agreement with the theoretical result.

Task

1. In the above experiment, $x[n]$ is a rectangular pulse signal. What will be the impact of increasing the width of the signal $x[n]$ in the magnitude and phase responses?

5.4.4 Analysis of Discrete-Time LTI System Using DTFT

The block diagram of discrete-time LTI system with the input signal $x[n]$, impulse response $h[n]$ and the output signal $y[n]$ is shown in Fig. 5.24.

The relationship between the input and output of the system if it is LTI is given by

$$y[n] = x[n] * h[n] \tag{5.52}$$

Upon taking DTFT on both sides of the above equation, we get

$$Y(e^{j\omega}) = X(e^{j\omega})H(e^{j\omega}) \tag{5.53}$$

Equation (5.53) is obtained using the fact that convolution in time domain is equivalent to multiplication in the Fourier domain.

The frequency response of the system from Eq. (5.53) can be expressed as

$$H(e^{j\omega}) = \frac{Y(e^{j\omega})}{X(e^{j\omega})} \tag{5.54}$$

The frequency response of the system is a combination of magnitude and phase responses. This is expressed as

$$H(e^{j\omega}) = |H(e^{j\omega})|e^{j\phi(e^{j\omega})} \tag{5.55}$$

The frequency response defines how a complex exponential is changed in amplitude and phase by a system.

Experiment 5.14 Computation of the Magnitude and Phase Responses of Discrete-Time System Using DTFT

This experiment discusses the computation of magnitude and phase responses of DT system using DTFT. Let us consider two discrete-time systems and its impulse responses given by $h_1[n] = \{\frac{1}{2}, \frac{1}{2}\}$ and $h_2[n] = \{\frac{1}{2}, -\frac{1}{2}\}$. The python code, which obtains the magnitude and phase responses of the two systems, is shown in Fig. 5.25, and the corresponding output is shown in Fig. 5.26.

Inferences

From Fig. 5.26, the following inferences can be made:

1. The magnitude response of system-1 shows that the system behaves like a lowpass filter.
2. The magnitude response of system-2 shows that the system behaves like a highpass filter.
3. The phase responses of both these systems reveal that both systems exhibit linear phase characteristics.
4. The response of the two systems is in agreement with the theoretical result.

Task

1. From the magnitude response, it is possible to observe that the roll-off rate is not sharp? What has to be done to improve the roll-off rate?

```
import numpy as np
import matplotlib.pyplot as plt
from scipy import signal
#Step 1: Impulse response of the two systems
h1=[0.5,0.5]
h2=[0.5,-0.5]
#Step 2: Obtaining the frequency response
w1, H1 = signal.freqz(h1,1)
w2, H2 = signal.freqz(h2,1)
angle_1 = np.unwrap(np.angle(H1))
angle_2 = np.unwrap(np.angle(H2))
#Step3 : Plotting the responses
plt.subplot(2,3,1),plt.stem(h1),plt.xlabel('n-->'),plt.ylabel('h$_1$[n]')
plt.title('Impulse response'),plt.subplot(2,3,2),plt.plot(w1, 10 * np.log10(abs(H1)))
plt.xlabel('$\omega$-->'),plt.ylabel('|X($j\omega$)|'),plt.title('Magnitude response')
plt.subplot(2,3,3),plt.plot(w1,(angle_1)),plt.xlabel('$\omega$--
>'),plt.ylabel('$\phi(j\omega})$')
plt.title('Phase response'),plt.subplot(2,3,4),plt.stem(h2),plt.xlabel('n-->'),
plt.ylabel('h$_2$[n]'),plt.title('Impulse response'),plt.subplot(2,3,5),
plt.plot(w1, 10 * np.log10(abs(H2))),plt.xlabel('$\omega$-->'),
plt.ylabel('|X($j\omega}$)|'),plt.title('Magnitude response')
plt.subplot(2,3,6),plt.plot(w1,(angle_2)),plt.xlabel('$\omega$-->'),
plt.ylabel('$\phi(j\omega})$'),plt.title('Phase response')
plt.tight_layout()
```

Fig. 5.25 Python code to obtain the magnitude and phase response of the systems

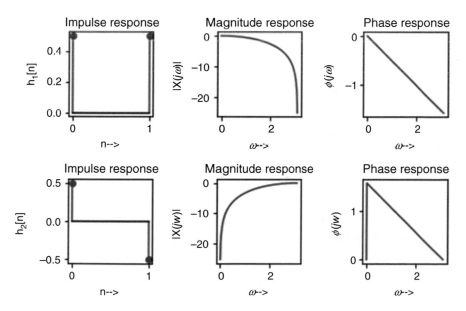

Fig. 5.26 Responses of the two systems

5.4.5   Discrete Fourier Transform

Discrete Fourier transform (DFT) represents a signal in terms of sinusoids. For a discrete-time signal of length N, the basis functions are sinusoids of length N. Discrete Fourier transform is used when the signal is discrete-time and periodic only. In practice, it calculates the frequency domain representation of aperiodic signals in a given time interval, by assuming their periodic extension. The discrete Fourier transform (DFT) of the signal $x[n]$ of length N is given by

$$X[k] = \sum_{n=0}^{N-1} x[n] e^{-j\frac{2\pi}{N}kn}, \quad k = 0, 1, 2, \ldots, N-1 \tag{5.56}$$

The above expression can be written in the form of

$$X[k] = \sum_{n=0}^{N-1} x[n] W_N^{kn} \tag{5.57}$$

where

$$W_N = e^{-j\frac{2\pi}{N}} \tag{5.58}$$

The signal is reconstructed by using the inverse discrete Fourier transform, which is defined as

$$x[n] = \frac{1}{N} \sum_{k=0}^{N-1} X[k] e^{j\frac{2\pi}{N}kn} \tag{5.59}$$

The forward transform is generally known as 'analysis', and the inverse transform is called as 'synthesis'.

Experiment 5.15 Plotting the Twiddle Factor for $N = 8$
The aim of this experiment is to plot the twiddle factor of DFT with the length $N = 8$. The python code, which plots the twiddle factor or phase factor for $N = 8$, is shown in Fig. 5.27, and the corresponding output is shown in Fig. 5.28.

Inference
From Fig. 5.28, it is possible to observe that for the choice of $N = 8$, the unit circle is divided into eight equal portions.

(a) **DFT matrix**
    The DFT matrix of order N is given by

Fig. 5.27 Python code to plot the twiddle factor for $N = 8$

```
import numpy as np
import matplotlib.pyplot as plt
n=8
for k in range(0,n):
    z=np.exp(2*np.pi*1j*k/n)
    plt.plot([0,np.real(z)],[0,np.imag(z)])
    x=np.linspace(0,2*np.pi,100)
    plt.plot(np.cos(x),np.sin(x),color='gray')
    plt.axis('square')
```

Fig. 5.28 Result of python code shown in Fig. 5.27

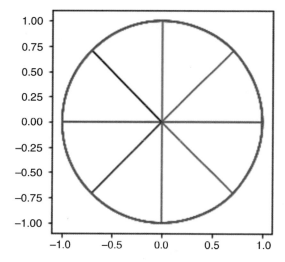

$$W_N = \begin{bmatrix} 1 & 1 & 1 & \cdots & 1 \\ 1 & W_N^{1 \cdot 1} & W_N^{1 \cdot 2} & \cdots & W_N^{1 \cdot (N-1)} \\ 1 & W_N^{2 \cdot 1} & W_N^{2 \cdot 2} & \cdots & W_N^{2 \cdot (N-1)} \\ \vdots & \vdots & \vdots & \cdots & \vdots \\ 1 & W_N^{(N-1) \cdot 1} & W_N^{(N-1) \cdot 2} & \cdots & W_N^{(N-1) \cdot (N-1)} \end{bmatrix} \qquad (5.60)$$

Substituting $N = 2$, the DFT matrix of order 2 is given by

$$W_2 = \begin{bmatrix} 1 & 1 \\ 1 & -1 \end{bmatrix} \qquad (5.61)$$

Substituting $N = 4$, the DFT matrix of order 4 is given by

Fig. 5.29 Python code to obtain 4 point DFT matrix

```
import numpy as np
np.set_printoptions(precision=2, suppress=True)
N=4;
n=np.arange(0,N,1)
k1=np.outer(n, n)
D=np.exp(-1j*2*np.pi*k1/N)
print('{} point DFT Matrix'.format(N))
print(D)
```

Fig. 5.30 4 point DFT matrix

4 point DFT Matrix

[[1.+0.j 1.+0.j 1.+0.j 1.+0.j]

 [1.+0.j 0.-1.j -1.-0.j -0.+1.j]

 [1.+0.j -1.-0.j 1.+0.j -1.-0.j]

 [1.+0.j -0.+1.j -1.-0.j 0.-1.j]]

$$W_4 = \begin{bmatrix} 1 & 1 & 1 & 1 \\ 1 & -j & -1 & j \\ 1 & -1 & 1 & -1 \\ 1 & j & -1 & -j \end{bmatrix} \tag{5.62}$$

Experiment 5.16 Computation of 4 Point DFT Matrix

This experiment deals with the computation of 4 point DFT matrix using python. The python code of 4 point DFT matrix generation is given in Fig. 5.29, and its corresponding simulation output is shown in Fig. 5.30.

The built-in function *dft* available in *scipy.linalg* can be used to obtain the DFT matrix. The python code to obtain the DFT matrix of order $N = 4$ is shown in Fig. 5.31a, and the corresponding output is shown in Fig. 5.31b.

Inferences

The python code to compute the 4 point DFT matrix is shown in Fig. 5.29. From this figure, it is possible to observe that the DFT computation formula is implemented in python. Also, it is possible to see that 'np.outer' python command is used to generate outer product of two vectors (1 for time index (n) and other for frequency index (k)). From this Fig. 5.31, it is possible to observe that *dft* python command called from the *linalg* library package to compute the DFT matrix. In both these methods, the simulation result is in agreement with the theoretical result.

```
from scipy.linalg import dft
import numpy as np
np.set_printoptions(precision=2, suppress=True)
N=4
W= dft(N)
print(W)
```

```
[[ 1.+0.j  1.+0.j  1.+0.j  1.+0.j]

 [ 1.+0.j  0.-1.j -1.-0.j -0.+1.j]

 [ 1.+0.j -1.-0.j  1.+0.j -1.-0.j]

 [ 1.+0.j -0.+1.j -1.-0.j  0.-1.j]]
```

(a) Python code   (b) Simulation result

Fig. 5.31 Python code to obtain the DFT matrix and its result. (**a**) Python code. (**b**) Simulation result

```
from scipy.linalg import dft
from numpy.linalg import matrix_rank
import numpy as np
np.set_printoptions(precision=2, suppress=True)
N=4
W= dft(N)
rank=matrix_rank(W)
print("Rank of the matrix is {}".format(rank))
```

```
Rank of the matrix is 4
```

(a) Python code   (b) Simulation result

Fig. 5.32 Python code to compute the rank of the matrix and its result. (**a**) Python code. (**b**) Simulation result

Experiment 5.17 Computation of the Rank of DFT Matrix of Order 4

The python code to obtain the rank of 4×4 DFT matrix is shown in Fig. 5.32a, and the corresponding output is shown in Fig. 5.32b.

Inference

From Fig. 5.32b, it is possible to observe that the rank of 4×4 DFT matrix is 4. Similarly, it is possible to prove that the rank of $N \times N$ DFT matrix is N. Full rank of the matrix indicates that all rows and columns are linearly independent.

Task

1. Investigate the nature of DFT matrix. Find whether the DFT matrix is unitary or not.

Experiment 5.18 Computattion of DFT of a Sequence Using DFT Matrix

This experiment deals with the computation of DFT of a input sequence using DFT matrix. Let us consider the input sequence $x[n] = \{1, 1, 1, 1\}$. The python code to obtain 4 point DFT of the sequence $x[n]$ is given in Fig. 5.33, and the corresponding output obtained is $\{4, 0, 0, 0\}$.

Fig. 5.33 Python code to obtain DFT of a sequence using DFT matrix

```
from scipy.linalg import dft
import numpy as np
np.set_printoptions(precision=2, suppress=True)
N=4
x=np.ones(N)
W= dft(N)
X=W@x # matrix multiplication
print(np.abs(X))
```

Fig. 5.34 Python code to compute inverse DFT using DFT matrix

```
from scipy.linalg import dft
import numpy as np
np.set_printoptions(precision=2, suppress=True)
N=4
X=[4,0,0,0]
W= dft(N)
x=(W@X)/N
print(np.abs(x))
```

Inference

The signal $x[n]$ is a DC signal, which is given by $x[n] = \{1, 1, 1, 1\}$. The DFT of the sequence $x[n]$ is obtained as $X[k] = \{4, 0, 0, 0\}$. The maximum energy of the sequence $x[n]$ is packed into one transform coefficient. This is through energy compaction property of DFT. In Fig. 5.33, the symbol ('@') denotes matrix multiplication in python.

Experiment 5.19 Computation of Inverse DFT Through DFT Matrix

Let us consider the DFT coefficients $X[k] = \{4, 0, 0, 0\}$. The python code to compute the inverse DFT of $X[k] = \{4, 0, 0, 0\}$ is given in Fig. 5.34. After executing this code, we get the result of $\{1, 1, 1, 1\}$, which is in agreement with the theoretical result.

Inference

The inverse DFT of $X[k] = \{4, 0, 0, 0\}$ is $x[n] = \{1, 1, 1, 1\}$. The experimental result is in agreement with the fact that DFT is invertible. This is to inform that the inverse DFT computation is done by the forward DFT matrix, which can be seen in Fig. 5.34.

5.4.6 Properties of DFT

Discrete Fourier transform is applied to discrete-time signal $x[n]$ that are zero for $n < 0$ and $n \geq N$. However, the discrete-time signal $x[n]$ must be considered a periodic signal. Therefore, some of the DFT properties are based on *modulo N* or

```
#Linearity property of DFT
import numpy as np
import matplotlib.pyplot as plt
x1=[1,1,1,1,1,1,1,1]
x2=[1,-1,1,-1,1,-1,1,-1]
a,b=5,10
x3=np.add(np.multiply(a,x1),np.multiply(b,x2))
N=len(x2)
n=np.arange(0,N,1)
k=np.arange(0,N,1)
k1=np.outer(n, k)
D=np.exp(-1j*2*np.pi*k1/N)#DFT matrix
X1=np.dot(D,x1)
X2=np.dot(D,x2)
X3=np.dot(D,x3)
X4=a*X1+b*X2
plt.subplot(2,2,1),plt.stem(n,x1),plt.xlabel('n-->'),plt.ylabel('x$_1$[n]')
plt.title('x$_1$[n]'), plt.subplot(2,2,2),plt.stem(n,x2),plt.xlabel('n-->'),
plt.ylabel('x$_2$[n]'),plt.title('x$_2$[n]')
plt.subplot(2,2,3), plt.stem(k,X1),plt.xlabel('k-->'),plt.ylabel('X$_1$[k]')
plt.tltle('X$_1$[k]'),plt.subplot(2,2,4),plt.stem(k,X2),plt.xlabel('k-->'),
plt.ylabel('X$_2$[k]'),plt.title('X$_2$[k]'),plt.tight_layout()
plt.figure(2),plt.subplot(3,1,1),plt.stem(n,x3),plt.xlabel('n-->'),
plt.ylabel('x$_3$[n]'),plt.title('x$_3$[n]=a*x$_1$[n]+b*x$_2$[n]')
plt.subplot(3,1,2),plt.stem(k,X3),plt.xlabel('k-->'),plt.ylabel('X$_3$[k]')
plt.title('X$_3$[k]'),plt.subplot(3,1,3),plt.stem(k,X4),plt.xlabel('k-->'),
plt.ylabel('X$_4$[k]'),plt.title('X$_4$[k]=a*X$_1$[k]+b*X$_2$[k]')
plt.tight_layout()
```

Fig. 5.35 Python code for linearity property

mod N operation. The *modulo* operation yields a division's remainder or signed remainder after dividing one number by another. For example, (5 mod 2), the result will be '1' (i.e. remainder value getting 5 divided by 2).

Experiment 5.20 Verification of the Linearity Property of DFT
The DFT of a linear combination of two sequences is the linear combination of the DFT of the individual sequences. The DFT property is given by

$$\text{DFT}\{\alpha x_1[n] + \beta x_2[n]\} = \alpha X_1[k] + \beta X_2[k] \tag{5.63}$$

The python code, which verifies the linearity property of DFT, is shown in Fig. 5.35, and the corresponding output is shown in Fig. 5.36.

Inferences
The following inferences can be made from this experiment:

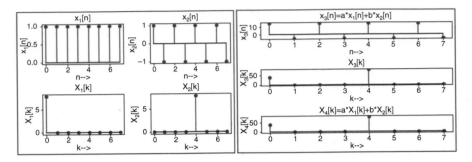

Fig. 5.36 Verification of the linearity property of DFT

1. From Fig. 5.35, the input signals chosen to prove the linearity property of DFT
   are $x_1[n] = \{1, 1, 1, 1, 1, 1, 1\}$, which is a DC signal, and $x_2[n] = \{1, -1, 1, -1, -1, 1, -1\}$ which is an AC signal. The scaling factors chosen in this example are
   $a = 5, b = 10$.
2. From Fig. 5.36, it is possible to observe the following facts:

   (a) DFT of the DC signal $x_1[n]$ is represented as $X_1[k]$, which exhibits peak at
       $k = 0$. DFT of the signal $x_2[n]$ shows the peak at $k = 4$.
   (b) The DFT of $ax_1[n] + bx_2[n]$ is equal to $aX_1[k] + bX_2[k]$. This implies that DFT
       obeys homogeneity and additivity properties; hence, it is a linear transform.

Experiment 5.21 Verification of Circular Shift Property of DFT
The circular shift property of DFT is expressed as

$$x[(n-m) \bmod N] \overset{\text{DFT}}{\longleftrightarrow} e^{-j\frac{2\pi}{N}km} X[k] \qquad (5.64)$$

The python code to illustrate the circular shift property of DFT is shown in
Fig. 5.37, and the corresponding output is shown in Fig. 5.38.

Inferences
After executing the python code given in Fig. 5.37, the result of input signal $x[n]$ is
$\{4, 3, 2, 1\}$, and the circularly shifted sequence with $k = 2$ is obtained as $\{2, 1, 4, 3\}$.
The DFT of the input sequence $x[n]$ and its circularly shifted version is shown in
Fig. 5.38. From this figure, it is possible to confirm that the magnitude spectrum of
both sequences is the same, whereas the phase spectrum is different. This indicates
that 'time shift in the time domain corresponds to phase shift in the frequency
domain'.

Experiment 5.22 Verification of the Parseval's Relationship of DFT
According to Parseval's relation, energy in time domain is equivalent to energy in
frequency domain.

```
# Python code for circular shift property
import numpy as np
import matplotlib.pyplot as plt
x=[4,3,2,1]
z=[]
k=2# Circular shifting factor
for i in range(len(x)):
    m=np.mod((i-k),len(x))
    y=x[m]
    z=np.append(z,y)
N=len(x)
n=np.arange(0,N,1)
k=np.arange(0,N,1)
k1=np.outer(n, k)
D=np.exp(-1j*2*np.pi*k1/N)#DFT matrix
X1=np.dot(D,x) # DFT computation of x[n]
X2=np.dot(D,z) # DFT computation of x[n-k]
plt.figure(1),plt.subplot(3,1,1),plt.stem(n,x),plt.xlabel('n-->')
plt.ylabel('x[n]'),plt.title('Input sequence')
plt.subplot(3,1,2),plt.stem(k,np.abs(X1)),plt.xlabel('k-->')
plt.ylabel('|X[k]|'),plt.title('Magnitude Response')
plt.subplot(3,1,3),plt.stem(k,(np.angle(X1))),plt.xlabel('k-->')
plt.ylabel('$\phi$[k]'),plt.title('Phase Response'),plt.tight_layout()
plt.figure(2),plt.subplot(3,1,1),plt.stem(n,z),plt.xlabel('n-->')
plt.ylabel('z[n]'),plt.title('Circularly shifted Sequence')
plt.subplot(3,1,2),plt.stem(k,np.abs(X2)),plt.xlabel('k-->')
plt.ylabel('|Z[k]|'),plt.title('Magnitude Response')
plt.subplot(3,1,3),plt.stem(k,(np.angle(X2))),plt.xlabel('k-->')
plt.ylabel('$\phi$[k]'),plt.title('Phase Response'),plt.tight_layout()
```

Fig. 5.37   Python code to illustrate circular shifting property of DFT

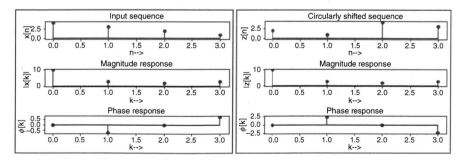

Fig. 5.38   Simulation result of Python code given in Fig. 5.37

Fig. 5.39 Python code for
Parseval's relation

```
# Python code for Parseval's relation
import numpy as np
np.set_printoptions(precision=2, suppress=True)
x=[1,2,3,4];
N=len(x);
n=np.arange(0,N,1);
k=np.arange(0,N,1);
k1=np.outer(n, k)
D=np.exp(-1j*2*np.pi*k1/N)# DFT matrix
print('Input Sequence x[n]: ', x)
y=np.sum((np.abs(x)**2))
print('\u03A3|x[n]|^2: ', y)
Y1=np.dot(D,x) # DFT Computation
print('DFT output X[k]: ', Y1)
Y=np.sum((np.abs(Y1)**2))/N
print('\u03A3|X[k]|^2: ', Y)
```

Fig. 5.40 Result of
Parseval's relation property

```
Input Sequence x[n]:  [1, 2, 3, 4]
Σ|x[n]|^2: 30
DFT output X[k]:  [10.+0.j  -2.+2.j -2.-0.j -2.-2.j]
Σ|X[k]|^2: 30.0
```

$$\sum_{n=0}^{N-1} |x[n]|^2 = \frac{1}{N} \sum_{k=0}^{N-1} |X[k]|^2 \tag{5.65}$$

The python code to verify the Parseval's relation of DFT is given in Fig. 5.39, and its simulation output is shown in Fig. 5.40.

Inference

From Fig. 5.40, it is possible to confirm that the energy in time domain and the energy in the frequency domain are the same. Hence, converting the time domain signal into the frequency domain using DFT always preserves energy. Therefore, the perfect reconstruction of the original signal from the frequency components is possible for the DFT.

5.4.7 Limitations of Fourier Transform

The basis function of Fourier transform is a complex exponential, which oscillates for all the time. This means that Fourier transform describes frequency components in the signal averaged over all the time. It is difficult for the Fourier transform to represent signals that are localized in time. Hence, Fourier transform is not an

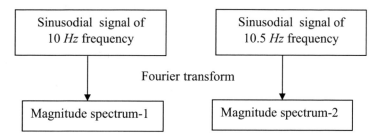

Fig. 5.41 Block diagram of problem statement

effective tool to analyse non-stationary signals. To overcome this drawback, time localization in Fourier transform can be achieved by windowing the signal over which the signal is nearly stationary, which leads to the development of short-time Fourier transform (STFT).

Experiment 5.23 Limitation of Fourier Transform

The objective of this experiment is to prove that Fourier transform cannot estimate fractional frequencies. Fourier transform of signal with fractional frequencies results in the spreading of the spectrum to other frequencies, which are not present in the signal. This fact is verified in this experiment. In this experiment, two sinusoidal signals of frequency 10 and 10.5 Hz are generated, and the magnitude spectrum of the generated signals is obtained by taking the Fourier transform of the generated signals. This is shown in Fig. 5.41.

The python code, which performs this task, is shown in Fig. 5.42, and the corresponding output is shown in Fig. 5.43.

Inference

From Fig. 5.43, the following inference can be drawn:

1. Fourier transform of 10 Hz sinusoidal signal has a peak exactly at 10 Hz.
2. Fourier transform of 10.5 Hz frequency component sinusoidal signal does not show peak at 10.5 Hz. Instead, it resulted in spreading of the spectrum to other frequencies. To avoid the spreading of the spectrum to other frequencies, the value of N has to be increased.

Task

1. Increase the value of N in the python code shown in Fig. 5.42 from 100 to 256 and 512 and comment on the observed output.

```
#Fourier transform of fractional frequency component
import numpy as np
import matplotlib.pyplot as plt
from scipy.fft import fft,fftfreq
# Step 1: Signal generation
fs=100
T=1/fs
f1=10  #10 Hz frequency component
f2=10.5 #10.5 Hz frequency component
N=100
t=np.linspace(0,N*T,N)
x1=np.sin(2*np.pi*f1*t)
x2=np.sin(2*np.pi*f2*t)
#Step 2: Spectrum of the signals
faxis=fftfreq(N,T)[0:N//2]
X1=fft(x1)
X2=fft(x2)
#Step 3: Ploting the result
plt.subplot(2,1,1),plt.stem(faxis,2/N*np.abs(X1)[0:N//2])
plt.xlabel('Frequency (Hz)'),plt.ylabel('Magnitude'),
plt.title('Spectrum of 10 Hz sine wave')
plt.subplot(2,1,2),plt.stem(faxis,2/N*np.abs(X2)[0:N//2])
plt.xlabel('Frequency (Hz)'),plt.ylabel('Magnitude'),
plt.title('Spectrum of 10.5 Hz sine wave')
plt.tight_layout()
```

Fig. 5.42 Fourier transform of fractional frequency component signal

5.5   Discrete Cosine Transform (DCT)

Discrete cosine transform was developed by Ahmed, Rao and Natarajan in the year 1974. DCT is a unitary transform, and it is not a discrete version of the cosine functions. The DCT has better energy compaction than DFT; hence, it is widely used in signal compression. DCT is employed in JPEG compression standard. DCT is based on the DFT with imposed even symmetry through reflection; hence, DCT is a real-valued transform.

The formula to compute forward discrete cosine transform is given by

$$X[k] = \alpha(k) \sum_{n=0}^{N-1} x[n] \cos \frac{(2n+1)\pi k}{2N} \tag{5.66}$$

where

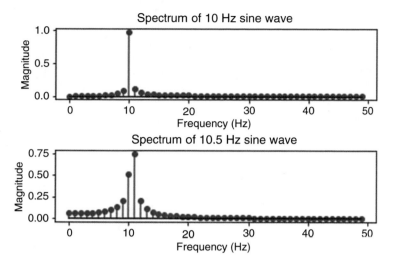

Fig. 5.43 Result of python code shown in Fig. 5.42

$$\alpha(k) = \begin{cases} \sqrt{\dfrac{1}{N}}, & \text{for } k = 0 \\[2ex] \sqrt{\dfrac{2}{N}}, & \text{Otherwise} \end{cases} \tag{5.67}$$

The formula to compute inverse discrete cosine transform is given by

$$x[n] = \alpha(k) \sum_{k=0}^{N-1} X[k] \cos \frac{(2n+1)\pi k}{2N} \tag{5.68}$$

where

$$\alpha(k) = \begin{cases} \sqrt{\dfrac{1}{N}}, & \text{for } k = 0 \\[2ex] \sqrt{\dfrac{2}{N}}, & \text{Otherwise} \end{cases} \tag{5.69}$$

Experiment 5.24 Computation of Forward and Inverse Discrete Cosine Transform of a Given Signal

This experiment deals with the computation of the forward and inverse DCT of a given input signal. The python code to compute the forward and inverse DCT of signal is given in Fig. 5.44, and its simulation result is shown in Fig. 5.45. From this figure, it is possible to infer that the DCT output is always real value, and there is no phase component in it.

```
# DCT and IDCT python implementation
import numpy as np
import matplotlib.pyplot as plt
from scipy import fft
np.set_printoptions(precision=2, suppress=True)
n=np.arange(0,40,1)
k=np.arange(0,40,1)
x=np.sin(2*np.pi*(5/100)*(n))+np.sin(2*np.pi*(15/100)*(n));
plt.figure(1),plt.subplot(3,1,1),plt.stem(n,x),plt.xlabel('n--
>'),plt.ylabel('x[n]'),plt.title('Input Signal')
y1=np.zeros(len(k))
alpha=np.zeros(len(k));
for i in range(len(n)):
   if k[i]==0:
      alpha[i]=np.sqrt(1/len(n));
   else:
      alpha[i]=np.sqrt(2/len(n));
   y=alpha[i]*x[i]*np.cos(((2*i)+1)*np.pi*k/(2*len(n)))
   y1=y1+y
plt.subplot(3,1,2),plt.stem(k,y1),plt.xlabel('k-->'),plt.ylabel('X[k]'),plt.title('DCT output')
y2=np.zeros(len(n))
for i in range(len(k)):
   if k[i]==0:
      alpha[i]=np.sqrt(1/len(n));
   else:
      alpha[i]=np.sqrt(2/len(n));
   x1=alpha[i]*y1[i]*np.cos(((2*i)+1)*np.pi*k/(2*len(n)))
   y2=y2+x1
y2=(1/2)*y2
plt.subplot(3,1,3),plt.stem(n,y2),plt.xlabel('n-->'),plt.ylabel('y[n]'),plt.title('IDCT output')
plt.tight_layout()
y3=fft.dct(x);# Built in command for DCT
z=fft.idct(y3);#Built in Command for IDCT
plt.figure(2),plt.subplot(3,1,1),plt.stem(n,x),plt.xlabel('n-->'),plt.ylabel('x[n]'),
plt.title('Input Signal'),plt.subplot(3,1,2),plt.stem(k,y1),plt.xlabel('k-->'),plt.ylabel('X[k]'),
plt.title('DCT output Using in-built'),plt.subplot(3,1,3),plt.stem(n,y2),plt.xlabel('n-->'),
plt.ylabel('y[n]'),plt.title('IDCT output using in-built')
```

Fig. 5.44 Python code for forward and inverse DCT

Inferences

The following inferences can be made from this experiment:

1. The input is real valued mulitiple sinusoidal signal, and the DCT output of the input signal shows that most of the DCT coefficients are zero, which indicates that the signal is highly correlated, and then the few DCT coefficients are used to represent the signal.

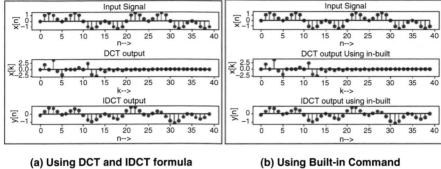

(a) Using DCT and IDCT formula   **(b) Using Built-in Command**

Fig. 5.45 Simulation result. (**a**) Using DCT and IDCT formula. (**b**) Using built-in command

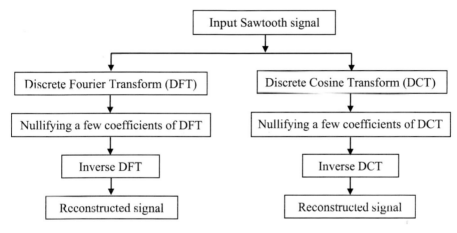

Fig. 5.46 Problem statement

2. Inverse DCT is used to reconstruct the original signal from the DCT coefficients. From Fig. 5.45, it is possible to observe that the reconstructed signal is exactly the same as the original signal.
3. The same result is verified with the built-in commands (*fft.dct* and *fft.idct*).

Task

1. Investigate whether DCT matrix entries are all real.

Experiment 5.25 Comparison Between Discrete Fourier Transform and Discrete Cosine Transform

The objective of this experiment is to compare the performance of discrete Fourier transform (DFT) with discrete cosine transform (DCT). This is done by taking DFT and DCT of the sawtooth signal. After taking both DFT and DCT, the last ten coefficients are nullified. Then, inverse DFT and inverse DCT of the modified coefficients are taken to obtain the reconstructed signal. This is depicted in Fig. 5.46.

```
#Comparison of Fourier and DCT
from scipy.fftpack import dct, idct
from scipy.fftpack import fft, ifft
import numpy as np
import matplotlib.pyplot as plt
from scipy import signal
#Step 1: Generation of sawtooth signal
t=np.linspace(0,1,100)
x=signal.sawtooth(2*np.pi*5*t)
#Step 2: Modifying DFT coefficients
X1=fft(x)
X1[89:99]=0
y1=ifft(X1)
#Step 3: Modifying DCT coefficients
X=dct(x)
X[89:99]=0
y2=idct(X)
#Step 4: Plotting the results
plt.subplot(3,1,1),plt.stem(x),plt.xlabel('n-->'),plt.ylabel('x[n]')
plt.title('Input Sawtooth signal')
plt.subplot(3,1,2),plt.stem(y1),plt.xlabel('n-->'),plt.ylabel('y$_1$[n]')
plt.title('Reconstructed signal using DFT')
plt.subplot(3,1,3),plt.stem(y2),plt.xlabel('n-->'),plt.ylabel('y$_2$[n]')
plt.title('Reconstructed signal using DCT')
plt.tight_layout()
```

Fig. 5.47 Python code to compare DCT with Fourier transform

The python code, which performs this task, is shown in Fig. 5.47, and the corresponding output is shown in Fig. 5.48.

Inferences
From Fig. 5.48, it is possible to interpret that the reconstructed signal obtained using DCT is better than DFT. DCT has better energy compaction than the DFT. This means that DCT can pack signal energy into a few coefficients.

5.6   Short-Time Fourier Transform

The short-time Fourier transform of the signal $x(t)$ is given by

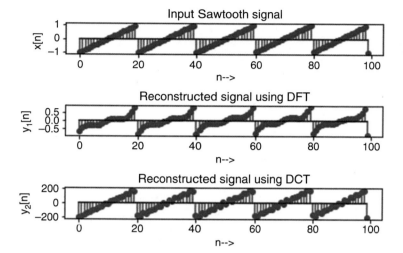

Fig. 5.48 Result of python code shown in Fig. 5.47

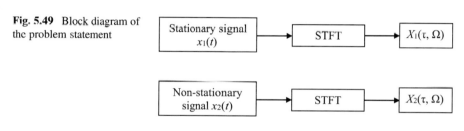

Fig. 5.49 Block diagram of the problem statement

$$X(\tau, \Omega) = \int_{-\infty}^{\infty} x(t)w^*(t-\tau)e^{-j\Omega t}dt \qquad (5.70)$$

where '$w(t)$' is the window function and 'τ' is the centre of the window. Equation (5.70) can be interpreted as 'STFT provides two-dimensional representation of the one-dimensional signal $x(t)$'. Narrow window provides good time resolution but poor frequency resolution, whereas wider window provides good frequency resolution but poor time resolution. According to the Heisenberg uncertainty principle, it is difficult to obtain both good time and frequency resolutions at the same resolution.

Experiment 5.26 STFT of Stationary and Non-stationary Signal

In this experiment, STFT of stationary and non-stationary signals are obtained, and their results are interpreted. The stationary signal $x_1(t)$ is generated using the formula $x_1(t) = \sin(2\pi ft)$; the non-stationary signal $x_2(t)$ is generated using the formula $x_2(t) = \sin(2\pi ft^2)$. The built-in function available in scipy library *stft* is used to obtain time-frequency representation of the two signals. The block diagram of the problem statement is given in Fig. 5.49.

```
#STFT of stationary and non-stationary signals
import numpy as np
import matplotlib.pyplot as plt
from scipy import signal
#Step 1: Generation of a stationary signal
fs=100
T,N,f=1/fs, 100, 5
t=np.linspace(0,N*T,N)
x1=np.sin(2*np.pi*f*t)#Step 2: Generation of non-stationary signal
x2=np.sin(2*np.pi*f*t**2)#Step 3: Obtaining the STFT of signals
f1,t1,z1=signal.stft(x1,fs,'hamming',64)
f2,t2,z2=signal.stft(x2,fs,'hamming',64)
#Step 4: Plotting the results
plt.subplot(2,2,1),plt.plot(t,x1),plt.xlabel('t-->'),plt.ylabel('x$_1$(t)'),
plt.title('Signal-1'),plt.subplot(2,2,2),plt.plot(t,x2)
plt.xlabel('t-->'),plt.ylabel('x$_2$(t)'),plt.title('Signal-2')
plt.subplot(2,2,3),plt.pcolormesh(t1,f1,np.abs(z1),shading='gouraud')
plt.xlabel('Time (t-->)'),plt.ylabel('Frequency ($\omega$-->)'),plt.title('STFT of Signal-1')
plt.subplot(2,2,4),plt.pcolormesh(t2,f2,np.abs(z2),shading='gouraud')
plt.xlabel('Time (t-->)'),plt.ylabel('Frequency ($\omega$-->)'),plt.title('STFT of Signal-2')
plt.tight_layout()
```

Fig. 5.50 Python code to obtain time-frequency representation of stationary and non-stationary signals

The python code to implement the task is done in four steps. First step deals with the generation of stationary signals, and second step deals with the generation of non-stationary signals. Obtaining the STFT of the two signals is done in the third step. Finally, the results are plotted in the fourth step. The python code, which performs the abovementioned task, is given in Fig. 5.50, and the corresponding output is shown in Fig. 5.51.

Inferences

From Fig. 5.51, the following inferences can be drawn:

1. Signal-1 and Signal-2 are stationary and non-stationary signals, respectively. For Signal-1, the frequency does not change with respect to time; hence, it is stationary, for Signal-2, the frequency increases with an increase in time; hence, it is non-stationary.
2. STFT provides time-frequency representation of the signal. STFT of Signal-1 is a horizontal line indicating that Signal-1 has one frequency component at all times. The STFT of Signal-2 shows the gradual variation of frequency with respect to time. With respect to time, the frequency changes, which is depicted in the spectrogram plot.

Experiment 5.27 Impact of Choice of Window Length

In this experiment, the built-in function available in 'matplot' library *plt.specgram* is used to analyse the impact of the choice of width of the window in STFT. The

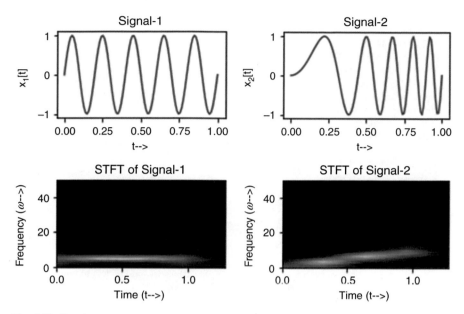

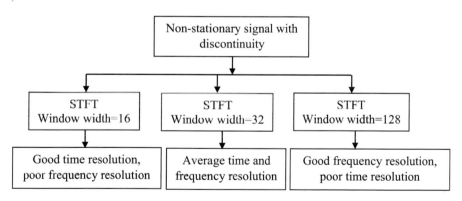

Fig. 5.51 Result of python code shown in Fig. 5.50

Fig. 5.52 Problem statement

objective of this experiment is to verify the fact that a shorter window gives a good time resolution and a wider window gives a good frequency resolution. To demonstrate this fact, a non-stationary signal is generated. This signal has 5, 0 and 10 Hz frequency components. In 0 Hz or DC, a discontinuity is introduced. The discontinuity is the increase in the amplitude of the signal from 1 to 2 V. This non-stationary signal is analysed using a spectrogram of different window widths, namely, 16, 32 and 128. The problem statement is depicted in Fig. 5.52.

The python code, which performs this task, is shown in Fig. 5.53, and the corresponding output is shown in Fig. 5.54.

```
#Effect of window length of STFT
import numpy as np
import matplotlib.pyplot as plt
#Step1: Signal generation
fs=100
T,N=1/fs, 100;
#Frequency components of the signal
f1,f2,f3=5, 0, 10;
t1=np.linspace(0,N*T,N)
t=np.linspace(0,N*T,3*N)
x1=np.sin(2*np.pi*f1*t1)
x2=np.sin(2*np.pi*f2*t1)
x3=np.sin(2*np.pi*f3*t1)
x=np.concatenate([x1,x2,x3])
x[150:160]=2  #Discontinuity
#Step 2: Plotting the signal and its spectrogram
plt.subplot(2,2,1),plt.plot(t,x),plt.xlabel('t-->'),plt.ylabel('x(t)')
plt.title('Signal')
plt.subplot(2,2,2),plt.specgram(x, Fs=fs, NFFT=16, noverlap=1,window =None)
plt.xlabel('Time (t-->)'),plt.ylabel('Frequency ($\omega$-->)'),
plt.title('Window length=16')
plt.subplot(2,2,3),plt.specgram(x, Fs=fs, NFFT=32, noverlap=1,window =None)
plt.xlabel('Time (t-->)'),plt.ylabel('Frequency ($\omega$-->)'),
plt.title('Window length=32')
plt.subplot(2,2,4),plt.specgram(x, Fs=fs, NFFT=128, noverlap=1,window =None)
plt.xlabel('Time (t-->)'),plt.ylabel('Frequency ($\omega$-->)'),
plt.title('Window length=128')
plt.tight_layout()
```

Fig. 5.53 Python code to analyse the impact of window width in STFT

Inferences

From Fig. 5.54, the following inferences can be drawn:

1. The signal is a non-stationary signal with three frequency components 5 Hz, 0 Hz (DC component) and 10 Hz, respectively. There is a discontinuity in the signal in the DC component. The discontinuity refers to an abrupt change in amplitude from 1 to 2 V.
2. The STFT of the signal is obtained for different window widths, namely, 16, 32 and 128.
3. The spectrogram corresponding to window width 16 gives good time information. The occurrence of discontinuity at a particular instant is clearly visible in the spectrogram with a window width of 16. But the frequency resolution is poor. The two frequency components present in the signal, namely, 5 and 10 Hz, are not visible in the spectrogram with a window width of 16. That is, a shorter window gives good time resolution but poor frequency resolution.

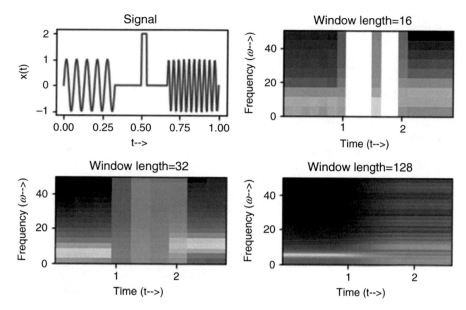

Fig. 5.54 Result of python code shown in Fig. 5.53

4. When the window width is 32, the frequency components present in the signal, namely, 5 and 10 Hz, are partially visible. When the window width is 32, the average time and frequency resolution are obtained.
5. When the window width is 128, the frequency resolution is good. From this spectrogram, it is possible to identify 5 and 10 Hz frequency components. But it is not possible to locate the discontinuity present in the signal. This means that time resolution is poor.
6. In a nut shell, shorter window gives good time resolution but poor frequency resolution, whereas a wider window gives good frequency resolution but poor time resolution.

Task
1. Repeat this experiment by choosing different types of window functions like Bartlett and Kaiser for specified value of 'β'.

Experiment 5.28 Choice of Window Function in Resolving Two Close Frequency Components

The objective of this experiment is to analyse the choice of window function in resolving two close frequency components of the input signal. The input signal is the addition of two sinusoidal signals of frequencies 5 and 8 Hz. The spectrogram of this signal is obtained for different choices of window functions like rectangular window, Blackman window and Kaiser window. The impact of window choices in frequency resolution is analysed in this experiment. The python code, which performs this task, is shown in Fig. 5.55, and the corresponding output is shown in Fig. 5.56.

```
#Choice of window function
import numpy as np
import matplotlib.pyplot as plt
#Step1: Signal generation
fs,f1,f2=100, 5, 8;
t=np.linspace(0,1,100)
x1=np.sin(2*np.pi*f1*t)
x2=np.sin(2*np.pi*f2*t)
x=x1+x2
#Step 2: Generation of window functions
NFFT=64
win1 =np.ones((NFFT)) #Rectangular window
win2=np.blackman(NFFT) #Blackman window
beta=1
win3 = np.kaiser(NFFT,beta)
#Step 3: Plotting the results
plt.subplot(2,2,1),plt.plot(t,x),plt.xlabel('t-->'),plt.ylabel('x(t)')
plt.title('Signal'),plt.subplot(2,2,2),
plt.specgram(x, Fs=fs, NFFT=64, noverlap=1,window = win1)
plt.xlabel('Time (t-->)'),plt.ylabel('Frequency ($\omega$-->)'),plt.title('Rectangular
window')
plt.subplot(2,2,3),plt.specgram(x, Fs=fs, NFFT=64, noverlap=1,window = win2)
plt.xlabel('Time (t-->)'),plt.ylabel('Frequency ($\omega$-->)'),plt.title('Blackman window')
plt.subplot(2,2,4),plt.specgram(x, Fs=fs, NFFT=64, noverlap=1,window = win3)
plt.xlabel('Time (t-->)'),plt.ylabel('Frequency ($\omega$-->)'),plt.title('Kaiser window')
plt.tight_layout()
```

Fig. 5.55 Window function and frequency resolution

Inferences

The following inference can be made from this experiment:

1. From the python code, it is possible to observe that the signal consists of two frequency components, 5 and 8 Hz, that are added to obtain the input signal whose time-frequency representation for different windows is obtained.
2. Rectangular window is able to resolve two closely spaced frequency components.
3. Blackman window has a wider main lobe; hence, it could not resolve the frequency components present in the signal.
4. Kaiser window successfully resolves the frequency components present in the signal for the choice of $\beta = 1$.
5. If the main lobe width of the window is small, then good frequency resolution could be obtained. Side lobes affect the extent to which adjacent frequency components leak into the adjacent frequency bins.

Experiment 5.29 Comparison of FT with STFT

This experiment aims to compare Fourier transform with short-time Fourier transform in analysing non-stationary signal. The non-stationary signal considered in this example has three frequency components, namely, 5, 0 and 15 Hz. In non-stationary

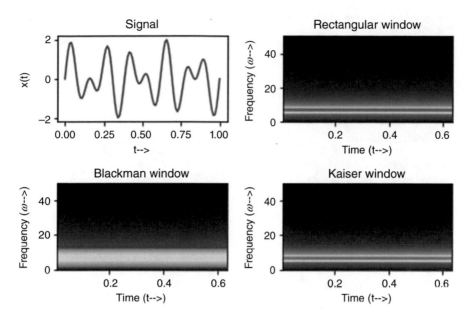

Fig. 5.56 Choice of window and its impact in frequency resolution

Fig. 5.57 Comparison of FT and STFT

Signal-1, 5 Hz signals appear first, followed by the DC and 15 Hz frequency components. In non-stationary Signal-2, 15 Hz frequency components appear first followed by DC and 5 Hz frequency components. For these two signals, Fourier transform and short-time Fourier transform are taken. This objective is illustrated in Fig. 5.57.

The python code, which implements this task, is given in Fig. 5.58 and the corresponding output is shown in Fig. 5.59.

Inference

From Fig. 5.59, the following inferences can be drawn:

1. Signal-1 and Signal-2 are non-stationary signals, because the frequency of these two signals changes with respect to time.
2. The frequency components present in Signal-1 and Signal-2 are 5, 0 and 10 Hz. In Signal-1, the 5 Hz frequency component appears first, followed by the 0 Hz frequency component and the 15 Hz frequency component. In Signal-2, 15 Hz

```
#Comparison of FT and STFT
import numpy as np
import matplotlib.pyplot as plt
from scipy.fft import fft,fftfreq
from scipy import signal
#Step1: Signal generation
fs=100
T=1/fs
N,f1,f2,f3=100,5,0,15
t1=np.linspace(0,N*T,N)
t=np.linspace(0,N*T,3*N)
x1=np.sin(2*np.pi*f1*t1)
x2=np.sin(2*np.pi*f2*t1)
x3=np.sin(2*np.pi*f3*t1)
x=np.concatenate([x1,x2,x3])
y=np.concatenate([x3,x2,x1])
plt.subplot(3,2,1),plt.plot(t,x),plt.xlabel('t-->'),plt.ylabel('x$_1$(t)'),plt.title('Signal-1')
plt.subplot(3,2,2),plt.plot(t,y),plt.xlabel('t-->'),plt.ylabel('x$_2$(t)'),plt.title('Signal-2')
#Step 2: Obtain the spectrum
faxis=fftfreq(3*N,T)[0:3*N//2]
X=fft(x)
Y=fft(y)
#Step 3: Plotting the result
plt.subplot(3,2,3),plt.plot(faxis,2/N*np.abs(X)[0:3*N//2])
plt.xlabel('Frequency ($\omega$-->)'),plt.ylabel('|X$_1$($\omega$)|'),plt.title('Spectrum-1')
plt.subplot(3,2,4),plt.plot(faxis,2/N*np.abs(Y)[0:3*N//2])
plt.xlabel('Frequency ($\omega$-->)'),plt.ylabel('|X$_2$($\omega$)|'),plt.title('Spectrum-2')
#Step 4: STFT of the signals
f1,t1,z1=signal.stft(x,fs,'hamming',1024)
f2,t2,z2=signal.stft(y,fs,'hamming',1024)
plt.subplot(3,2,5),plt.pcolormesh(t1, f1, np.abs(z1),shading='gouraud')
#plt.pcolormesh(t1,f1,np.abs(z1),shading='flat')
plt.xlabel('Time (t-->)'),plt.ylabel('Freq($\omega$-->)'),plt.title('STFT of Signal-1')
plt.subplot(3,2,6),plt.pcolormesh(t2,f2,np.abs(z2),shading='gouraud')
plt.xlabel('Time (t-->)'),plt.ylabel('Freq ($\omega$-->)'),plt.title('STFT of Signal-2')
plt.tight_layout ()
```

Fig. 5.58 Python code to compare FT with STFT

frequency component appears first, then followed by 0 Hz and finally by 5 Hz frequency component.

3. From Fig. 5.59, it is possible to interpret that spectrum-1 and spectrum-2 are alike. That is magnitude spectrum of Fourier transform cannot distinguish Signal-1 and Signal-2. The reason is Fourier transform is an effective tool for the frequency representation of the stationary signal, but it does not provide time information.

4. STFT of Signal-1 shows time-frequency representation of the signal. STFT of Signal-1 indicates that 5 Hz frequency component appears first, followed by 15 Hz frequency component. In STFT of Signal-2, it is possible to observe that

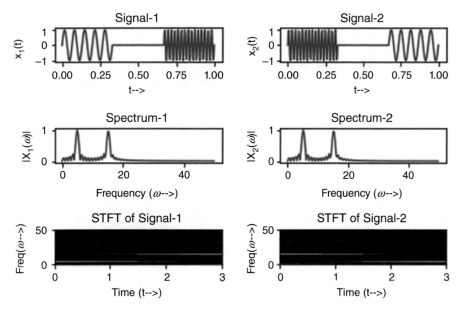

Fig. 5.59 Result of comparison of FT with STFT

15 Hz frequency component appears first, followed by 5 Hz frequency component.

5. It is possible to interpret that STFT is effective in analysing non-stationary signals.

5.6.1   Shortcoming of STFT

The width of the window cannot be changed once it is fixed. This implies that STFT provides fixed resolution. For the multi-resolution representation of the signal, wavelet transform is employed.

5.7   Continuous Wavelet Transform (CWT)

Wavelets are oscillatory functions of finite duration. Wavelet transform provides timescale relationship of the signal. The continuous wavelet transform of the signal $f(t)$ is expressed as

$$W_f(a,b) = \frac{1}{\sqrt{a}} \int\limits_{-\infty}^{\infty} f(t)\psi^*\left(\frac{t-b}{a}\right) dt \qquad (5.71)$$

In the above equation, $f(t)$ represents signal of interest, $\psi(t)$ denotes 'mother wavelet', b is the shifting parameter and a is the scaling parameter. The above equation can be written as

$$W_f(a,b) = \langle f(t), \psi_{a,b}(t) \rangle \qquad (5.72)$$

The above equation indicates that wavelet transform is basically taking inner product of the function $f(t)$ with the 'daughter wavelet' $\psi_{a,b}(t)$. The daughter wavelets are derived from the mother wavelet $\psi(t)$ using the relation

$$\psi_{a,b}(t) = \frac{1}{\sqrt{a}} \psi\left(\frac{t-b}{a}\right) \qquad (5.73)$$

5.7.1 Continuous Wavelets Family

A variety of continuous wavelets filter are currently in use. They are (1) Haar, (2) Mexican Hat, (3) Morlet, (4) Complex Morlet, (5) Gaussian, (6) Shannon and (7) Daubechies. The wavelet family and its mathematical expression are given in Table 5.3.

Experiment 5.30 Detection of Discontinuity in the Signal Using CWT
The objective of this experiment is to detect the discontinuity present in the signal using continuous wavelet transform (CWT). The built-in function *cwt* available in the library *pywt* is utilized in this experiment. The three steps followed in this experiment are the following: Step 1: generating signal with discontinuity; Step 2: obtaining timescale relationship using CWT, in which the wavelet chosen for this study is Gaussian wavelet; and Step 3: plotting the signal and the corresponding scalogram. The python code which performs this task is shown in Fig. 5.60, and the corresponding output is shown in Fig. 5.61.

Inferences
From Fig. 5.61, the following inferences can be drawn:

1. The input signal is a smooth sinusoidal signal with a sharp discontinuity at a particular location.
2. Upon observing the CWT result, it is possible to interpret that discontinuity occurs at 90th sample of the sinusoidal signal, which has 200 samples of data.
3. Thus, CWT is capable of detecting the discontinuity present in the signal.

Table 5.3 List of wavelet family

Name of wavelet	Mathematical expression	Python command pywt.Wavelet ('wavelet_name')
Haar	$\psi(t) = \begin{cases} 1, & 0 \le t < \frac{1}{2} \\ -1 & \frac{1}{2} \le t < 1 \\ 0, & \text{otherwise} \end{cases}$	'haar'
Mexican Hat	$\psi(t) = \frac{2}{\sqrt{3}\sqrt[4]{\pi}}(1 - t^2)\exp\left(-\frac{t^2}{2}\right)$	'mexh'
Morlet	$\psi(t) = \exp\left(-\frac{t^2}{2}\right)\cos(5t)$	'morl'
Complex Morlet	$\psi(t) = \frac{2}{\sqrt{\pi B}}\exp\left(-\frac{t^2}{B}\right)\exp(j2\pi Ct)$ Where B is Bandwidth and C is centre frequency	'cmor'
Gaussian wavelet	$\psi(t) = C\exp(-t^2)$ where C is an order-dependent normalization constant	'gaus'
Shannon wavelet	$\psi(t) = \sqrt{B}\frac{\sin(\pi Bt)}{\pi Bt}\exp(j2\pi Ct)$ Where B is Bandwidth and C is centre frequency	'shan'

Task
1. Repeat the experiment for different choices of mother wavelet and comment on the observed result.

5.7.2   Drawback of CWT

CWT is a redundant representation because of continuous values taken by scaling and shifting parameters. Overcoming the problem of redundant representation, a discrete wavelet transform was proposed.

5.8   Discrete Wavelet Transform

The discrete wavelet transform decomposes the signal into approximation and detail. The process is further iterated by decomposing the approximation with the assumption that much of the signal energy is in approximation. This idea is illustrated in Fig. 5.62.

In Fig. 5.62, L_1 corresponds to first-level decomposition, where the signal is decomposed into approximation and detail. In the second-level of decomposition (L_2), the approximation obtained in L_1 is further decomposed into approximation and detail. In the third-level of decomposition (L_3), the approximation of level L_2 is

```
#Discontinuity detection using CWT
import pywt
import numpy as np
import matplotlib.pyplot as plt
#Step 1: Signal generation
#t=np.linspace(0,1,200)
t=np.arange(0,200,1);
x=np.sin(2*np.pi*5*t/len(t))
x[90]=10
#Step 2: CWT of the signal
scale=np.arange(1,5)
coef,freqs=pywt.cwt(x,scale,'gaus1')
plt.subplot(2,1,1),plt.plot(t,x),plt.xlabel('t-->'),plt.ylabel('x(t)')
plt.title('Signal with discontinuity')
#Step 3: Plotting the reslt
plt.subplot(2,1,2),
plt.imshow(abs(coef),extent=[0,200,30,1],interpolation='bilinear',cmap='winter',
        aspect='auto',vmax=abs(coef).max(),vmin=-abs(coef).max())
plt.gca().invert_yaxis()
plt.xticks(np.arange(0,201,20))
plt.xlabel('Time (t-->)'),plt.ylabel('Freq Scale ($\omega$-->)'),
plt.title('CWT of the signal')
plt.tight_layout()
plt.show()
```

Fig. 5.60 Discontinuity detection using CWT

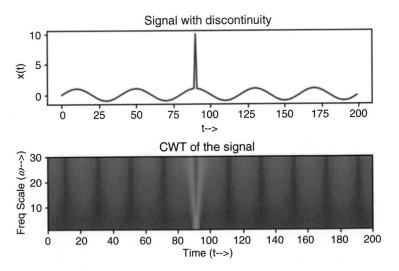

Fig. 5.61 CWT of signal with discontinuity

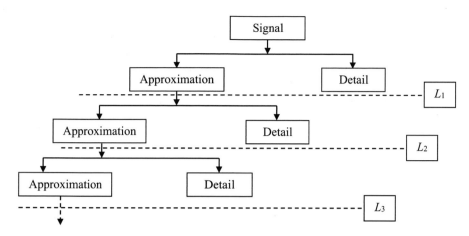

Fig. 5.62 Wavelet decomposition

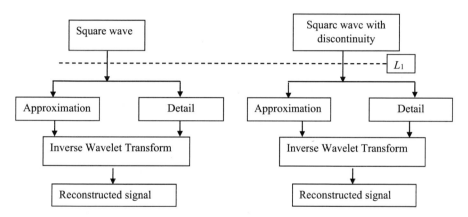

Fig. 5.63 Pictorial representation of problem statement

decomposed further into approximation and detail. This is done assuming that most of the signal energy is in approximation.

Experiment 5.31 Detection of Discontinuity in Signal Using DWT

The objective of this experiment is to compare the first-level approximation and detail of DWT coefficient of a signal with the coefficient of the signal with discontinuity. Here discontinuity refers to sudden changes in the amplitude of the signal. The problem statement is depicted in Fig. 5.63.

From Fig. 5.63, it is possible to observe that two signals are considered in this experiment, Signal-1 is a square wave, whereas Signal-2 is a square wave with discontinuity. L_1 in the figure represents the first-level of decomposition. Upon first-level of decomposition, the signal is split into approximation and detail. Upon taking an inverse discrete wavelet transform, it is possible to reconstruct the signal. The

```
#DWT of a signal with discontinuity
import pywt
import numpy as np
import matplotlib.pyplot as plt
from scipy import signal
#Step 1: Signal generation
f=5
n=np.arange(0,100,1)
x1=signal.square(2*np.pi*f*n/len(n)) #Square wave
x2=signal.square(2*np.pi*f*n/len(n)) #Square wave with discontinuity
x2[50]=5
#Step 2: DWT of the signal
cA,cD=pywt.dwt(x1,'db1')
cA1,cD1=pywt.dwt(x2,'db1')
#Step 3: Inverse DWT
y1=pywt.idwt(cA,cD,'db1')
y2=pywt.idwt(cA1,cD1,'db1')
#Step 4: Plotting the result
plt.subplot(3,2,1),plt.stem(n,x1),
plt.xlabel('n-->'),plt.ylabel('x$_1$[n]'),plt.title('Signal-1')
plt.subplot(3,2,2),plt.stem(n,x2),
plt.xlabel('n-->'),plt.ylabel('x$_2$[n]'),plt.title('Signal-2')
WC=np.concatenate([cA,cD])
WC1=np.concatenate([cA1,cD1])
plt.subplot(3,2,3),plt.stem(n,WC),plt.title('First level Decomposition')
plt.subplot(3,2,4),plt.stem(n,WC1),plt.title('First level Decomposition')
plt.subplot(3,2,5),plt.stem(n,y1),
plt.xlabel('n-->'),plt.ylabel('y$_1$[n]'),plt.title('Reconstructed signal-1')
plt.subplot(3,2,6),plt.stem(n,y2),
plt.xlabel('n-->'),plt.ylabel('y$_2$[n]'),plt.title('Reconstructed signal-2')
plt.tight_layout()
```

Fig. 5.64 Python code to compute the DWT CWT of signal with discontinuity

python code that performs this task mentioned above is shown in Fig. 5.64, and the corresponding output is in Fig. 5.65.

Inferences

From Fig. 5.65, the following inferences are drawn:

1. Signal-1 is a square wave with 5 Hz fundamental frequency; Signal-2 is a square wave with discontinuity.
2. The first-level decomposition of the signal gives approximation and detail coefficients. For Signal-1, the approximation coefficient is similar to the signal, whereas the detail coefficient is almost zero. For Signal-2, the discontinuity is captured in detail coefficient.
3. Upon taking inverse DWT, the reconstructed signals are obtained, which resembles the input signal. Thus, DWT is a reversible transform.

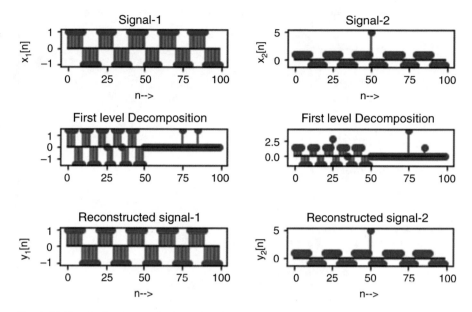

Fig. 5.65   Result of python code shown in Fig. 5.64

4. In this experiment, it is possible to observe that discontinuity is captured in the detail coefficient.

Task

1. Repeat this experiment for different choices of mother wavelet and comment on the observed result.

Experiment 5.32 Denoising of Signal Using DWT and Thresholding Approach
The objective of this experiment is to denoise the signal using discrete wavelet transform and inverse discrete wavelet transform. The input signal (sawtooth signal) is corrupted by white noise, which follows normal distribution of zero mean and variance of 0.125. The wavelet decomposition of the noisy signal is performed using the built-in function '*wavedec*' available in 'pywavelet' library. The wavelet chosen for decomposition is *db2*, and the level of decomposition chosen is 3. After wavelet decomposition, the detail coefficients are thresholded using the built-in function '*pywt.threshold*'. The choice of threshold is *soft 'thresholding'*. After thresholding, the modified wavelet coefficients are reconstructed using the built-in function *wavedec* to obtain the reconstructed (filtered) signal. The python code, which performs this task, is shown in Fig. 5.66, and the corresponding output is shown in Fig. 5.67.

Inferences
The following inferences can be made from this experiment:

1. The input signal (clean signal) is a sawtooth signal of 5 Hz fundamental frequency.

```
#Denoising of signals using DWT
import pywt
import numpy as np
import matplotlib.pyplot as plt
from scipy import signal
#Step 1: Signal generation
f=5
t=np.arange(0,50,1)
x1=signal.sawtooth(2*np.pi*f*t/len(t))
#Step 2: Adding noise to the clean signal
n=np.random.normal(0,0.125,len(x1))
x=x1+n
wavelet = 'db2'
level =3
# Step 3: Perform wavelet decomposition
coeffs = pywt.wavedec(x, wavelet, level=level)
# Step 4: Define threshold for filtering
threshold = 0.75 * np.max(coeffs[-1])
# Step 5: Perform wavelet thresholding
coeffs_filtered = [pywt.threshold(c, threshold, mode='soft') for c in coeffs]
# Step 6: Reconstruct filtered signal
y= pywt.waverec(coeffs_filtered, wavelet)
#Step 7: Plotting the result
plt.subplot(3,1,1),plt.plot(t,x1),plt.xlabel('n-->'),plt.ylabel('x[n]')
plt.title('Clean signal')
plt.subplot(3,1,2),plt.plot(t,x),plt.xlabel('n-->'),plt.ylabel('z[n]')
plt.title('Noisy signal')
plt.subplot(3,1,3),plt.plot(t,y),plt.xlabel('n-->'),plt.ylabel('y[n]')
plt.title('Filtered signal')
plt.tight_layout()
```

Fig. 5.66 Python code to perform denoising of the signal

2. The noisy signal is obtained by adding white noise, which follows normal distribution to the input signal.
3. The noisy signal is decomposed using *db2* wavelet. The level of decomposition is three.
4. After wavelet decomposition, the detail coefficients are thresholded using soft thresholding to minimize the impact of noise. It is generally believed that much of the signal energy will be in low-frequency regions and noise will reside in high-frequency regions.
5. The inverse wavelet transform of the modified wavelet coefficients is performed to obtain the filtered signal.
6. From Fig. 5.67, it is possible to interpret that the impact of noise is less in filtered signal when compared to noisy signal.

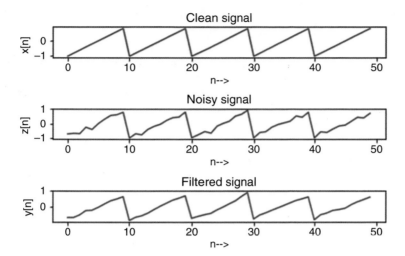

Fig. 5.67 Denoising of sawtooth signal

Tasks

1. The experiment can be repeated by the following: (a) Choose different wavelet family other than '*db2*'. (b) The level of decomposition can be changed. (c) Instead of soft thresholding, hard thresholding can be tried.

Exercises

1. Write a python code to obtain the Z-transform of the following sequences:
   (a) $x_1[n] = \delta[n - 5]$ (b) $x_2[n] = u[n] - u[n - 1]$ (c) $x_3[n] = nu[n]$
   (d) $x_4[n] = \sin(\omega_0 n)$.
2. Write a python code to compute the inverse Z-transform of (a) $X_1(z) = z^{-2}$
   (b) $X_2(z) = \frac{1}{(1-z^{-1})^2}$.
3. Write a python code to compute the magnitude and phase responses of the system, whose transfer function is given by $H(z) = \frac{1}{1-z^{-1}}$.
4. Let the signal $x[n]$ represent 100 samples of 5 Hz sine wave. Now increase the length of the signal by padding 50 sample values of zeros to $x[n]$. Zero padding is done at the end of 100 samples of $x[n]$. Let the zero padded signal be denoted as $y[n]$. Write a python code to plot the spectrum of the signal $x[n]$ and $y[n]$. Comment on the observed result.
5. Obtain the DFT of the sequences $x_1[n] = \{1, 1, 1, 1\}$ and $x_2[n] = \{1, -1, 1, -1\}$. Plot their magnitude responses and comment on the observed result.
6. Generate a square wave of 5 Hz fundamental frequency. Write a python code to plot the spectrum of the square wave and comment on the observed result.
7. Let $x[n]$ represent 100 samples of 5 Hz sine wave. Let $y[n]$ represent 100 samples of 5 Hz cosine wave. Take Fourier transform of $x[n]$ and $y[n]$ to obtain $X[k]$ and $Y[k]$. Extract the magnitude and phase components of $X[k]$ and $Y[k]$. Now interchange the phase of $X[k]$ with $Y[k]$. After phase interchange, take inverse

Fourier transform to obtain $x'[n]$ and $y'[n]$. Use subplot to plot the signals $x[n]$, $y[n]$, $x'[n]$ and $y'[n]$ and comment on the observed result.

8. Generate a linear chirp signal whose frequency varies from 10 to 1 Hz in 10 s. Plot the spectrum and spectrogram of this chirp signal and comment on the observed result.

9. Write a python code to verify the fact that a shorter window gives good time resolution and a wider window gives good frequency resolution in short-time Fourier transform.

10. Generate sinusoidal signal with momentary interruption. Apply CWT to identify the momentary interruption present in the signal.

Objective Questions

1. The region of convergence of unit sample signal ($\delta[n]$) is

   A. Entire Z-plane except $z = 0$
   B. Entire Z-plane except $z = $ infinity
   C. Entire Z-plane
   D. Entire Z-plane except $z = 0$ and $z = $ infinity

2. Convolution in time domain is equivalent to

   A. Addition in Z-domain
   B. Subtraction in Z-domain
   C. Multiplication in Z-domain
   D. Division in Z-domain

3. For a discrete-time system to be stable

   A. Pole should lie inside the unit circle.
   B. Pole should lie outside the unit circle.
   C. Pole should lie on the unit circle.
   D. Pole can lie anywhere in the z-plane.

4. Z-transform of $x[n] = nu[n]$ is

   A. $X(z) = \frac{z}{z-1}$
   B. $X(z) = \frac{z}{(z-1)^2}$
   C. $X(z) = \left(\frac{z}{z-1}\right)^2$
   D. $X(z) = \frac{z}{(z-1)^3}$

5. Let $X(z)$ be the Z-transform of the signal $x[n]$. If $X(z) = \frac{z}{z-1}$, then $\lim\limits_{n \to \infty} x[n]$ is

   A. 0
   B. 1
   C. -1
   D. Infinite

6. Let $x[n] = \left(\frac{2}{5}\right)^n u[n] - \left(\frac{5}{2}\right)^n u[-n-1]$. Let $X(z)$ be the Z-transform of the given signal $x[n]$; then, the region of convergence of its Z-transform is

   A. $\frac{2}{5} < |z| < \infty$
   B. $\frac{5}{2} < |z| < \infty$
   C. $\frac{2}{5} < |z| < \frac{5}{2}$
   D. $-\infty < |z| < \frac{5}{2}$

7. The transfer function of an LTI system is

   A. Linear function of 'z'
   B. Rational function of 'z'
   C. Logarithmic function of 'z'
   D. Exponential function of 'z'

8. The basis function of Fourier transform is

   A. Triangular function
   B. Rectangular function
   C. Complex exponential function
   D. Prolate spheroidal function

9. Fourier transform of a Gaussian function will result in

   A. Triangular function
   B. Rectangular function
   C. Sinc function
   D. Gaussian function

10. Fourier transform of a rectangular function will result in

   A. Triangular function
   B. Rectangular function
   C. Sinc function
   D. Gaussian function

11. The 2×2 DFT matrix is given by

   A. $\begin{bmatrix} 1 & 1 \\ 1 & 1 \end{bmatrix}$

   B. $\begin{bmatrix} 1 & 1 \\ 1 & -1 \end{bmatrix}$

   C. $\begin{bmatrix} 1 & 0 \\ 0 & 1 \end{bmatrix}$

   D. $\begin{bmatrix} 1 & 0 \\ 0 & -1 \end{bmatrix}$

Bibliography

1. Ronald N. Bracewell, "Fourier Transform and its Applications", McGraw Hill, 1978.
2. Alexander D. Poularikas, and Richard C. Dorf, "Transforms and Applications Handbook", Wiley, 2021.
3. Martin Vetterli, and Jelena Kovacevic, "Wavelets and Subband Coding", CreateSpace Independent Publishing Platform, 2013.
4. Ronald L. Allen, Duncan W. Mills, "Signal Analysis: Time, Frequency, Scale, and Structure", Wiley-IEEE Press, 2004.
5. B. P. Lathi, "Signals, Systems and Communication", B.S publication, 2001.

Chapter 6
Filter Design Using Pole-Zero Placement Method

Learning Objectives

After reading this chapter, the reader is expected to

- Design, implement and analyse first-order infinite impulse response filter.
- Design, implement and analyse the moving average filter.
- Design and analyse digital resonator.
- Design and analyse notch filter and comb filter.
- Design and analyse all-pass filter.

Roadmap of the Chapter

Digital filters can be considered as a linear time-invariant system that accepts input and gives modified input as the output. Based on the input-output relation, discrete-time systems can be classified as autoregressive system (AR), moving average system (MA) and autoregressive moving average (ARMA) system. If the current output is a function of the current input and past outputs, then the system is autoregressive. If the current output of the system is a function of current input and past inputs, the system is a moving average system. If the current output is a function of both past input and past output, the system is autoregressive moving average system. An example of autoregressive system is the IIR filter and notch filter. Digital resonator and M-point moving average systems are examples of MA system. All-pass filter is an example of an ARMA system. This is depicted below.

S. Esakkirajan et al., *Digital Signal Processing*,
https://doi.org/10.1007/978-981-99-6752-0_6

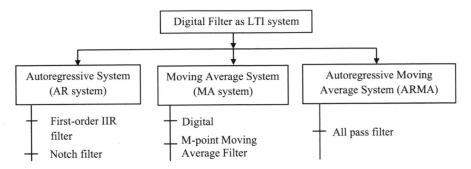

The filters discussed in this chapter include the IIR filter, moving average filter, digital resonator, notch filter, comb filter and all-pass filter.

PreLab Questions

1. If $h[n]$ represents the impulse response of a lowpass filter, what would be the behaviour of the filter whose impulse response is $(-1)^n h[n]$?
2. When a discrete-time system is said to be a minimum phase system?
3. What do you understand by the term 'poles' and 'zeros' of a system?
4. How is the stability of the discrete-time system related to (a) location of poles of the system and (b) impulse response of the system?
5. What is the basic principle involved in the design of digital filter using pole-zero placement method?
6. What do you understand by the term 'delay equalizer' or 'phase equalizer'?
7. Mention two applications of notch filter.
8. A square wave is fed as input to M-point moving average filter. What would be the output of M-point moving average filter?
9. The relationship between the input and output of a digital filter is given by $y[n] = \alpha x[n] + \beta x[n-1] + \gamma x[n-2]$. Is this a finite impulse response filter (FIR) or infinite impulse response filter (IIR)? Justify your choice.
10. What is a pole-zero plot? What information one gets by interpreting the pole-zero plot?

6.1   First-Order IIR Filter

This section begins with the design of a first-order IIR filter. The transfer function of a first-order IIR filter is given by

$$H(z) = \frac{1}{1 - p_1 z^{-1}} \tag{6.1}$$

If the pole lies on the unit circle ($p_1 = 1$), the transfer function of the filter is given by

$$H(z) = \frac{1}{1 - z^{-1}} \tag{6.2}$$

Step 1: Magnitude response of the filter

From the magnitude response of the filter, it is possible to observe the filter behaviour at low and high frequencies. The frequency response of the system is obtained by substituting $z = e^{j\omega}$ in Eq. (6.2), we get

$$H(e^{j\omega}) = \frac{1}{1 - e^{-j\omega}} \tag{6.3}$$

The above equation can be expressed as

$$H(e^{j\omega}) = \frac{1}{1 - \cos(\omega) + j\sin(\omega)} \tag{6.4}$$

The magnitude response is obtained as

$$\left| H(e^{j\omega}) \right| = \frac{1}{\sqrt{(1 - \cos(\omega))^2 + \sin^2(\omega)}} \tag{6.5}$$

Upon simplifying the above equation

$$\left| H(e^{j\omega}) \right| = \frac{1}{\sqrt{2 - 2\cos(\omega)}} \tag{6.6}$$

When $\omega = 0$, the magnitude response tends to infinity, and when $\omega = \pi$, the magnitude response is given by $|H(e^{j\omega})| = \frac{1}{2}$. The filter passes low-frequency components and attenuates high-frequency components. The filter behaves like a lowpass filter.

Step 2: Impulse response of the filter

The impulse response of the filter is obtained by taking inverse Z-transform of the transfer function. The impulse response of the filter is given by

$$h[n] = Z^{-1}\{H(z)\} \tag{6.7}$$

Upon substituting Eq. (6.2) in Eq. (6.7), we get

$$h[n] = Z^{-1}\left\{ \frac{1}{1 - z^{-1}} \right\} = u[n] \tag{6.8}$$

The impulse response of the filter is the unit step function. The unit step function is not absolutely summable; hence, the filter is not stable.

Table 6.1 Built-in function in the design of first-order IIR filter

Built-in function	Library	Use
freqz	*signal. scipy*	To obtain the frequency response of the filter, which is a combination of magnitude and phase responses
tf2zpk	*signal. scipy*	To obtain the poles, zeros and gain of the filter

Experiment 6.1 Characteristics of First-Order IIR Filter

The experiment is about obtaining the characteristics of the filter like (a) impulse response, (b) pole-zero plot and (c) magnitude and phase responses of the filter. The built-in functions used to obtain the responses of the filter are given in Table 6.1

The python code which obtains the characteristics of the first-order IIR filter is given in Fig. 6.1, and the corresponding output is shown in Fig. 6.2.

The python code consists of five steps which are given as S_1 to S_5. Step 1 (S_1) generation of an impulse to obtain the impulse response of the filter. Step 2 (S_2) defines the system in terms of the transfer function of the system. Step 3 (S_3) deals with obtaining the impulse response of the filter for which the input to the filter is unit sample signal. Step 4 (S_4) deals with obtaining the frequency response of the filter, and Step 5 (S_5) deals with plotting the characteristics of the filter. The result obtained upon execution of the code shown in Fig. 6.1 is given in Fig. 6.2.

Inferences

From Fig. 6.2, the following inferences can be obtained:

1. The impulse response of the filter is obtained as a unit step function, which is in agreement with the theoretical result.
2. From the pole-zero plot, it is possible to observe that the pole lies on the unit circle.
3. From the frequency response, it is possible to observe that the filter behaves like a lowpass filter.
4. For a discrete-time system to be stable, the impulse response should be absolutely summable. From Fig. 6.2, the impulse response is a unit step function that is not absolutely summable; hence, the given filter is not stable.
5. For discrete-time system to be stable, the poles should lie within the unit circle. From the pole-zero plot, it is possible to observe that the pole lies on the unit circle; hence, the filter is BIBO stable.

Task

1. Repeat this experiment for $H(z) = \frac{1}{1+z^{-1}}$, and comment on the observed result. Will the pole-position change the nature of the system?

Experiment 6.2 Input-Output of First-Order IIR Filter

In order to understand the behaviour of the filter, two types of inputs are fed to the filter. Input 1 is a DC signal, whereas input 2 is a high-frequency signal. The python

```
#Characteristics of first-order IIR filter
import numpy as np
import matplotlib.pyplot as plt
from scipy import signal
#S1: Generation of impulse input
x=np.zeros(100)
x[0]=1
#S2: Define the system
num,den=[1],[1,-1]
#S3: To obtain the impulse response
h=signal.lfilter(num,den,x)
#S4: Characteristics of the first order IIR filter
fs=100
w,H=signal.freqz(num,den)
z,p,k=signal.tf2zpk(num,den)
#S5: Plotting the result
plt.figure(1),plt.subplot(2,2,1),plt.stem(h),plt.xlabel('n-->'),plt.ylabel('h[n]'),
plt.title('Impulse Response')
plt.subplot(2,2,2),plt.plot((w/np.pi)*fs/2,20*np.log10(np.abs(H))),
plt.xlabel('$\omega$-->'),plt.ylabel('|H($j\omega$)|'),
plt.title('Magnitude Response'),plt.subplot(2,2,3),plt.xlabel('$\sigma$'),
plt.ylabel('$j\omega$'),plt.title('Pole Zero Plot')
plt.plot(np.real(z),np.imag(z),'ko'),plt.plot(np.real(p),np.imag(p),'rx')
theta=np.linspace(0,2*np.pi,100)
plt.plot(np.cos(theta),np.sin(theta))
plt.subplot(2,2,4),plt.plot(w,np.angle(H)),plt.xlabel('$\omega$-->'),
plt.ylabel('$\u2220H(j\omcga)$'),plt.title('Phase Response')
plt.tight_layout()
```

Fig. 6.1 Python code to obtain the characteristics of first order IIR filter

code which deals with the response of the filter for these two different types of inputs is shown in Fig. 6.3, and the corresponding output is shown in Fig. 6.4.

Inferences

From Fig. 6.4, the following inferences can be drawn:

1. $y_1[n]$ is the output signal corresponding to the input signal $x_1[n]$. Here $x1[n]$ is a DC signal. The signal $x_1[n]$ is generated from the expression $x_1[n] = e^{j\omega n}$ by substituting $\omega = 0$. From the output signal $y_1[n]$, it is possible to interpret that the filter amplifies $x_1[n]$, which is a DC signal.
2. $y_2[n]$ is the output signal corresponding to the input signal $x_2[n]$. Here $x_2[n]$ is an AC signal. The signal $x_2[n]$ is generated from the expression $x_2[n] = e^{j\omega n}$ by substituting $\omega = \pi$. From the output signal $y_2[n]$, it is possible to interpret that the filter blocks $x_2[n]$, which is a high-frequency signal.

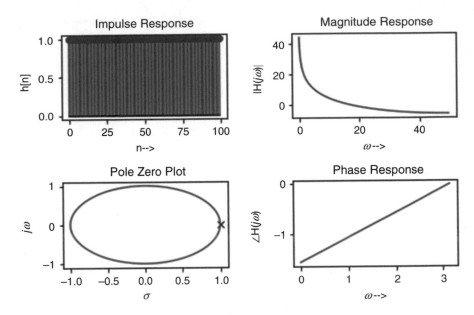

Fig. 6.2 Characteristics of first-order IIR filter

```
#Input-Output of first-order IIR filter
import numpy as np
import matplotlib.pyplot as plt
from scipy import signal
#Step 1: Signal generation
n=np.arange(-5,6)
omega1=0
omega2=np.pi
x1=np.exp(1j*omega1*n)
x2=np.exp(1j*omega2*n)
#S2: Define the system
num,den=[1],[1,-1]
#S3: To obtain the output
y1=signal.lfilter(num,den,x1)
y2=signal.lfilter(num,den,x2)
#S4: Plotting the input and output of the filter
plt.subplot(2,2,1),plt.stem(n,x1),plt.xticks(n),plt.xlabel('n-->'),plt.ylabel('Amplitude'),
plt.title('$x_1[n]$'),plt.subplot(2,2,2),plt.stem(n,y1),plt.xticks(n),
plt.xlabel('n-->'),plt.ylabel('Amplitude'), plt.title('$y_1[n]$')
plt.subplot(2,2,3),plt.stem(n,x2),plt.xticks(n),plt.xlabel('n-->'),
plt.ylabel('Amplitude'),plt.title('$x_2[n]$'),plt.subplot(2,2,4),plt.stem(n,y2),plt.xticks(n),
plt.xlabel('n-->'),plt.ylabel('Amplitude'),plt.title('$y_2[n]$')
plt.tight_layout()
```

Fig. 6.3 Python code to obtain the output of the filter

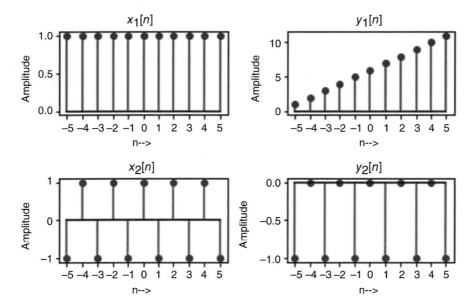

Fig. 6.4 Input and output of first-order IIR filter

3. When the poles are at position $p = 1$, the filter amplifies the DC signal and blocks the high-frequency signal. The filter behaves like a lowpass filter.

Experiment 6.3 Impact of Pole Position on the Magnitude and Impulse Responses of First-Order IIR Filter

The objective of this experiment is to analyse the impact of pole position on the impulse and magnitude response of first-order IIR filter whose transfer functions are given by $H_1(z) = \frac{1}{1 - 0.25z^{-1}}$ and $H_2(z) = \frac{1}{1 + 0.25z^{-1}}$. The python code which obtains the impulse response and the magnitude response of the two filters is given in Fig. 6.5, and the corresponding output is shown in Fig. 6.6.

Inferences

From Fig. 6.6, the following inferences can be drawn:

1. The pole of System-1 whose transfer function is given by $H_1(z) = \frac{1}{1 - 0.25z^{-1}}$ lies on the positive half of the Z-plane.
2. The impulse response of System-1 is observed to be an exponentially decreasing function.
3. From the magnitude response of System-1, it is possible to observe that System-1 behaves like a lowpass filter.
4. The pole of System-2 whose transfer function is given by $H_2(z) = \frac{1}{1 + 0.25z^{-1}}$ lies on the negative half of Z-plane.
5. From the magnitude response of System-2, it is possible to infer that the System-2 behaves like a highpass filter.

```
#Impact of pole position on the behaviour of the system
import numpy as np
import matplotlib.pyplot as plt
from scipy import signal
#Defining systems
num,den1,den2=[1],[1,-0.25],[1,0.25]
#Obtaining the magnitude response
w1,H1=signal.freqz(num,den1)
z1,p1,k1=signal.tf2zpk(num,den1)
w2,H2=signal.freqz(num,den2)
z2,p2,k2=signal.tf2zpk(num,den2)
#To generate imupluse input
x=np.zeros(15)
x[0]=1
h1=signal.lfilter(num,den1,x)
h2=signal.lfilter(num,den2,x)
plt.subplot(2,3,1),plt.plot(np.real(z1),np.imag(z1),'ko'),plt.xlabel('$\sigma$'),
plt.ylabel('$j\omega$'),plt.title('Pole Zero Plot')
plt.plot(np.real(p1),np.imag(p1),'rx')
theta=np.linspace(0,2*np.pi,100)
plt.plot(np.cos(theta),np.sin(theta))
plt.subplot(2,3,2),plt.stem(h1),plt.xlabel('n-->'),plt.ylabel('Amplitude'),plt.title('$h_1[n]$')
plt.subplot(2,3,3),plt.plot(w1,20*np.log10(np.abs(H1)))
plt.xlabel('$\omega$-->'),plt.ylabel('|$H_1(j\omega)$|'),plt.title('Magnitude response')
plt.subplot(2,3,4),plt.plot(np.real(z2),np.imag(z2),'ko')
plt.plot(np.real(p2),np.imag(p2),'rx'),plt.xlabel('$\sigma$'),
plt.ylabel('$j\omega$'),plt.title('Pole Zero Plot')
theta=np.linspace(0,2*np.pi,100)
plt.plot(np.cos(theta),np.sin(theta))
plt.subplot(2,3,5),plt.stem(h2),plt.xlabel('n-->'),plt.ylabel('Amplitude'),plt.title('$h_2[n]$')
plt.subplot(2,3,6),plt.plot(w2,20*np.log10(np.abs(H2)))
plt.xlabel('$\omega$-->'),plt.ylabel('|$H_2(j\omega)$|'),plt.title('Magnitude response')
plt.tight_layout()
```

Fig. 6.5 Python code to analyse the impact of pole position on the behaviour of the filter

6. The impulse response of System-1 is given by $h_1[n] = (0.25)^n u[n]$, whereas the impulse response of System-2 is given by $h_2[n] = (-1)^n h_1[n]$. If $h_1[n]$ acts as a lowpass filter, then $h_2[n]$ behaves like a highpass filter.
7. This experiment concludes that the pole position changes the behaviour of the filter from lowpass to highpass filter.

Task

1. Repeat the above experiment for the system whose transfer function is given by $H(z) = \frac{1}{1 - 2z^{-1}}$, and comment on the observed result.

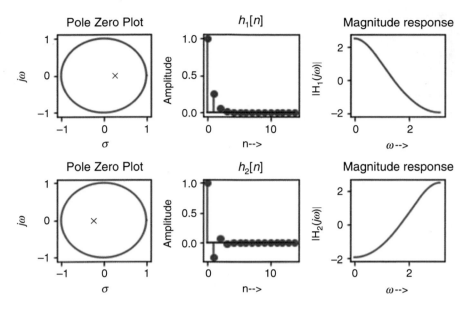

Fig. 6.6 Result of python code shown in Fig. 6.5

6.2 Moving Average filter

The relationship between the input and output of M-point moving average filter is given by

$$y[n] = \frac{1}{M} \sum_{k=0}^{M-1} x[n-k] \tag{6.9}$$

For a three-point moving average filter, $M = 3$, substituting $M = 3$ in Eq. (6.9), we get the input-output relationship as

$$y[n] = \frac{1}{3}\{x[n] + x[n-1] + x[n-2]\} \tag{6.10}$$

From Eq. (6.10), it is possible to interpret that equal weightage is given to $x[n]$, $x[n-1]$ and $x[n-2]$. This type of system is termed as '*moving average system*'. This system performs the weighted average of three input samples $x[n]$, $x[n-1]$ and $x[n-2]$; hence, it is termed as '*moving average filter*'.

Experiment 6.4 Characteristics of Moving Average Filter

This experiment tries to obtain the characteristics of a moving average filter using python. The python code, which obtains the characteristics of three-point moving average filter, is shown in Fig. 6.7, and the corresponding output is shown in Fig. 6.8.

```
#Characteristics of Moving average filter
import numpy as np
import matplotlib.pyplot as plt
from scipy import signal
#S1: Generation of impulse input
x=np.zeros(100)
x[0]=1
#S2: Define the three-point Moving average system
num,den=[1/3,1/3,1/3],[1]
#S3: To obtain the impulse response
h=signal.lfilter(num,den,x)
#S4: Characteristics of the first-order IIR filter
fs=100
w,H=signal.freqz(num,den)
z,p,k=signal.tf2zpk(num,den)
#S5: Plotting the result
plt.figure(1),plt.subplot(2,2,1),plt.stem(h),plt.xlabel('n-->'),plt.ylabel('Amplitude'),
plt.title('h[n]'),plt.subplot(2,2,2),plt.plot((w/np.pi)*fs/2,20*np.log10(np.abs(H))),
plt.xlabel('$\omega$-->'),plt.ylabel('|H($j\omega$)|'),plt.title('Magnitude response')
plt.subplot(2,2,3),plt.xlabel('$\sigma$'),plt.ylabel('$j\omega$'),plt.title('Pole Zero Plot')
plt.plot(np.real(z),np.imag(z),'ko'),plt.plot(np.real(p),np.imag(p),'rx')
theta=np.linspace(0,2*np.pi,100)
plt.plot(np.cos(theta),np.sin(theta)),plt.subplot(2,2,4),plt.plot(w,np.angle(H)),
plt.xlabel('$\omega$-->'),plt.ylabel('$\u2220H(j\omega)$'),plt.title('Phase response')
plt.tight_layout()
```

Fig. 6.7 Python code to obtain the characteristics of three-point moving average filter

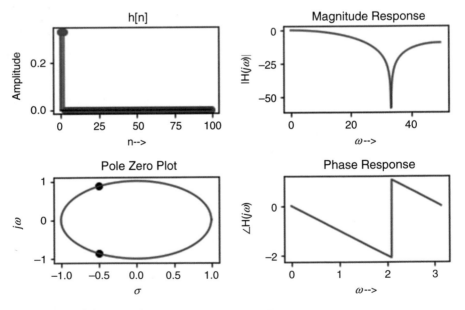

Fig. 6.8 Characteristics of three-point moving average filter

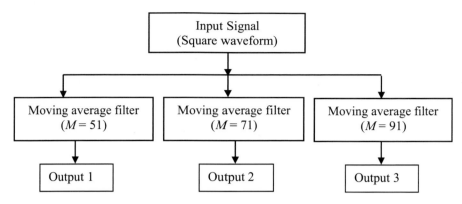

Fig. 6.9 Problem statement illustration

Inferences

From Fig. 6.8, the following inferences can be drawn:

1. The impulse response of a moving average filter is finite. If the impulse response of the system is absolutely summable, then the discrete-time system is a stable system. Thus, three-point moving average system is inherently stable.
2. Moving average filter is an all-zero filter.
3. From the magnitude response, it is possible to observe that three-point moving average filter act as a lowpass filter.
4. From the phase response, it is possible to conclude that a moving average filter exhibits linear phase characteristics in the pass band.

Task

1. Repeat the above experiment for a five-point moving average filter and six-point moving average filter, and comment on the observed output. What change do you observe in the pole-zero plot for $M = 5$ and $M = 6$?

Experiment 6.5 Impact of the Order of Moving Average Filter

The objective of this experiment is to observe the impact of the order of the moving average filter with respect to the extent of filtering. This objective is shown in Fig. 6.9. From Fig. 6.9, it is possible to interpret that the input signal to the three moving average filters of orders 51, 71 and 91 is a square wave. The reason for choosing square wave as input is that it exhibits sharp transition between '*ON*' and '*OFF*' state. The python code, which implements the task shown in Fig. 6.9, is given in Fig. 6.10, and the corresponding output is shown in Fig. 6.11.

Inferences

The following inferences can be drawn from Fig. 6.11:

1. The input signal to the moving average filter is a square wave. The input signal exhibits sudden transitions between states '0' and '1'.
2. The input signal is passed through 3 moving average filters of order 51, 71 and 91.

```
#Impact of the order of Moving average filter
import numpy as np
import matplotlib.pyplot as plt
from scipy import signal
#Step 1: Generating the input square waveform
t=np.linspace(0,1,1000)
x=signal.square(2*np.pi*5*t)
#Step 2: Defining the MA filters
num1=1/51*np.ones(51)
num2=1/71*np.ones(71)
num3=1/91*np.ones(91)
den=[1]
#Step 3: Obtaining the outputs
y1=signal.lfilter(num1,den,x)
y2=signal.lfilter(num2,den,x)
y3=signal.lfilter(num3,den,x)
#Step 4: Plotting the results
plt.subplot(2,2,1),plt.plot(t,x),plt.xlabel('Time'),plt.ylabel('Amplitude'),
plt.title('Input signal'),plt.subplot(2,2,2),plt.plot(t,y1)
plt.xlabel('Time'),plt.ylabel('Amplitude'),plt.title('Filtered signal (M=51)')
plt.subplot(2,2,3),plt.plot(t,y2),plt.xlabel('Time'),plt.ylabel('Amplitude'),
plt.title('Filtered signal (M=71)'),plt.subplot(2,2,4),plt.plot(t,y3)
plt.xlabel('Time'),plt.ylabel('Amplitude'),plt.title('Filtered signal (M=91)')
plt.tight_layout()
```

Fig. 6.10 Python code to obtain the results of moving average filter

3. The square wave is transformed into a triangular wave for the moving average filter of order 91. The square wave, when passed through an integrator (lowpass filter), results in a triangular wave. The triangular wave exhibits gradual variation between the states '0' and '1'.
4. The extent of smoothing increases with an increase in the order of the moving average filter.

6.3   *M*-Point Exponentially Weighted Moving Average Filter (EWMA)

The relationship between the input and output of *M*-point exponentially weighted moving average filter is given by

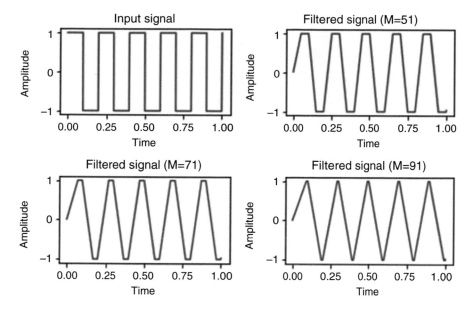

Fig. 6.11 Results of moving average filter

$$y[n] = C \sum_{k=0}^{M-1} \alpha^k x[n-k] \tag{6.11}$$

In the above expression, '*C*' is the normalization constant, and '*α*' is the exponential weighting factor, where $0 < \alpha < 1$.

To Find the Expression for Normalization Constant (*C*) One way to obtain the value of '*C*' is that it should preserve the DC gain. If the input is a constant signal (*x*[*n*] = *K*), if the filter preserves the DC component of the signal, then the output is also expected to be '*K*'. Substituting $x[n] = K$ and $y[n] = K$ in Eq. (6.11), we get

$$K = C \sum_{k=0}^{M-1} \alpha^k K \tag{6.12}$$

The above equation can be written as

$$K = CK \sum_{k=0}^{M-1} \alpha^k \tag{6.13}$$

From the above expression, the expression for the constant '*C*' can be written as

$$C \sum_{k=0}^{M-1} \alpha^k = 1 \tag{6.14}$$

The expression for the constant 'C' is written as

$$C = \frac{1}{\displaystyle\sum_{k=0}^{M-1} \alpha^k} \tag{6.15}$$

Using summation formula

$$\sum_{k=0}^{M-1} \alpha^k = \frac{1 - \alpha^M}{1 - \alpha} \tag{6.16}$$

Substituting Eq. (6.16) in Eq. (6.15), we have

$$C = \frac{1 - \alpha}{1 - \alpha^M} \tag{6.17}$$

The expression for M-point exponentially weighted moving average filter is given by

$$y[n] = \frac{1 - \alpha}{1 - \alpha^M} \sum_{k=0}^{M-1} \alpha^k x[n - k] \tag{6.18}$$

If 'α' value is closer to 1, then M-point exponentially weighted moving average filter will behave like an M-point moving average filter.

Experiment 6.6 Comparing the Impulse Response of MA and EWMA Filter
The objective of this experiment is to compare the impulse responses of the moving average filter with the exponentially weighted moving average filter for $M = 3$ and 5. The python code, which plots the impulse response of the moving average filter and exponentially weighted average filter, is shown in Fig. 6.12, and the corresponding output is shown in Fig. 6.13.

Inferences
From Fig. 6.13, it is possible to interpret the following:

1. The MA filter gives equal weightage to all the input sample values.
2. The EWMA filter gives more weightage to the current input sample and less weightage to the past input samples.

```
#Impulse response of MA and EWMA
import numpy as np
import matplotlib.pyplot as plt
M1, M2=3,5
alpha=0.5
h1=1/M1*np.ones(M1) #MA filter for M=3
h2=1/M2*np.ones(M2) #MA filter for M=5
C1=(alpha-1)/(alpha**M1-1)
C2=(alpha-1)/(alpha**M2-1)
h3=C1*np.array([1,alpha,alpha**2]) #EWMA filter for M=3
h4=C2*np.array([1,alpha,alpha**2,alpha**3,alpha**4])#EWMA filter for M=5
plt.subplot(2,2,1),plt.stem(h1),plt.xlabel('n-->'),plt.ylabel('$h_1[n]$'),
plt.title('$h_1[n]$ of MA filter for M = 3')
plt.subplot(2,2,2),plt.stem(h3),plt.xlabel('n-->'),plt.ylabel('$h_3[n]$'),
plt.title('$h_3[n]$ of EWMA filter for M = 3')
plt.subplot(2,2,3),plt.stem(h2),plt.xlabel('n-->'),plt.ylabel('$h_2[n]$'),
plt.title('$h_2[n]$ of MA filter for M = 5')
plt.subplot(2,2,4),plt.stem(h4),plt.xlabel('n-->'),plt.ylabel('$h_4[n]$'),
plt.title('$h_4[n]$ of EWMA filter for M = 5'),plt.tight_layout()
```

Fig. 6.12 Python code which obtains the impulse response of MA and EWMA filter

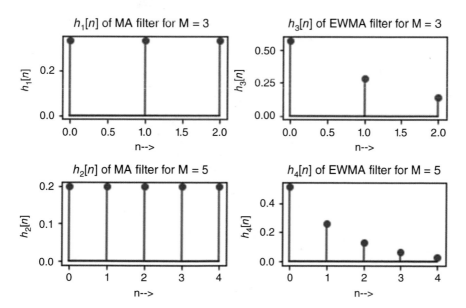

Fig. 6.13 Impulse response of MA and EWMA filter

6.4   Digital Resonator

A resonator is designed to have its strongest response to match certain input signal. Resonators find application in communication receivers, AM/FM demodulators, etc. A digital resonator is a two-pole bandpass filter with a pair of complex-conjugate poles near the unit circle to create a resonant peak at the desired frequency. The digital resonator has a large magnitude response in the vicinity of the pole location. If one pole is located at $p_1 = re^{j\omega}$, then the other pole will be at $p_2 = re^{-j\omega}$, where $0 < r < 1$. The expression for the transfer function of the system is given by

$$H(z) = \frac{1}{(1 - p_1 z^{-1})(1 - p_2 z^{-1})} \tag{6.19}$$

Substituting $p_1 = re^{j\omega}$ and $p_2 = re^{-j\omega}$ in the above expression, we get

$$H(z) = \frac{1}{(1 - re^{j\omega} z^{-1})(1 - re^{-j\omega} z^{-1})}$$

Simplifying the above expression, we get

$$H(z) = \frac{1}{1 - re^{-j\omega} z^{-1} - re^{j\omega} z^{-1} + r^2 z^{-2}}$$

The above equation can be expressed as

$$H(z) = \frac{1}{1 - rz^{-1}[e^{-j\omega} + e^{j\omega}] + r^2 z^{-2}}$$

Simplifying the above equation, we get

$$H(z) = \frac{1}{1 - 2rz^{-1} \cos(\omega) + r^2 z^{-2}} \tag{6.20}$$

Experiment 6.7 Digital Resonator
This experiment analyses the concept of digital resonator using python. The python illustration of digital resonator with two complex conjugate poles occurring at $r = 0.98$ is shown in Fig. 6.14, and the corresponding output is shown in Fig. 6.15.

Inferences
The following inferences can be made from this experiment:

1. From the input and output signals, it is possible to observe from Fig. 6.15 that the input is a unit sample signal. The system is excited with an impulse signal. The output of the system produces an oscillation.

```
#Digital resonator
import numpy as np
import matplotlib.pyplot as plt
from scipy import signal
#Defining the system
r,fs=0.98,100  #Sampling frequency
fn,fc=fs/2, 5   #Cutoff frequency
w=2*np.pi*(fc/fn)
b=[1]
a=[1,-2*r*np.cos(w),r**2]
#Generating the input and obtaining the response
x=np.zeros(25)  #Input to the resonator
x[0]=1
y=signal.lfilter(b,a,x) #Output of resonator
plt.subplot(3,2,1),plt.stem(x),plt.xlabel('n-->'),plt.ylabel('x[n]'),plt.title('Input signal')
plt.subplot(3,2,2),plt.stem(y),plt.xlabel('n-->'),plt.ylabel('y[n]'),plt.title('Output signal')
#Impulse response of the system
h_1=np.zeros(25)
h_1[0]=1
h=signal.lfilter(b,a,h_1)
plt.subplot(3,2,3),plt.stem(h),plt.xlabel('n-->'),plt.ylabel('y1[n]'),plt.title('Impulse response')
# Pole-zero plot
z, p, k = signal.tf2zpk(b, a)
theta = np.linspace(0, np.pi*2, 500)
circle = np.exp(1j*theta)
plt.subplot(3,2,5),plt.plot(circle.real, circle.imag, 'k--')
plt.plot(z.real, z.imag, 'ro', ms=7.5),plt.plot(p.real, p.imag, 'rx',ms=7.5)
plt.xlabel('Real part'),plt.ylabel('Imaginary part'),plt.title('Pole-zero plot'),plt.grid()
#Magnitude and phase response
w, h = signal.freqz(b,a)
plt.subplot(3,2,4),plt.plot(w, 10 * np.log10(abs(h))),plt.xlabel('$\omega$ [rad/sample]'),
plt.ylabel('$|H(e^{j\omega})|$ in [dB]'),plt.title('Magnitude response'),plt.subplot(3,2,6),
plt.plot(w,np.unwrap(np.angle(h))),plt.xlabel('$\omega$ [rad/sample]'),plt.ylabel('Degree')
plt.title('Phase response'),plt.tight_layout()
```

Fig. 6.14 Python code to implement digital resonator

2. From the magnitude response, it is possible to observe that the system behaves like a narrow bandpass filter.
3. From the pole-zero plot, it is possible to observe that two complex conjugate poles occur very closer to the unit circle.
4. From the phase response, it is possible to observe that the system exhibits non-linear phase characteristics.

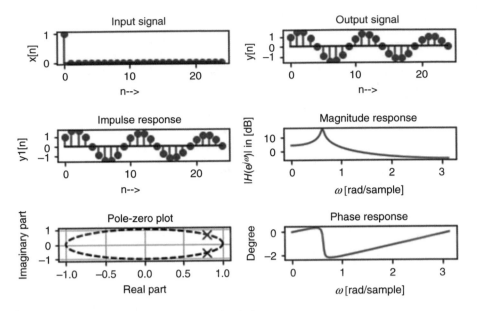

Fig. 6.15 Result of python code shown in Fig. 6.14

Task

1. Repeat the above experiment for $r = 1$ and comment on the observed result.

6.5   Notch Filter

Notch filter has two complex conjugate zeros placed on the unit circle to create a null at a desired frequency. A notch filter has the ability to reject one particular frequency. A pair of complex conjugate zeros on the unit circle produces a null in the frequency response, which results in the rejection of one particular frequency. Let the conjugate zeros be represented as $z_1 = re^{j\omega}$ and $z_2 = re^{-j\omega}$. If the zeros occur on the unit circle, then $r = 1$. The transfer function of such a system is given by

$$H(z) = \left(1 - z_1 z^{-1}\right)\left(1 - z_2 z^{-1}\right) \tag{6.21}$$

Substituting $z_1 = e^{j\omega}$ and $z_2 = e^{-j\omega}$ in the above expression, we get

$$H(z) = \left(1 - e^{j\omega} z^{-1}\right)\left(1 - e^{-j\omega} z^{-1}\right)$$

The above equation can be written as

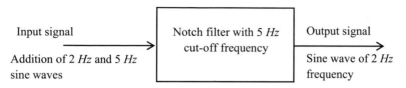

Fig. 6.16 Problem illustration

$$H(z) = 1 - e^{-j\omega}z^{-1} - e^{j\omega}z^{-1} + z^{-2}$$

The transfer function of the system is expressed as

$$H(z) = 1 - z^{-1}\left(e^{-j\omega} + e^{j\omega}\right) + z^{-2}$$

The above equation can be expressed as

$$H(z) = 1 - 2z^{-1}\cos(\omega) + z^{-2} \tag{6.22}$$

Experiment 6.8 Notch Filter

This experiment discusses the design of a notch filter to eliminate one particular frequency component. The input signal to the notch filter is the addition of two sine waves of frequency components, 2 and 5 Hz. The notch filter cut-off frequency is 5 Hz. It is expected that the notch filter will eliminate 5 Hz frequency component so that the filtered signal will have only 2 Hz frequency component. This is illustrated in Fig. 6.16.

The python code which performs the above-mentioned task is shown in Fig. 6.17, and the corresponding output is shown in Fig. 6.18.

Inferences

From Fig. 6.18, the following inferences can be drawn

1. The input signal to the notch filter is an addition of 2 and 5 Hz sine waves.
2. The output of the notch filter clearly shows that it is a 2 Hz sine wave. This means that the notch filter has filtered 5 Hz sinusoidal component only.
3. The impulse response shows that the filter designed has a finite impulse response.
4. From the magnitude response, it is possible to observe that the notch occurs at 5 Hz.
5. From the pole-zero plot, it is possible to observe two conjugate zeros.
6. From the phase response, it is possible to observe that the phase response varies linearly with respect to frequency. Therefore, the designed filter exhibits linear phase characteristics.

Experiment 6.9 Design of Notch Filter Using Built-In Function

The built-in function '*iirnotch*' available in '*scipy*' library can be used to design a notch filter. The python code to design a notch filter for cut-off frequency of 50 Hz

```
#Notch filter
import numpy as np
import matplotlib.pyplot as plt
from scipy import signal
#Step 1: Generating the input signal
f1,f2=2,5 # 1Hz and 5Hzfrequency component
n=np.arange(0,100)
x1=np.sin(2*np.pi*f1*n/100)
x2=np.sin(2*np.pi*f2*n/100)
x=x1+x2  #Input signal has 1 Hz and 5 Hz component
#Step 2: Design of notch filter
r,fs,fc=0.99,100,5 # Sampling, Cutoff frequencies
w=2*np.pi*(fc/fs)
b=[1,-2*np.cos(w),1]
a=[1]
#Step 3: Obtaining the output
y=signal.lfilter(b,a,x) #Output of resonator
plt.subplot(3,2,1),plt.stem(x),plt.xlabel('n-->'),plt.ylabel('x[n]'),plt.title('Input signal'),
plt.subplot(3,2,2),plt.stem(y),plt.xlabel('n-->'),plt.ylabel('y[n]'),plt.title('Output signal')
#Impulse response of the system
h_1=np.zeros(25)
h_1[0]=1
h=signal.lfilter(b,a,h_1)
plt.subplot(3,2,3),plt.stem(h),plt.xlabel('n-->'),plt.ylabel('y1[n]'),plt.title('Impulse response')
# Pole-zero plot
z, p, k = signal.tf2zpk(b, a)
theta = np.linspace(0, np.pi*2, 500)
circle = np.exp(1j*theta)
plt.subplot(3,2,5),plt.plot(circle.real, circle.imag, 'k--'),plt.plot(z.real, z.imag, 'ro', ms=7.5),
plt.plot(p.real, p.imag, 'rx',ms=7.5)
plt.xlabel('$\sigma$'),plt.ylabel('$j\omega$'),plt.title('Pole-zero plot'),plt.grid()
#Magnitude and phase response
w, h = signal.freqz(b,a)
plt.subplot(3,2,4),plt.plot(0.5*fs*w/np.pi, 10 * np.log10(abs(h)))
plt.xlabel('$\omega$ [rad/sample]'),plt.ylabel('$|H(j\omega)|$ in [dB]')
plt.title('Magnitude response'),plt.subplot(3,2,6),
plt.plot(0.5*fs*w/np.pi,np.unwrap(np.angle(h))),plt.xlabel('$\omega$ [rad/sample]'),
plt.ylabel('$\u2220H(j\omega)$'),plt.title('Phase response'),plt.tight_layout()
```

Fig. 6.17 Python code which performs notch filtering of the input signal

and a sampling frequency of $f_s = 1000$ Hz is shown in Fig. 6.19, and the corresponding characteristics are shown in Fig. 6.20.

Inferences

From Fig. 6.20, it is possible to observe the following facts:

1. The impulse response is of infinite duration. The impulse response is not symmetric.

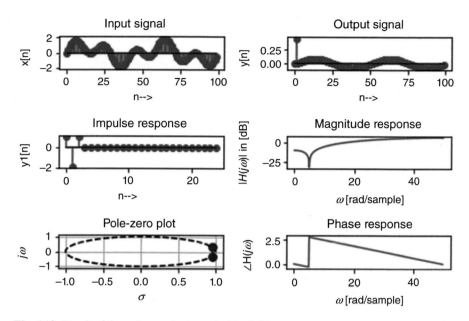

Fig. 6.18 Result of the python code shown in Fig. 6.17

2. The pole-zero plot shows two complex conjugate poles and zeros on the unit circle. The presence of poles on the unit circle indicates that the stability of the filter is not guaranteed.
3. The magnitude response shows that the 50 Hz notch frequency is the cut-off frequency.
4. The phase response indicates that it is non-linear. Since the impulse response is not symmetric, the phase response is not linear.

6.6   All-Pass Filter

The transfer function of first-order all-pass filter is given by

$$H(z) = \frac{z^{-1} - a}{1 - az^{-1}}, \quad \text{where } |a| < 1 \tag{6.23}$$

The frequency response of the system is obtained by substituting $z = e^{j\omega}$ in the above equation, we get

$$H(e^{j\omega}) = \frac{e^{-j\omega} - a}{1 - ae^{-j\omega}} \tag{6.24}$$

```
#Characteristics of notch filter
import matplotlib.pyplot as plt
import numpy as np
from scipy import signal
#Step 1: Design of notch filter
fs, fc, QF = 1000, 50, 10 # Sampling, Cut-off frequency and Quality factor
w0=fc/(fs/2)
b, a = signal.iirnotch(w0, QF)
#Step 2: Plotting the characteristics
h_1=np.zeros(25)
h_1[0]=1
h=signal.lfilter(b,a,h_1)
plt.subplot(2,2,1),plt.stem(h),plt.xlabel('n-->'),plt.ylabel('h[n]'),plt.title('Impulse response')
# Pole-zero plot
z, p, k = signal.tf2zpk(b, a)
theta = np.linspace(0, np.pi*2, 100)
circle = np.exp(1j*theta)
plt.subplot(2,2,2),plt.plot(circle.real, circle.imag, 'k--')
plt.plot(z.real, z.imag, 'ro', ms=7.5),plt.plot(p.real, p.imag, 'gx',ms=7.5)
plt.xlabel('$\sigma$'),plt.ylabel('$j\omega$'),plt.title('Pole-zero plot'),plt.grid()
#Magnitude and phase response
w, H = signal.freqz(b,a)
plt.subplot(2,2,3),plt.plot(w/np.pi*fs/2, 10*np.log(abs(H)))
plt.xlabel('$\omega$ [rad/sample]'),plt.ylabel('$|H(j\omega)|$ in [dB]')
plt.title('Magnitude response'),plt.subplot(2,2,4),plt.plot(w/np.pi*fs/2,np.unwrap(np.angle(H)))
plt.xlabel('$\omega$ [rad/sample]'),plt.ylabel('$\u2220H(j\omega)$')
plt.title('Phase response'),plt.tight_layout()
```

Fig. 6.19 Python code to design notch filter using built-in function

Using Euler's formula $e^{-j\omega} = \cos(\omega) - j\sin(\omega)$, the above expression can be written as

$$H\left(e^{j\omega}\right) = \frac{\cos(\omega) - j\sin(\omega) - a}{1 - a[\cos(\omega) - j\sin(\omega)]} \tag{6.25}$$

Now the expression for squared magnitude response is given by

$$\left|H\left(e^{j\omega}\right)\right|^2 = \frac{(\cos(\omega) - a)^2 + \sin^2(\omega)}{(1 - a\cos(\omega))^2 + a^2\sin^2(\omega)}$$

The above equation can be simplified as

$$\left|H\left(e^{j\omega}\right)\right|^2 = \frac{\cos^2(\omega) + a^2 - 2a\cos(\omega) + \sin^2(\omega)}{1 + a^2\cos^2(\omega) - 2a\cos(\omega) + a^2\sin^2(\omega)}$$

Using the fact that $\sin^2(\omega) + \cos^2(\omega) = 1$, the above equation can be simplified as

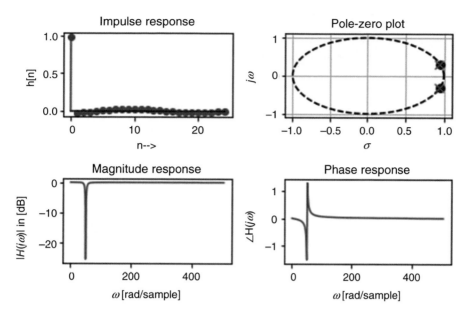

Fig. 6.20 Characteristics of a notch filter

$$|H(e^{j\omega})|^2 = \frac{1 + a^2 - 2a\cos(\omega)}{1 + a^2[\cos^2(\omega) + \sin^2(\omega)] - 2a\cos(\omega)}$$

Upon simplifying the above expression, we get

$$|H(e^{j\omega})|^2 = \frac{1 + a^2 - 2a\cos(\omega)}{1 + a^2 - 2a\cos(\omega)} = 1 \tag{6.26}$$

Thus, the magnitude response of the all-pass filter is unity. This means all-pass filters pass all frequency components of the input signal.

Experiment 6.10 All-Pass Filter

This experiment discusses the python implementation of all-pass filter. The python implementation of first-order all-pass filter with the value of '$a = 0.5$' is shown in Fig. 6.21, and the corresponding output is shown in Fig. 6.22.

Inferences

From Fig. 6.22, the following inferences can be drawn:

1. The input to the all-pass filter is a sine wave of 5 Hz frequency.
2. The output of the all-pass filter is almost the same as the input signal. Thus, all-pass filter passes all the frequency components of the input signal.
3. The impulse response is not finite. It slowly reaches the value of zero. Hence, it is an IIR filter.

```
#First order all-pass filter
import numpy as np
import matplotlib.pyplot as plt
from scipy import signal
#Step 1: Generating the input signal
f=5 # frequency
n=np.arange(0,50)
x=np.sin(2*np.pi*f*n/50)
#Step 2: Design of all-pass filter
a1=0.5
b,a=[-a1,1],[1,-a1]
#Step 3: Obtaining the output
y=signal.lfilter(b,a,x) #Output of resonator
plt.subplot(3,2,1),plt.stem(n,x),plt.xlabel('n-->'),plt.ylabel('x[n]'),plt.title('Input signal')
plt.subplot(3,2,2),plt.stem(n,y),plt.xlabel('n-->'),plt.ylabel('y[n]'),plt.title('Output signal')
#Impulse response of the system
h_1=np.zeros(25)
h_1[0]=1
h=signal.lfilter(b,a,h_1)
plt.subplot(3,2,3),plt.stem(h),plt.xlabel('n-->'),plt.ylabel('h[n]'),plt.title('Impulse
response')
# Pole-zero plot
z, p, k = signal.tf2zpk(b, a)
theta = np.linspace(0, np.pi*2, 500)
circle = np.exp(1j*theta)
plt.subplot(3,2,4),plt.plot(circle.real, circle.imag, 'k--')
plt.plot(z.real, z.imag, 'ro', ms=7.5),plt.plot(p.real, p.imag, 'gx',ms=7.5)
plt.xlabel('$\sigma$'),plt.ylabel('$j\omega$'),plt.title('Pole-zero plot'),plt.grid()
#Magnitude and phase response
w, h = signal.freqz(b,a)
plt.subplot(3,2,5),plt.plot(w,np.abs(h))
plt.xlabel('$\omega$ [rad/sample]'),plt.ylabel('$|H(j\omega)|$')
plt.title('Magnitude response'),plt.subplot(3,2,6),
plt.plot(w,np.unwrap(np.angle(h))),plt.xlabel('$\omega$ [rad/sample]'),
plt.ylabel('$\u2220H(j\omega)$'),plt.title('Phase response'),plt.tight_layout()
```

Fig. 6.21 First-order all-pass filter

4. The magnitude response indicates that the filter gain is one for all frequency components.
5. From the pole-zero plot, a pole lies at 0.5, whereas a zero lies at 2. The given system is not a minimum phase system.
6. The phase response of the system is non-linear. The phase is not varying linearly with respect to frequency.

Task

1. Repeat the above experiment for $a = 0.25$ and $a = 0.75$, and comment on the observed results.

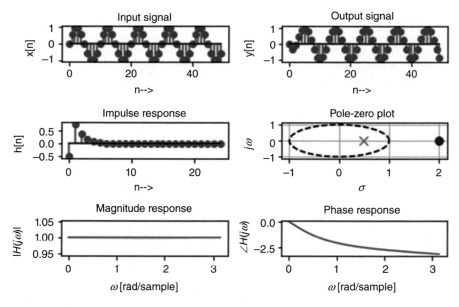

Fig. 6.22 Result of first-order all-pass filter

Fig. 6.23 Block diagram representation of comb filter

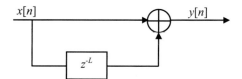

6.7   Comb Filter

A comb filter is a notch filter with a number of equally spaced nulls. The block diagram representing the comb filter is shown in Fig. 6.23.

The relationship between the input and output of the comb filter is expressed as

$$y[n] = x[n] + x[n - L] \tag{6.27}$$

Taking Z-transform on both sides of the above equation, we get

$$Y(z) = X(z) + z^{-L}X(z) \tag{6.28}$$

The above equation can be expressed as

$$Y(z) = X(z)\left[1 + z^{-L}\right]$$

The transfer function can be expressed as

$$H(z) = \frac{Y(z)}{X(z)} = 1 + z^{-L} \tag{6.29}$$

The frequency response of the system is obtained by substituting $z = e^{j\omega}$ in the above expression, we get

$$H(e^{j\omega}) = 1 + e^{-j\omega L} \tag{6.30}$$

6.7.1   Location of Poles and Zeros of Comb Filter

From the expression of the transfer function given in Eq. (6.30), it is possible to interpret; there is a pole of multiplicity 'L' at the origin. The location of zeros is obtained by equating the numerator of the transfer function to zero, which results in

$$z^{-L} = -1$$

The above equation can be expressed as

$$e^{-j\omega L} = e^{j(2k+1)\pi}$$

From the above expression

$$\omega_k = \frac{(2k+1)\pi}{L} \tag{6.31}$$

The zeros of the FIR filter are uniformly spaced $\frac{2\pi}{L}$ radians apart around the unit circle starting at $\omega = \frac{\pi}{L}$. For odd 'L', there is a zero at $\omega = \pi$.

Experiment 6.11 Comb Filter
The objective of this experiment is to plot the pole-zero pattern of Comb filter for even and odd values of L. The odd value is chosen as $L = 5$, and the even value is chosen as $L = 6$. The python code which plots the pole-zero plot of comb filters is shown in Fig. 6.24, and the corresponding output is shown in Fig. 6.25.

Inferences
From the pole-zero plot, which is shown in Fig. 6.25, the following inferences can be made:

1. The zeros are uniformly spaced $\frac{2\pi}{L}$ radians apart around the unit circle.
2. For odd values of 'L', there is a zero at $\omega = \pi$.
3. The poles lie at the origin, which implies that the filters are inherently stable.

```
#Pole-zero plot of comb filter
import numpy as np
import matplotlib.pyplot as plt
from scipy import signal
h1=[1,0,0,0,0,1] # Comb filter for N=5 (Odd)
h2=[1,0,0,0,0,0,1] #Comb filter for N=6 (Even)
z1, p1, k1 = signal.tf2zpk(h1,1) #Pole-zero for N=5
z2, p2, k2 = signal.tf2zpk(h2,1) #Pole-zero for N=6
#Plotting the pole-zero plot
theta = np.linspace(0, np.pi*2, 500)
circle = np.exp(1j*theta)
plt.subplot(2,1,1),plt.plot(circle.real, circle.imag, 'k--')
plt.plot(z1.real, z1.imag, 'ro', ms=7.5)
plt.plot(p1.real, p1.imag, 'gx',ms=7.5)
plt.xlabel('$\sigma$'),plt.ylabel('$j\omega$'),
plt.title('Pole-zero plot for L = 5'),plt.grid()
plt.subplot(2,1,2),plt.plot(circle.real, circle.imag, 'k--')
plt.plot(z2.real, z2.imag, 'ro', ms=7.5)
plt.plot(p2.real, p2.imag, 'gx', ms=7.5)
plt.xlabel('$\sigma$'),plt.ylabel('$j\omega$'),
plt.title('Pole-zero plot for L = 6'),plt.grid(),plt.tight_layout()
```

Fig. 6.24 Pole-zero plot of comb filter

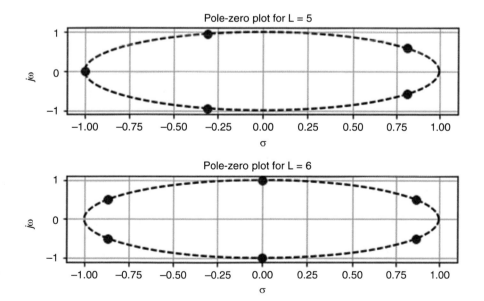

Fig. 6.25 Pole-zero plot of comb filters for odd and even values of 'L'

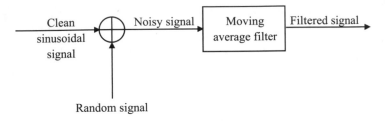

Fig. 6.26 Block diagram representation of problem statement

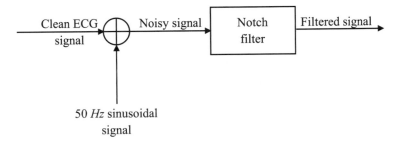

Fig. 6.27 Block diagram of problem statement

Exercises

1. Generate sinusoidal signal of 5 Hz frequency. Add random noise which follows uniform distribution to the clean sinusoidal signal. Pass the noisy signal to the moving average filter and comment on the observed result. The block diagram of the problem statement is shown in Fig. 6.26.
2. Generate three tones of frequencies 500, 1000 and 1500 Hz. Append the three tones together as one signal. Now pass this signal to a notch filter, which will block the frequency component of a specific frequency (say 1000 Hz). Hear the input and output signals and comment on your observation.
3. Design a notch filter to minimize 50 Hz powerline interference in ECG signal. Read an ECG signal which is stored in 'mat' file format. Add 50 Hz powerline interference to the clean ECG signal to generate noisy signal. Pass the noisy signal to the notch filter to minimize the powerline interference. Plot the clean, noisy and filtered signals and comment on the observed result. The problem statement is depicted in the form of a block diagram and is shown in Fig. 6.27.
4. Generate 10 Hz square waveform. Design a comb filter to eliminate 10 Hz frequency component in this signal and its odd harmonics.
5. Generate sine wave of frequencies 5 and 10 Hz. Add these two waveforms. Now pass this signal through a notch filter, which should eliminate the 5 Hz frequency component, so that the output signal contains a 10 Hz frequency component.

Objective Questions

1. The filter which is used to reject one particular frequency is

   A. Lowpass filter
   B. Highpass filter
   C. All-pass filter
   D. Notch filter

2. The filter which can be used as a delay equalizer is

   A. Lowpass filter
   B. Highpass filter
   C. All-pass filter
   D. Notch filter

3. Cascading of lowpass and highpass filter will result in

   A. Lowpass filter
   B. Highpass filter
   C. Band pass filter
   D. All-pass filter

4. The filter which is used to minimize the impact of power line interference is

   A. Lowpass filter
   B. Highpass filter
   C. All-pass filter
   D. Notch filter

5. The impulse response of three-point moving average filter is given by

   A. $h[n] = \frac{1}{3}\{ -\delta[n] - \delta[n-1] - \delta[n-2]\}$
   B. $h[n] = \frac{1}{3}\{ -\delta[n] - \delta[n-1] + \delta[n-2]\}$
   C. $h[n] = \frac{1}{3}\{\delta[n] - \delta[n-1] + \delta[n-2]\}$
   D. $h[n] = \frac{1}{3}\{\delta[n] + \delta[n-1] + \delta[n-2]\}$

6. The filter which has the ability to remove fundamental frequency and its harmonics is

   A. Notch filter
   B. Comb filter
   C. Lowpass filter
   D. Highpass filter

7. The impulse response of a digital filter is given by $h[n] = \delta[n] + \delta[n-8]$. The filter behaves like a

   A. All-pass filter
   B. Highpass filter

    C. Comb filter

    D. Notch filter

8. Statement 1: Stable filters are always causal

    Statement 2: Causal filters are always stable:

    A. Both statements are wrong.

    B. Both statements are true.

    C. Statement 1 is true, and Statement 2 is wrong.

    D. Statement 1 is wrong, and Statement 2 is true.

9. Assertion: Moving average filter attenuates quick change in the signal.

    Reason: Moving average filter is a lowpass filter.

    A. Both assertion and reason are wrong.

    B. Assertion is true, reason is wrong.

    C. Assertion is wrong, reason may be true.

    D. Both assertion and reason are true.

10. The frequency response of lowpass filter is given by $H(e^{j\omega}) = \frac{1+e^{-j\omega}+e^{-j2\omega}+e^{-j3\omega}}{4}$.
Using frequency shift, the lowpass filter can be converted to a highpass filter.
The impulse response of the highpass filter is

    A. $h[n] = 0.25\{\delta[n] + \delta[n-1] + \delta[n-2] + \delta[n-3]\}$

    B. $h[n] = 0.25\{\delta[n] + \delta[n-1] - \delta[n-2] - \delta[n-3]\}$

    C. $h[n] = 0.25\{-\delta[n] - \delta[n-1] - \delta[n-2] - \delta[n-3]\}$

    D. $h[n] = 0.25\{\delta[n] - \delta[n-1] + \delta[n-2] - \delta[n-3]\}$

11. Statement 1: Digital resonator generates sinusoidal signal of specific frequency.

    Statement 2: Digital resonator has complex conjugate pole located on the unit circle.

    A. Statement 1 is correct, and Statement 2 is wrong.

    B. Statement 1 is wrong, and Statement 2 is correct.

    C. Both Statements 1 and 2 are correct.

    D. Both Statements 1 and 2 are wrong.

12. In the design of a simple digital filter using pole-zero placement:

    Statement 1: To suppress a frequency component, locate a zero at this frequency on the unit circle.

    Statement 2: To amplify a frequency, locate a pole at this frequency inside the unit circle.

    A. Statement 1 is correct, and Statement 2 is wrong.

    B. Statement 1 is wrong, and Statement 2 is correct.

    C. Both Statements 1 and 2 are wrong.

    D. Both Statements 1 and 2 are correct.

13. The transfer function of a linear time-invariant system is expressed as
$H(z) = \frac{B(z)}{A(z)}$.

Statement 1: Zeros are roots of the polynomial $B(z)$.
Statement 2: Poles are roots of the polynomial $A(z)$.

A. Statement 1 is correct, and Statement 2 is wrong.
B. Statement 1 is wrong, and Statement 2 is correct.
C. Both Statements 1 and 2 are wrong.
D. Both Statements 1 and 2 are correct.

Bibliography

1. Ashok Ambardar, "Digital Signal Processing: A Modern Introduction", Cengage Learning India, 2007.
2. Sanjit K. Mitra, "Digital Signal Processing: A Computer-Based Approach", McGraw Hill Education, 2013.
3. Boaz Porat, "A Course in Digital Signal Processing", John Wiley and Sons, 1996.
4. Dimitris G. Manolakis, Vinay K. Ingle, "Applied Digital Signal Processing: Theory and Practice", Cambridge University Press, 2011.
5. L. B. Jackson, "Signals, Systems and Transforms", Addison-Wesley, 1991.

Chapter 7
FIR Filter Design

Learning Objectives

After completing this chapter, the reader is expected to

- Analyse the characteristics of Type-I, Type-II, Type-III and Type-IV FIR filters.
- Design and analyse window-based finite impulse response filter.
- Design and analyse frequency sampling based finite impulse response filter.
- Design and analyse optimal finite impulse response filter.

Roadmap of the Chapter

This chapter discusses the type of FIR filters and its characteristic. Also, it gives detail about the designs of the FIR filter. The roadmap of this chapter is given in the form of flowchart below.

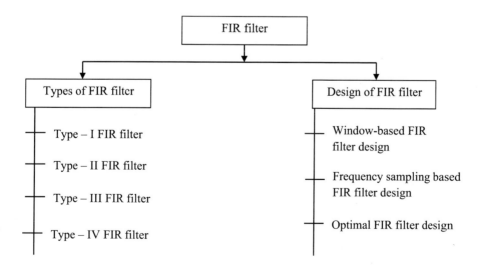

© The Author(s), under exclusive license to Springer Nature Singapore Pte Ltd. 2024
S. Esakkirajan et al., *Digital Signal Processing*,
https://doi.org/10.1007/978-981-99-6752-0_7

PreLab Questions

1. What is the difference equation relating the input and output of a finite impulse response filter? What are the inferences that could be made from the difference equation?
2. On what basis are FIR filters classified as Type-I, Type-II, Type-III and Type-IV FIR filters?
3. When is a FIR filter coefficient said to exhibit (a) even symmetry and (b) odd symmetry?
4. What is the condition for the digital filter to exhibit linear phase characteristics?
5. What is the advantage of 'linear phase' characteristics of digital filter?
6. What is the relationship between the group delay and the phase response of the FIR filter?
7. What is the relationship between the order (M) and the number of coefficients (N) of FIR filters?
8. Why FIR filter is considered as an 'inherently stable' filter?
9. List four advantages of FIR filter.
10. Mention different methods of design of FIR filter.

7.1 FIR Filter

FIR stands for finite impulse response. The coefficients of FIR filter are either symmetric or anti-symmetric. Due to the symmetric nature of FIR filter coefficients, it exhibits linear phase characteristics. Because of linear phase characteristics, the FIR filter has no phase distortion. Also, FIR filter exhibits constant group delay. In FIR filter, the current output is a function of the current and previous inputs. This implies that FIR filters are non-recursive filters; hence, they are inherently stable. FIR filter is an all-zero filter, and the zeros occur in conjugate reciprocal pair.

7.2 Classification of FIR Filter

Based on nature of symmetry and the number of coefficients, FIR filter can be classified as Type-I, Type-II, Type-III and Type-IV. The classification is given in Table 7.1.

Table 7.1 Classification of FIR filter

S. No.	Nature of symmetry	Number of coefficients	Type of FIR filter
1	Even symmetry	Odd	Type-I
2	Even symmetry	Even	Type-II
3	Odd symmetry	Odd	Type-III
4	Odd symmetry	Even	Type-IV

```
#Type-I FIR filter
import numpy as np
import matplotlib.pyplot as plt
from scipy import signal
#Step 1 : Type-I FIR filter
b,a=[1,2,3,2,1],[1]
#Impulse response of the system
h_1=np.zeros(25)
h_1[0]=1
h=signal.lfilter(b,a,h_1)
plt.subplot(2,2,1),plt.stem(h),plt.xlabel('n-->'),
plt.ylabel('Amplitude'),plt.title('Impulse response')
# Pole-zero plot
z, p, k = signal.tf2zpk(b, a)
theta = np.linspace(0, np.pi*2, 500)
circle = np.exp(1j*theta)
plt.subplot(2,2,2),plt.plot(circle.real, circle.imag, 'k--'),plt.plot(z.real, z.imag, 'ro', ms=7.5)
plt.xlabel('$\sigma$'),plt.ylabel('$j\omega$'),plt.title('Pole-zero plot'),plt.grid()
#Magnitude and phase response
w, h = signal.freqz(b,a)
plt.subplot(2,2,3),plt.plot(w, np.abs(h)),plt.xlabel('$\omega$-->'),
plt.ylabel('|H(j$\omega$)|'),plt.title('Magnitude response')
plt.subplot(2,2,4),plt.plot(w,np.unwrap(np.angle(h))), plt.xlabel('$\omega$-->'),
plt.ylabel('$\u2220H(j\omega)$'),plt.title('Phase response'),plt.tight_layout()
```

Fig. 7.1 Python code to obtain the characteristics of Type-I FIR filter

Experiment 7.1 Characteristics of Type-I FIR Filter
The objective of this experiment is to plot the characteristics of Type-I FIR filter. Here, the impulse response of Type-I FIR filter is chosen as $h[n] = \{1, 2, 3, 2, 1\}$. The filter coefficients satisfy even symmetry, and the number of coefficients is odd; hence, it belongs to Type-I FIR filter. The python code, which obtains the impulse response, magnitude response, phase response and pole-zero plot, is shown in Fig. 7.1, and the corresponding output is shown in Fig. 7.2. The built-in functions used in the program are given in Table 7.2.

Inferences
From Fig. 7.2, the following inferences can be drawn:

1. From the impulse response, it is possible to observe that the impulse response is of finite duration, and the filter coefficients exhibit even symmetry.
2. From the pole-zero plot, the filter is an all-zero filter. The zeros occur in conjugate pair.
3. From the magnitude response, it is possible to observe that the filter behaves like a lowpass filter.
4. From the phase response, it is possible to observe that the filter exhibits a linear phase characteristic in the passband.

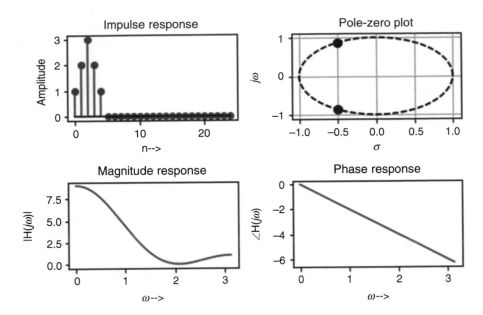

Fig. 7.2 Characteristics of Type-I FIR filter

Table 7.2 Built-in functions used in the program

S. No.	Built-in function	Library	Application
1	signal. lfilter	Scipy	To obtain the output of LTI system
2	signal. tf2zpk	Scipy	To obtain the zeros, poles and gain of the LTI system
3	signal. freqz	Scipy	To obtain the frequency response of the LTI system. Frequency response is a combination of magnitude and phase responses
4	abs	Numpy	To obtain the magnitude response of the system
5	angle	Numpy	To obtain the phase response of the system

Task

1. Write a python code to illustrate that Type-I FIR filter is versatile; (i.e.) it can be used as a lowpass, highpass, bandpass and band reject filter.

Experiment 7.2 Characteristics of Type-II FIR Filter

Type-II FIR filter exhibits even symmetry with an even number of coefficients. The Type-II FIR filter coefficients considered in this experiment is $h[n] = \{1, 2, 2, 1\}$. The python code to obtain the impulse response, pole-zero plot, magnitude and phase responses is shown in Fig. 7.3, and the corresponding output is shown in Fig. 7.4.

Inferences

From Fig. 7.4, the following inferences can be made:

```
#Type-II FIR filter
import numpy as np
import matplotlib.pyplot as plt
from scipy import signal
#Step 1 : Type-I FIR filter
b,a=[1,2,2,1],[1]
#Impulse response of the system
h_1=np.zeros(25)
h_1[0]=1
h=signal.lfilter(b,a,h_1)
plt.subplot(2,2,1),plt.stem(h),plt.xlabel('n-->'),
plt.ylabel('Amplitude'),plt.title('Impulse response')
# Pole-zero plot
z, p, k = signal.tf2zpk(b, a)
theta = np.linspace(0, np.pi*2, 500)
circle = np.exp(1j*theta)
plt.subplot(2,2,2),plt.plot(circle.real, circle.imag, 'k--'),plt.plot(z.real, z.imag, 'ro', ms=7.5)
plt.xlabel('$\sigma$'),plt.ylabel('$j\omega$'),plt.title('Pole-zero plot'),plt.grid()
#Magnitude and phase response
w, h = signal.freqz(b,a)
plt.subplot(2,2,3),plt.plot(w, np.abs(h)),plt.xlabel('$\omega--
>'),plt.ylabel('|H(j$\omega$)|'),plt.title('Magnitude response')
plt.subplot(2,2,4),plt.plot(w,np.unwrap(np.angle(h))), plt.xlabel('$\omega$-->'),
plt.ylabel('$\u2220H(j\omega)$'),plt.title('Phase response'),plt.tight_layout()
```

Fig. 7.3 Python code to obtain the characteristics of Type-II FIR filter

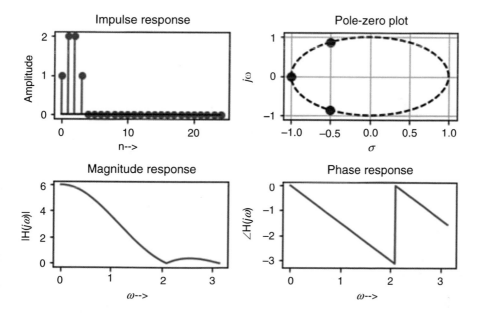

Fig. 7.4 Characteristics of Type-II FIR filter

```
#Type-III FIR filter
import numpy as np
import matplotlib.pyplot as plt
from scipy import signal
#Step 1 : Type-I FIR filter
b,a=[1,2,0,-2,-1],[1]
#Impulse response of the system
h_1=np.zeros(25)
h_1[0]=1
h=signal.lfilter(b,a,h_1)
plt.subplot(2,2,1),plt.stem(h),plt.xlabel('n-->'),
plt.ylabel('Amplitude'),plt.title('Impulse response')
# Pole-zero plot
z, p, k = signal.tf2zpk(b, a)
theta = np.linspace(0, np.pi*2, 500)
circle = np.exp(1j*theta)
plt.subplot(2,2,2),plt.plot(circle.real, circle.imag, 'k--'),plt.plot(z.real, z.imag, 'ro', ms=7.5)
plt.xlabel('$\sigma$'),plt.ylabel('$j\omega$'),plt.title('Pole-zero plot'),plt.grid()
#Magnitude and phase response
w, h = signal.freqz(b,a)
plt.subplot(2,2,3),plt.plot(w, np.abs(h)),plt.xlabel('$\omega$-->'),
plt.ylabel('|H(j$\omega$)|'),plt.title('Magnitude response')
plt.subplot(2,2,4),plt.plot(w,np.unwrap(np.angle(h))), plt.xlabel('$\omega$-->'),
plt.ylabel('$\u2220H(j\omega)$'),plt.title('Phase response'),plt.tight_layout()
```

Fig. 7.5 Python code to obtain the characteristics of Type-III FIR filter

1. From the impulse response, it is possible to observe that the impulse response is of finite duration, and it exhibits even symmetry.
2. From the pole-zero plot, it is possible to observe that Type-II FIR filter is an all-zero filter. The magnitude which is zero at $\omega = \pi$ indicates that Type-II FIR filter cannot be used as a highpass filter.
3. From the magnitude response, it is possible to conclude that the filter behaves like a lowpass filter.
4. From the phase response, it is possible to confirm that the Type-II FIR filter exhibits linear phase characteristics in the passband.

Task

1. Generate $x[n] = e^{j\pi n}$, $0 < n \leq 100$. Pass this signal through Type-II FIR filter whose impulse response is $h[n] = \{1, 2, 2, 1\}$. Use subplot to plot the input and output signals and comment on the observed result.

Experiment 7.3 Characteristics of Type-III FIR Filter

This experiment discusses the analysis of the characteristics of Type-III FIR filter using python. The python code, which obtains the characteristics of Type-III FIR filter, is shown in Fig. 7.5, and the corresponding output is shown in Fig. 7.6. The

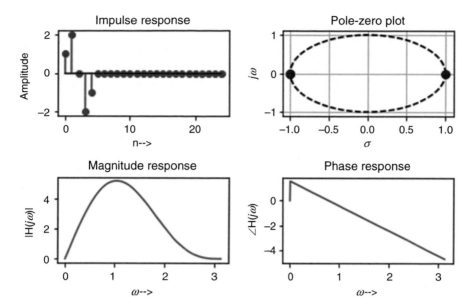

Fig. 7.6 Characteristics of Type-III FIR filter

impulse response chosen for Type-III FIR filter characteristics is {1, 2, 0, −2, −1}, and it satisfies both odd symmetry and the number of coefficients is odd.

Inferences
From Fig. 7.6, the following inferences can be drawn:

1. The impulse response shows odd symmetry with an odd number of coefficients. The duration of the impulse response is finite.
2. From the pole-zero plot, it is possible to observe that the magnitude value is zero at $\omega = 0$ and $\omega = \pi$. This implies that Type-III FIR filter cannot be used as a lowpass and a highpass filters.
3. From the magnitude response, it is possible to observe that the filter can act as a bandpass filter only.
4. From the phase response, it is possible to infer that the filter exhibits linear phase characteristics in the passband.

Task
1. Write a python code to illustrate the fact that cascading of lowpass and highpass filters will result in a bandpass filter.

Experiment 7.4 Characteristics of Type-IV FIR Filter
This experiment tries to analyse the characteristics of Type-IV FIR filter using python. The python code, which obtains the characteristics of Type-IV FIR filter, is shown in Fig. 7.7, and the corresponding output is shown in Fig. 7.8. The impulse response chosen for this illustration is $h[n] = \{1, 2, -2, -1\}$. The impulse response exhibits odd symmetry with an even number of filter coefficients.

```
#Type-IV FIR filter
import numpy as np
import matplotlib.pyplot as plt
from scipy import signal
#Step 1 : Type-I FIR filter
b,a=[1,2,-2,-1],[1]
#Impulse response of the system
h_1=np.zeros(25)
h_1[0]=1
h=signal.lfilter(b,a,h_1)
plt.subplot(2,2,1),plt.stem(h),plt.xlabel('n-->'),
plt.ylabel('Amplitude'),plt.title('Impulse response')
# Pole-zero plot
z, p, k = signal.tf2zpk(b, a)
theta = np.linspace(0, np.pi*2, 500)
circle = np.exp(1j*theta)
plt.subplot(2,2,2),plt.plot(circle.real, circle.imag, 'k--'),plt.plot(z.real, z.imag, 'ro', ms=7.5)
plt.xlabel('$\sigma$'),plt.ylabel('$j\omega$'),plt.title('Pole-zero plot'),plt.grid()
#Magnitude and phase response
w, h = signal.freqz(b,a)
plt.subplot(2,2,3),plt.plot(w, np.abs(h)),plt.xlabel('$\omega$-->'),
plt.ylabel('|H(j$\omega$)|'),plt.title('Magnitude response')
plt.subplot(2,2,4),plt.plot(w,np.unwrap(np.angle(h))), plt.xlabel('$\omega$-->'),
plt.ylabel('$\u2220H(j\omega)$'),plt.title('Phase response'),plt.tight_layout()
```

Fig. 7.7 Python code to obtain the characteristics of Type-IV FIR filter

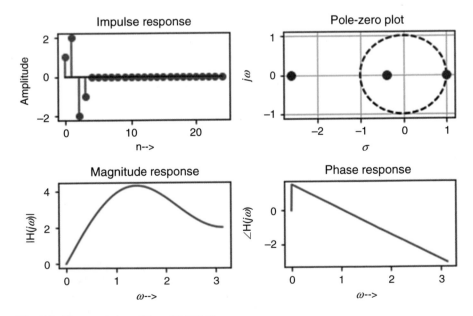

Fig. 7.8 Characteristics of Type-IV FIR filter

Inferences

From Fig. 7.8, the following inferences can be made:

1. The impulse response plot reveals that Type-IV FIR filter impulse response is anti-symmetric with an even number of coefficients.
2. From the pole-zero plot, it is possible to confirm that Type-IV FIR filter has a zero at $\omega = 0$; hence, it cannot be used as a lowpass filter.
3. The magnitude response resembles that of a bandpass filter.
4. The phase response plot reveals that Type-IV FIR filter exhibits linear phase characteristics in the passband.

Experiment 7.5 Comparison of Type-I, Type-II, Type-III and Type-IV FIR Filters with Respect to Their Location of Zeros

This experiment compares all four types of FIR filters with respect to their location of zeros using python. The python code used to plot the pole-zero plot of Type-I, Type-II, Type-III and Type-IV FIR filters is shown in Fig. 7.9, and the corresponding output is shown in Fig. 7.10. In this program, the impulse responses of four FIR filters are defined first. Then, the pole, zero and gain of each type of FIR filter are obtained using the built-in function '*tf2zpk*' available in 'scipy' package. Then, the extracted poles and zeros are plotted. The automatic location of zeros in Type-I, Type-II, Type-III and Type-IV FIR filters is given in Table 7.3.

Inferences

From Fig. 7.10, the following inferences can be drawn:

1. For Type-I FIR filter, there is no zero at $\omega = 0$ and $\omega = \pi$. It can be used as a versatile filter.
2. Type-II FIR filter has a zero at $\omega = \pi$. It cannot be used as a highpass filter.
3. Type-III FIR filter has zero at $\omega = 0$ and $\omega - \pi$. It cannot be used as both lowpass and highpass filters.
4. Type-IV FIR filter has zero at $\omega = 0$. It cannot be used as lowpass filter.
5. In general, zeros of FIR filter occur in conjugate, reciprocal pairs.

Task

1. Write a python code to illustrate the fact that all four types of FIR filters are inherently stable filters. Hint: For a discrete-time system to be stable, the impulse response should be absolutely summable.

7.3   Design of FIR Filter

The design of FIR filter starts with specification. The specification can be either in time domain or frequency domain. In time domain, the desired impulse response is given as specification. In frequency domain, the specification involves magnitude and phase response. Three prominent methods to design FIR filters are (1) window-based method, (2) frequency sampling method and (3) optimal method.

```
#Pole-zero plot of different types of FIR filter
import matplotlib.pyplot as plt
import numpy as np
from scipy import signal
#Defining four types of FIR filter
h1,h2,h3,h4=[1,2,5,2,1],[1,2,2,1],[1,2,0,-2,-1],[1,2,-2,-1]
#Poles and zeros of the filter
z1, p1, k1 = signal.tf2zpk(h1,1)
z2, p2, k2 = signal.tf2zpk(h2,1)
z3, p3, k3 = signal.tf2zpk(h3,1)
z4, p4, k4 = signal.tf2zpk(h4,1)
#Ploting the pole-zero plot
theta = np.linspace(0, np.pi*2, 500)
circle = np.exp(1j*theta)
plt.subplot(2,2,1),plt.plot(circle.real, circle.imag, 'k--'),plt.plot(z1.real, z1.imag, 'ro', ms=7.5)
plt.plot(p1.real, p1.imag, 'rx',ms=7.5)
plt.xlabel('$\sigma$'),plt.ylabel('$j\omega$'),plt.title('Pole-zero plot(Type-I)'),plt.grid()
theta = np.linspace(0, np.pi*2, 500)
circle = np.exp(1j*theta)
plt.subplot(2,2,2),plt.plot(circle.real, circle.imag, 'k--')
plt.plot(z2.real, z2.imag, 'ro', ms=7.5),plt.plot(p2.real, p2.imag, 'gx',ms=7.5)
plt.xlabel('$\sigma$'),plt.ylabel('$j\omega$'),plt.title('Pole-zero plot(Type-II)'),plt.grid()
theta = np.linspace(0, np.pi*2, 500)
circle = np.exp(1j*theta)
plt.subplot(2,2,3),plt.plot(circle.real, circle.imag, 'k--')
plt.plot(z3.real, z3.imag, 'ro', ms=7.5),plt.plot(p3.real, p3.imag, 'gx',ms=7.5)
plt.xlabel('$\sigma$'),plt.ylabel('$j\omega$'),plt.title('Pole-zero plot(Type-III)'),plt.grid()
theta = np.linspace(0, np.pi*2, 500)
circle = np.exp(1j*theta)
plt.subplot(2,2,4),plt.plot(circle.real, circle.imag, 'k--')
plt.plot(z4.real, z4.imag, 'ro', ms=7.5),plt.plot(p4.real, p4.imag, 'gx',ms=7.5)
plt.xlabel('$\sigma$'),plt.ylabel('$j\omega$'),plt.title('Pole-zero plot(Type-IV)'),plt.grid()
plt.tight_layout()
```

Fig. 7.9 Python code to obtain the pole-zero plots of different types of FIR filter

7.3.1 Steps in Window-Based FIR Filter Design

The steps followed in FIR filter design using Windows are summarized below:

1. The filter design starts with the specification of the filter in terms of desired frequency response.
2. The desired impulse response ($h_d[n]$) is obtained from the desired frequency response using inverse discrete-time Fourier transform.
3. Multiply the desired impulse response with the selected window function.
4. Delay the windowed impulse response by a factor of 'τ' to get the causal FIR filter coefficients.
5. The process is complete if the frequency response is satisfied as per the specification. If the frequency specifications are not satisfied, increase the filter order and repeat the steps.

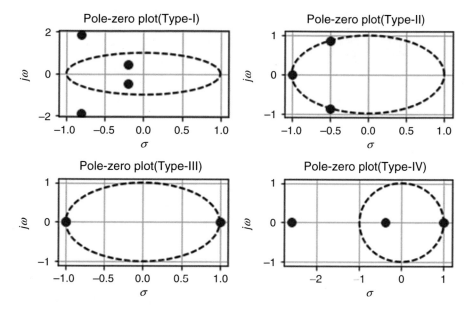

Fig. 7.10   Pole-zero plot of FIR filters

Table 7.3   Automatic location of zeros in different types of FIR filter

Type of FIR filter	Automatic zero location	Inference
Type-I	–	Type-I FIR filter is a versatile filter; it can be used to design lowpass, highpass, bandpass and band reject filters
Type-II	Zero at $\omega = \pi$	Type-II FIR filter cannot be used as a highpass filter
Type-III	Zero at $\omega = 0$ and $\omega = \pi$	Type-III FIR filter cannot be used as lowpass and highpass filters
Type-IV	Zero at $\omega = 0$	Type-IV FIR filter cannot be used as lowpass filter

7.3.2   Window-Based FIR Lowpass Filter

The expression for the impulse response of the ideal lowpass filter is given by

$$h_d[n] = \frac{\omega_c}{\pi} \sin c\left(\frac{\omega_c}{\pi} n\right) \tag{7.1}$$

From Eq. (7.1), it is possible to observe that the filter is neither causal nor finite in duration. To make it finite duration, the desired impulse response is multiplied with the window function. The mathematical expression for the impulse response multiplied with a rectangular window of length 'N' is given by

```
#Impulse response of ideal filter
import numpy as np
import matplotlib.pyplot as plt
from scipy import signal
N, omega = 50,1.0
n = np.arange(-N/2,N/2)
rect_win=np.ones(N)
hd = omega/np.pi * np.sinc(n*omega/np.pi)
h=hd*rect_win
plt.stem(n,h),plt.xlabel('n-->'),plt.ylabel('Amplitude'),plt.title('h[n]')
```

Fig. 7.11 Python code to obtain the impulse response of the ideal filter

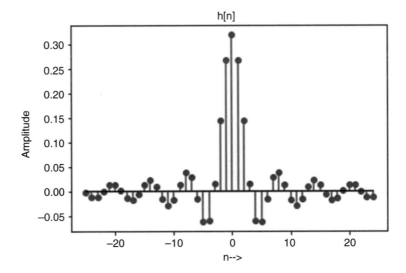

Fig. 7.12 Impulse response of an ideal filter

$$h[n] = h_d[n] \times \text{rect}_N[n] \tag{7.2}$$

Experiment 7.6 Window-Based Design of Ideal Lowpass Filter
This experiment discusses the ideal lowpass FIR filter design using windowing method. The python code, which obtains the impulse response of an ideal filter, is shown in Fig. 7.11, and the corresponding output is shown in Fig. 7.12.

Inferences
The following inferences can be drawn from Fig. 7.12:

1. The impulse response of the ideal filter is non-causal.
2. If the impulse response is non-causal, the filter is not physically realizable.

```
#Comparison of ideal and practical filter
import numpy as np
import matplotlib.pyplot as plt
from scipy import signal
N, omega = 32, 1.0
n = np.arange(0,N)
rect_win=np.ones(N)
hd = omega/np.pi * np.sinc(n*omega/np.pi)
hd1 = omega/np.pi * np.sinc((n-(N-1)/2)*omega/np.pi)
h1=hd1*rect_win
h=hd*rect_win
w,H=signal.freqz(h)
w1,H1=signal.freqz(h1)
plt.subplot(2,3,1),plt.stem(n,h),plt.xlabel('n-->'),plt.ylabel('Amplitude')
plt.title('h[n]:IF'),plt.subplot(2,3,2),plt.plot(w,20*np.log10(np.abs(H)))
plt.xlabel('$\omega$-->'),plt.ylabel('$|H(e^{j\omega})|$')
plt.title('Magnitude response:IF'),plt.subplot(2,3,3),plt.plot(w,np.unwrap(np.angle(H)))
plt.xlabel('$\omega$-->'),plt.ylabel('$\phi(e^{j\omega})$'),plt.title('Phase response:IF')
plt.subplot(2,3,4),plt.stem(n,h1),plt.xlabel('n-->'),plt.ylabel('Amplitude')
plt.title('h[n]:PF'),plt.subplot(2,3,5),plt.plot(w1,20*np.log10(np.abs(H1)))
plt.xlabel('$\omega$-->'),plt.ylabel('$|H(e^{j\omega})|$'),plt.title('Magnitude response:PF')
plt.subplot(2,3,6),plt.plot(w1,np.unwrap(np.angle(H1)))
plt.xlabel('$\omega$-->'),plt.ylabel('$\phi(e^{j\omega})$'),plt.title('Phase response:PF')
plt.tight_layout()
```

Fig. 7.13 Comparison of ideal and practical filter

3. Practically realizable filters have passband and stopband ripples and a non-zero transition band.
4. For practical filter, a delay is necessary to capture most of the signal energy in causal time.
5. Delay in time-domain is accomplished by multiplying the spectrum with a complex exponential. The magnitude response is multiplied by $e^{-j\omega\tau}$, which results in the time shift of the impulse response. This is discussed in the subsequent section.

Task

1. In the above experiment, increase the number of coefficients of the filter to 100, observe the filter's impulse response and comment on the observed result.

Experiment 7.7 Comparison of Ideal and Practical Lowpass Filter

The impulse response of an ideal filter is non-causal; hence, it is not physically realizable. To design a practical filter, the impulse response of the ideal filter has to be delayed to make it causal. Delay in the time-domain is accomplished by multiplying the spectrum with a complex exponential. The magnitude response is multiplied by $e^{-j\omega\tau}$, which results in the time shift of the impulse response.

This python illustration compares the ideal FIR filter with the practical FIR filter. The python code, which performs the comparison, is shown in Fig. 7.13, and the corresponding output is shown in Fig. 7.14.

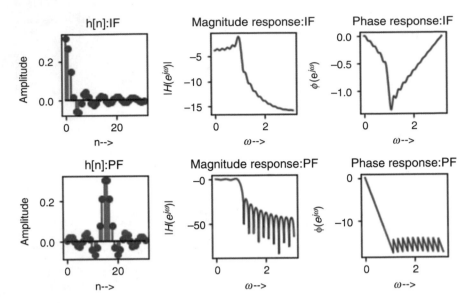

Fig. 7.14 Characteristics of Ideal and practical filters

In Fig. 7.14, IF and PF denote ideal and practical filters, respectively.

Inferences
The following inferences can be made from this experiment:

1. By comparing the impulse response of the ideal and practical filters, it is possible to observe that the practical filter exhibits a symmetric impulse response. In contrast, the impulse response of the ideal filter is not symmetric.
2. From the magnitude responses, it is possible to observe that the ideal filter exhibits ripples, whereas, in the practical filter, the ripples in the magnitude response are less.
3. Practical filter exhibits linear phase characteristics, whereas the ideal filter phase response is not linear.
4. From this experiment, it is possible to conclude that the impulse response should be symmetric for the phase response to be linear.

7.3.3   Window-Based FIR Highpass Filter

The desired impulse response of window-based highpass FIR filter is given by

```
#FIR high pass filter
import numpy as np
import matplotlib.pyplot as plt
from scipy import signal
#Step 1: Desired impulse response hd[n]
N,omega = 50,np.pi/4
n = np.arange(0,N)
rect_win=np.ones(N)
hd =np.sinc(n-(N-1)/2)-(omega/np.pi * np.sinc((n-(N-1)/2)*omega/np.pi))
h=hd*rect_win #Step 2: Windowed impulse response h[n]
w,H=signal.freqz(h) #Step 3: Frequency response of ideal filter
#Step 4:Impulse response of the filter
plt.subplot(2,2,1),plt.stem(n,h),plt.xlabel('n-->'),plt.ylabel('Amplitude'),plt.title('h[n]')
z, p, k = signal.tf2zpk(h,1) #Step 5: Pole-zero plot of the filter
theta = np.linspace(0, np.pi*2, 500)
circle = np.exp(1j*theta)
plt.subplot(2,2,2),plt.plot(circle.real, circle.imag, 'k--')
plt.plot(z.real, z.imag, 'ro', ms=7.5),plt.plot(p.real, p.imag, 'rx')
plt.xlabel('$\sigma$'),plt.ylabel('$j\omega$'),plt.title('Pole-zero plot')
#Step 6: Magnitude response of the filter
plt.subplot(2,2,3),plt.plot(w,20*np.log10(np.abs(H)))
plt.xlabel('$\omega$-->'),plt.ylabel('$|H(e^{j^\omega})|$'),plt.title('Magnitude response')
#Step 7: Phase response of the filter
plt.subplot(2,2,4),plt.plot(w,np.unwrap(np.angle(H)))
plt.xlabel('$\omega$-->'),plt.ylabel('$\phi(e^{j^\omega})$'),plt.title('Phase response')
plt.tight_layout()
```

Fig. 7.15 Python code to obtain the characteristics of highpass filter

$$h_{\mathrm{d}}[n] = \sin c(n-\tau) \quad \frac{\omega_c}{\pi} \sin c\left(\frac{\omega_c}{\pi}(n-\tau)\right) \qquad (7.3)$$

From Eq. (7.3), it is possible to know that the desired impulse response is a *sinc* function that is not of finite duration. The desired impulse response must be multiplied by the window function $w[n]$ to make the finite impulse response. This is expressed as

$$h[n] = h_{\mathrm{d}}[n] \times w[n] \qquad (7.4)$$

Experiment 7.8 Window-Based FIR Highpass Filter

This experiment discusses about the study of characteristics of FIR highpass filter design using the windowing method. The python code, which obtains the characteristics of a highpass filter, is given in Fig. 7.15, and the corresponding output is shown in Fig. 7.16.

Inferences

The following inferences can be made from this experiment:

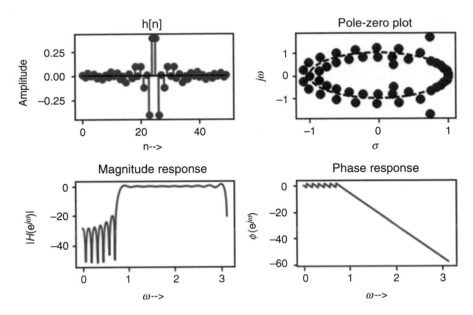

Fig. 7.16 Characteristics of highpass filter

1. Figure 7.16 shows that the code was written to simulate the desired impulse response, which is given in Eq. (7.3). The desired impulse response is multiplied by the window to get the actual response. The window chosen in this illustration is a rectangular window.
2. The '*scipy*' library is used here to obtain the frequency response and the pole-zero plot of the filter. The built-in function '*tf2zpk*' is utilized to obtain the pole-zero plot, whereas the built-in function '*freqz*' is used here to obtain the frequency response of the filter.
3. From the magnitude response shown in Fig. 7.16, it is possible to observe that the filter is a highpass filter that exhibits linear phase characteristics in the passband.
4. From the pole-zero plot, it is possible to observe that the zeros of FIR filter occur in conjugate reciprocal pair.

Task
1. In the python code given in Fig. 7.15, try to use windows like Hamming, Hanning, Bartlett and Blackman window, and observe the changes in magnitude and phase response.

7.3.4   Window-Based FIR Bandpass Filter

The expression for the desired impulse response of the bandpass filter is given by

```
#Characteristics of bandpass filter
import numpy as np
import matplotlib.pyplot as plt
from scipy import signal
#Step 1: Desired impulse response hd[n]
N,omega_1,omega_2 = 16,np.pi/4,np.pi/2
n = np.arange(0,N)
rect_win=np.ones(N)
hd =(omega_2/np.pi * np.sinc((n-(N-1)/2)*omega_2/np.pi))-(omega_1/np.pi * np.sinc((n-(N-
1)/2)*omega_1/np.pi))
h=hd*rect_win #Step 2: Windowed impulse response h[n]
w,H=signal.freqz(h) #Step 3: Frequency response of ideal filter
#Step 4:Impulse response of the filter
plt.subplot(2,2,1),plt.stem(n,h),plt.xlabel('n-->'),plt.ylabel('Amplitude'),plt.title('h[n]')
z, p, k = signal.tf2zpk(h,1) #Step 5: Pole-zero plot of the filter
theta = np.linspace(0, np.pi*2, 500)
circle = np.exp(1j*theta)
plt.subplot(2,2,2),plt.plot(circle.real, circle.imag, 'k--')
plt.plot(z.real, z.imag, 'ro', ms=7.5),plt.plot(p.real, p.imag, 'rx')
plt.xlabel('$\sigma$'),plt.ylabel('$j\omega$'),plt.title('Pole-zero plot')
#Step 6: Magnitude response of the filter
plt.subplot(2,2,3),plt.plot(w,20*np.log10(np.abs(H)))
plt.xlabel('$\omega$-->'),plt.ylabel('$|H(e^{j^\omega})|$'),plt.title('Magnitude response')
#Step 7: Phase response of the filter
plt.subplot(2,2,4),plt.plot(w,np.unwrap(np.angle(H)))
plt.xlabel('$\omega$-->'),plt.ylabel('$\phi(e^{j^\omega})$')
plt.title('Phase response'),plt.tight_layout()
```

Fig. 7.17 Python code to obtain the characteristics of the bandpass filter

$$h_d[n] = \frac{\omega_{c2}}{\pi} \sin c(\frac{\omega_{c2}}{\pi}(n-\tau)) - \frac{\omega_{c1}}{\pi} \sin c(\frac{\omega_{c1}}{\pi}(n-\tau)) \qquad (7.5)$$

Here 'ω_{c1}' and 'ω_{c2}' are pass band frequencies and $\omega_{c2} > \omega_{c1}$. The desired impulse response must be multiplied by the window function $w[n]$ to get a finite impulse response. This is expressed as

$$h[n] = h_d[n] \times w[n] \qquad (7.6)$$

Experiment 7.9 Window-Based FIR Bandpass Filter

This experiment discusses the FIR bandpass filter design using a windowing approach. The python code, which obtains the characteristics of bandpass filter with the cut-off frequencies $\omega_{c1} = \frac{\pi}{4}$ radians/sample and $\omega_{c2} = \frac{\pi}{2}$ radians/sample, is shown in Fig. 7.17, and the corresponding output is shown in Fig. 7.18.

Inferences

From Fig. 7.18, the following observations can be made:

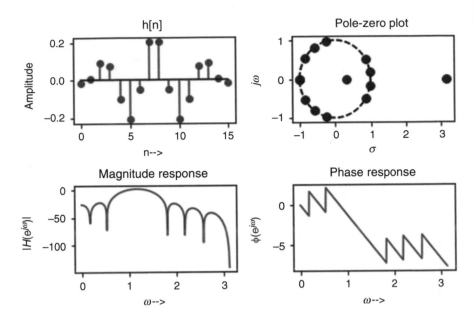

Fig. 7.18 Characteristics of the bandpass filter

1. The lower cut-off frequency chosen is $\omega_{c1} = \frac{\pi}{4}$ which is equal to 0.785 radians/sample, and the upper cut-off frequency chosen is $\omega_{c2} = \frac{\pi}{2}$ which is equal to 1.57 radians/sample. The magnitude response shows the passband between 0.785 and 1.57, and the gain drops beyond the cut-off frequency.
2. From the impulse response plot, it is possible to observe that the impulse response is symmetric in nature.
3. From the phase response, it is possible to observe that the filter exhibits linear phase characteristics in the passband. The linear phase is due to the symmetric nature of the impulse response.
4. The pole-zero plot shows that the filter is an all-zero filter with the zeros occurring in a conjugate reciprocal manner.

Task
1. In the above program, try to use windows like Hamming, Hanning, Bartlett and Blackman, and observe the change in magnitude and phase response.

7.3.5   Window-Based FIR Band Reject Filter

The expression for the desired impulse response of the band reject/stop filter is given by

```
#Characteristics of band-reject filter
import numpy as np
import matplotlib.pyplot as plt
from scipy import signal
#Step 1: Desired impulse response hd[n]
N,omega_1,omega_2 = 21,np.pi/4,np.pi/2
n = np.arange(0,N)
rect_win=np.ones(N)
hd =(omega_1/np.pi * np.sinc((n-(N-1)/2)*omega_1/np.pi))+np.sinc(n-(N-1)/2)-(omega_2/np.pi *
np.sinc((n-(N-1)/2)*omega_2/np.pi))
h=hd*rect_win#Step 2: Windowed impulse response h[n]
w,H=signal.freqz(h) #Step 3: Frequency response of ideal filter
#Step 4:Impulse response of the filter
plt.subplot(2,2,1),plt.stem(n,h),plt.xlabel('n-->'),plt.ylabel('Amplitude'),plt.title('h[n]')
z, p, k = signal.tf2zpk(h,1)#Step 5: Pole-zero plot of the filter
theta = np.linspace(0, np.pi*2, 500)
circle = np.exp(1j*theta)
plt.subplot(2,2,2),plt.plot(circle.real, circle.imag, 'k--')
plt.plot(z.real, z.imag, 'ro', ms=7.5),plt.plot(p.real, p.imag, 'rx')
plt.xlabel('$\sigma$'),plt.ylabel('$j\omega$'),plt.title('Pole-zero plot')
#Step 6: Magnitude response of the filter
plt.subplot(2,2,3),plt.plot(w,20*np.log10(np.abs(H))),plt.xlabel('$\omega$-->'),
plt.ylabel('$|H(c^{j^\omega})|$'),plt.title('Magnitude response')
#Step 7: Phase response of the filter
plt.subplot(2,2,4),plt.plot(w,np.unwrap(np.angle(H))),plt.xlabel('$\omega$-->'),
plt.ylabel('$\phi(e^{j^\omega})$'),plt.title('Phase response'),plt.tight_layout()
```

Fig. 7.19 Python code to obtain the characteristics of band reject filter

$$h_d[n] = \operatorname{sinc}(n-\tau) + \frac{\omega_{c1}}{\pi}\operatorname{sinc}\left(\frac{\omega_{c1}}{\pi}(n-\tau)\right) - \frac{\omega_{c2}}{\pi}\operatorname{sinc}\left(\frac{\omega_{c2}}{\pi}(n-\tau)\right) \quad (7.7)$$

Here 'ω_{c1}' and 'ω_{c2}' are stop band frequencies and $\omega_{c2} > \omega_{c1}$. To make the impulse response finite, the desired impulse response must be multiplied by the window function $w[n]$. This is expressed as

$$h[n] = h_d[n] \times w[n] \quad (7.8)$$

Experiment 7.10 Window-Based FIR Band Reject Filter
This experiment deals with the FIR band reject filter design using windowing method. The python code, which obtains the characteristics of band reject filter, is shown in Fig. 7.19, and the corresponding output is shown in Fig. 7.20. The cut-off frequencies chosen are $\omega_{c1} = \frac{\pi}{4}$ radians/sample and $\omega_{c2} = \frac{\pi}{2}$ radians/sample.

Inferences
The following inferences can be drawn from this experiment:

1. The lower cut-off frequency chosen is $\omega_{c1} = \frac{\pi}{4}$, which is equal to 0.785 radians/sample, and the upper cut-off frequency chosen is $\omega_{c2} = \frac{\pi}{2}$, which is equal to

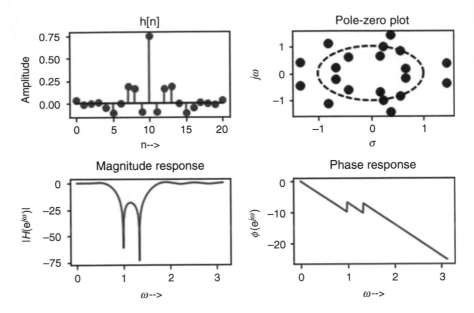

Fig. 7.20 Characteristics of band reject filter

   1.57 radians/sample. The magnitude response shows that the frequency between
   0.785 and 1.57 is attenuated.
2. From the impulse response plot, it is possible to observe that the impulse response
   is symmetric in nature.
3. From the phase response, it is possible to observe that the filter exhibits linear
   phase characteristics in the passband. The linear phase is due to symmetric nature
   of the impulse response.
4. The pole-zero plot shows that the filter is an all-zero filter with the zeros occurring
   in conjugate reciprocal manner.

Task
1. In the above program, try to use windows like Hamming, Hanning, Bartlett and
   Blackman, and observe the change in magnitude and phase responses.

7.3.6   Design of FIR Filter Using Built-In Function

The built-in function '*firwin*' available in '*scipy*' library is used here to generate FIR
filter coefficients using window-based method. The input to the built-in function is
the number of coefficients of the filter, cut-off frequency and window type.

```
#Characteristics of Low pass filter using firwin command
import numpy as np
import matplotlib.pyplot as plt
from scipy import signal
N,fs,LPF_cutoff=20,100,5
n=np.arange(N)
w_LPF=LPF_cutoff/(fs/2)
h=signal.firwin(N,w_LPF,window='hamming')
w,H=signal.freqz(h)
z, p, k = signal.tf2zpk(h,1)
plt.subplot(2,2,1),plt.stem(n,h),plt.xlabel('n-->'),plt.ylabel('Amplitude'),plt.title('h[n]')
theta = np.linspace(0, np.pi*2, 500)
circle = np.exp(1j*theta)
plt.subplot(2,2,2),plt.plot(circle.real, circle.imag, 'k--')
plt.plot(z.real, z.imag, 'ro', ms=7.5),plt.plot(p.real, p.imag, 'rx')
plt.xlabel('$\sigma$'),plt.ylabel('$j\omega$'),plt.title('Pole-zero plot')
plt.subplot(2,2,3),plt.plot((fs * 0.5 / np.pi) * w,20*np.log10(np.abs(H)))
plt.xlabel('Frequency (f) (Hz)'),plt.ylabel('$|H(jf)|$ in dB'),plt.title('Magnitude response')
#Step 7: Phase response of the filter
plt.subplot(2,2,4),plt.plot((fs * 0.5 / np.pi) * w,np.unwrap(np.angle(H)))
plt.xlabel('Frequency (f) (Hz)'),plt.ylabel('$\phi(jf)$')
plt.title('Phase response'),plt.tight_layout()
```

Fig. 7.21 Built-in function '*firwin*' to obtain the characteristics of lowpass filter

Experiment 7.11 Design of FIR Lowpass Filter Using a Built-In Function
This experiment intends to obtain the FIR filter coefficients using the built-in function '*firwin*'. After obtaining the coefficients, the characteristics of FIR filter, like impulse response, magnitude response, phase response and pole-zero plot, are plotted. The python code, which performs this task, is shown in Fig. 7.21, and the corresponding output is shown in Fig. 7.22.

Inferences
From Fig. 7.21, the following observations can be obtained:

1. The built-in function '*firwin*' available in '*scipy*' library is used here to obtain the filter coefficients.
2. The specifications of the lowpass filter are (a) number of coefficients = 20, cut-off frequency = 5 Hz, sampling frequency = 100 Hz and window chosen is '*Hamming*' window.

From Fig. 7.22, the following observations can be made:

1. The impulse response consists of 20 coefficients. From the impulse response, it is possible to observe that the coefficients are symmetric.
2. From the pole-zero plot, it is possible to observe that the zeros occur in conjugate reciprocal pair.

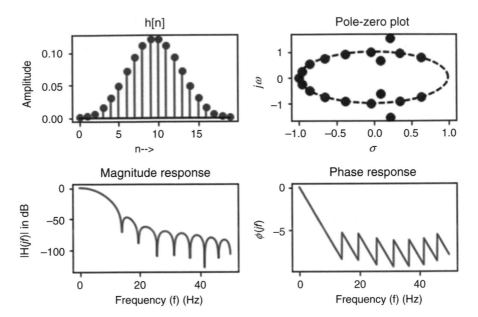

Fig. 7.22 Characteristics of lowpass filter

3. From the magnitude response, it is possible to observe that the gain drops after the cut-off frequency of 5 Hz.
4. From the phase response, it is possible to confirm that linear phase characteristic is obtained in the passband.

Experiment 7.12 Design of FIR Highpass Filter Using the Built-In Function
This experiment aims to obtain the highpass filter coefficients with a slight change in the code which generates the lowpass filter. The keyword '*pass-zero = false*' helps one to obtain the highpass filter. The python code, which performs this task, is shown in Fig. 7.23, and the corresponding output is shown in Fig. 7.24.

Inferences
The following inferences can be made from this experiment:

1. From Fig. 7.23, it is possible to observe that the keyword '*pass_zero = False*' allows one to obtain the coefficients of the highpass filter.
2. The characteristic of highpass filter is shown in Fig. 7.24. From this figure, it is possible to observe that beyond the cut-off frequency of 5 Hz, the gain reaches a value of 0 dB, and the phase response is linear curve.

Experiment 7.13 Design of FIR Bandpass Filter Using the Built-In Function
The objective of this experiment is to design a bandpass filter, which will pass signal in the frequency range 10–20 Hz. The sampling frequency chosen is 100 Hz. The order of the filter is 50, and the window chosen is 'Hamming window'.

```
#Characteristics of high pass filter using "firwin" command
import numpy as np
import matplotlib.pyplot as plt
from scipy import signal
N,fs,HPF_cutoff=21,100,5
n=np.arange(N)
w_HPF=HPF_cutoff/(fs/2)
h=signal.firwin(N,w_HPF,window='hamming',pass_zero=False)
w,H=signal.freqz(h)
z, p, k = signal.tf2zpk(h,1)
plt.subplot(2,2,1),plt.stem(n,h),plt.xlabel('n-->'),plt.ylabel('Amplitude'),plt.title('h[n]')
theta = np.linspace(0, np.pi*2, 500)
circle = np.exp(1j*theta)
plt.subplot(2,2,2),plt.plot(circle.real, circle.imag, 'k--')
plt.plot(z.real, z.imag, 'ro', ms=7.5),plt.plot(p.real, p.imag, 'rx')
plt.xlabel('$\sigma$'),plt.ylabel('$j\omega$'),plt.title('Pole-zero plot')
plt.subplot(2,2,3),plt.plot((fs * 0.5 / np.pi) * w,20*np.log10(np.abs(H)))
plt.xlabel('Frequency (f) (Hz)'),plt.ylabel('$|H(jf)|$ in dB'),plt.title('Magnitude response')
#Step 7: Phase response of the filter
plt.subplot(2,2,4),plt.plot((fs * 0.5 / np.pi) * w,np.unwrap(np.angle(H)))
plt.xlabel('Frequency (f) (Hz)'),plt.ylabel('$\phi(jf)$')
plt.title('Phase response'),plt.tight_layout()
```

Fig. 7.23 Python code to obtain the characteristics of highpass filter

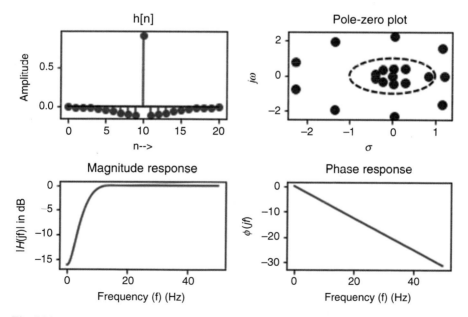

Fig. 7.24 Characteristics of highpass filter

```
#Characteristics of bandpass filter using "firwin" command
import numpy as np
import matplotlib.pyplot as plt
from scipy import signal
N,fs,lf,uf=20,100,10,20  #Specifications
n=np.arange(0,N)
w_LCF=lf/(fs/2)
w_UCF=uf/(fs/2)
h=signal.firwin(N,[w_LCF,w_UCF],window='hamming',pass_zero=False)
w,H=signal.freqz(h)
z, p, k = signal.tf2zpk(h,1)
plt.subplot(2,2,1),plt.stem(n,h),plt.xlabel('n-->'),plt.ylabel('Amplitude'),plt.title('h[n]')
theta = np.linspace(0, np.pi*2, 500)
circle = np.exp(1j*theta)
plt.subplot(2,2,2),plt.plot(circle.real, circle.imag, 'k--')
plt.plot(z.real, z.imag, 'ro', ms=7.5),plt.plot(p.real, p.imag, 'rx')
plt.xlabel('$\sigma$'),plt.ylabel('$j\omega$'),plt.title('Pole-zero plot')
plt.subplot(2,2,3),plt.plot((fs * 0.5 / np.pi) * w,20*np.log10(np.abs(H)))
plt.xlabel('Frequency (f) (Hz)'),plt.ylabel('$|H(jf)|$ in dB'),plt.title('Magnitude response')
#Step 7: Phase response of the filter
plt.subplot(2,2,4),plt.plot((fs * 0.5 / np.pi) * w,np.unwrap(np.angle(H)))
plt.xlabel('Frequency (f) (Hz)'),plt.ylabel('$\phi(jf)$')
plt.title('Phase response'),plt.tight_layout()
```

Fig. 7.25 Python code to obtain the characteristics of bandpass filter

The python code which generates the filter coefficient corresponding to the desired bandpass filter is shown in Fig. 7.25, and the corresponding filter characteristics are shown in Fig. 7.26.

Inferences
The following inferences can be drawn from this experiment:

1. From Fig. 7.26, it is possible to observe from the magnitude response that the filter passes a band of frequencies from 10 to 20 Hz.
2. From the impulse response, it is possible to observe that the impulse response of the filter is finite and symmetric.
3. The filter also exhibits linear phase characteristics in the passband. This is due to the symmetric nature of the impulse response.
4. From the pole-zero plot, it is possible to observe that the zeros occur in conjugate reciprocal pair.

Experiment 7.14 Design of FIR Band Reject Filter Using the Built-In Function
This experiment discusses the design of FIR band reject filter using built-in function. The band reject filter is obtained from bandpass filter design using the key term '*pass_zero = True*'. The python code which obtains the coefficient of the band reject filter is shown in Fig. 7.27, and the corresponding output is shown in Fig. 7.28.

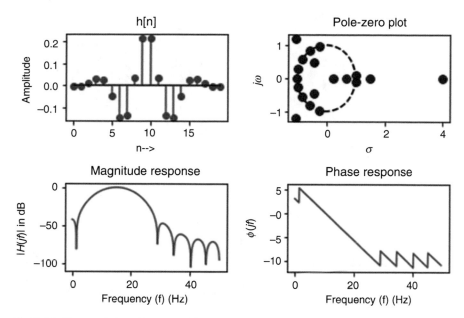

Fig. 7.26 Characteristics of bandpass filter

```
#Characteristics of band reject filter using "firwin" command
import numpy as np
import matplotlib.pyplot as plt
from scipy import signal
N,fs,lf,uf=25,100,10,30
n=np.arange(0,N)
w_LCF=lf/(fs/2)
w_UCF=uf/(fs/2)
h=signal.firwin(N,[w_LCF,w_UCF],window='hamming',pass_zero=True)
w,H=signal.freqz(h)
z, p, k = signal.tf2zpk(h,1)
plt.subplot(2,2,1),plt.stem(n,h),plt.xlabel('n-->'),plt.ylabel('Amplitude'),plt.title('h[n]')
theta = np.linspace(0, np.pi*2, 500)
circle = np.exp(1j*theta)
plt.subplot(2,2,2),plt.plot(circle.real, circle.imag, 'k--')
plt.plot(z.real, z.imag, 'ro', ms=7.5),plt.plot(p.real, p.imag, 'rx')
plt.xlabel('$\sigma$'),plt.ylabel('$j\omega$'),plt.title('Pole-zero plot')
plt.subplot(2,2,3),plt.plot((fs * 0.5 / np.pi) * w,20*np.log10(np.abs(H)))
plt.xlabel('Frequency (f) (Hz)'),plt.ylabel('$|H(jf)|$ in dB'),plt.title('Magnitude response')
#Step 7: Phase response of the filter
plt.subplot(2,2,4),plt.plot((fs * 0.5 / np.pi) * w,np.unwrap(np.angle(H)))
plt.xlabel('Frequency (f) (Hz)'),plt.ylabel('$\phi(jf)$')
plt.title('Phase response'),plt.tight_layout()
```

Fig. 7.27 Python code to generate the band reject filter coefficients and its characteristics

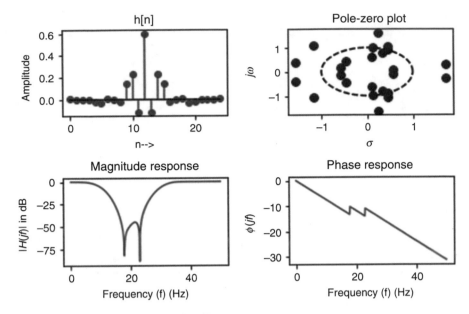

Fig. 7.28 Characteristics of band reject filter

Inferences

The following observations can be made from this experiment:

1. Figure 7.27 shows that the keyword '*pass-zero* = *True*' is used here to convert the bandpass filter to a band reject filter.
2. From the magnitude response shown in Fig. 7.28, it is possible to confirm that this filter blocks the frequency band from 10 to 30 Hz.

7.3.7 Window Functions

In this section, the window function is visualized in both time domain and frequency domain. Different types of window functions include rectangular, triangular, Hamming, Hanning, Kaiser, etc. The main lobe width of the window function controls the transition bandwidth, whereas the height of the side lobe controls the passband and stopband ripples. A rectangular window has the narrowest main lobe; hence, it gives sharpest transition. Compared to rectangular window, the Hamming and Hanning windows are smoother. By tapering the window smoothly to zero, the sidelobes can be reduced in amplitude, which will reduce the ripple, but the trade-off is a larger main lobe. A linear phase response can be achieved if the window function is symmetric. Some windows allow controlled trade-offs between sidelobe amplitude and main lobe width. One such window is the Kaiser window.

```
#Window functions in time and frequency domain
import numpy as np
import matplotlib.pyplot as plt
from scipy.fft import fft,fftshift
N = 51  #Length of the window
n =np.arange(-(N-1)/2, (N-1)/2)
#Defining different window functions
w_Rect = n-n+1; #Rectangular
w_Bart = 1 - 2* abs(n)/(N-1) #Bartlett
w_Han = 0.5 + 0.5 * np.cos(2*np.pi*n/(N-1)); #Hanning
w_Hamm = 0.54 + 0.46 * np.cos(2*np.pi*n/(N-1)); #Hamming
#Spectrum of window
W_Rect=fftshift(fft(w_Rect,1024)/len(w_Rect))
W_Bart =fftshift(fft(w_Bart,1024)/len(w_Bart))
W_Han=fftshift(fft(w_Han,1024)/len(w_Bart))
W_Hamm=fftshift(fft(w_Hamm,1024)/len(w_Bart))
plt.figure(1),plt.subplot(2,2,1),plt.stem(n,w_Rect),plt.xlabel('n-->'),plt.ylabel('w[n]'),
plt.title('Rectangular window'),plt.subplot(2,2,3),plt.stem(n,w_Bart),plt.xlabel('n-->')
plt.ylabel('w[n]'),plt.title('Bartlett window')
freq = np.linspace(-0.5, 0.5, len(W_Rect))
plt.subplot(2,2,2), plt.plot(freq, 20 * np.log10(W_Rect)),plt.axis([-0.5, 0.5, -120, 0]),
plt.xlabel('Normalized frequency'),plt.ylabel('Magnitude [dB]'),
plt.title('Spectrum of rectangular window'),
plt.subplot(2,2,4), plt.plot(freq, 20 * np.log10(W_Bart)),plt.xlabel('Normalized frequency'),
plt.ylabel('Magnitude [dB]'),plt.title('Spectrum of Bartlett window')
plt.axis([-0.5, 0.5, -120, 0]),plt.tight_layout()
plt.figure(2),plt.subplot(2,2,1),plt.stem(n,w_Han),plt.xlabel('n-->'),plt.ylabel('w[n]'),
plt.title('Hanning window'),plt.subplot(2,2,3),plt.stem(n,w_Hamm),plt.xlabel('n-->')
plt.ylabel('w[n]'),plt.title('Hamming window'),plt.subplot(2,2,2),
plt.plot(freq, 20 * np.log10(W_Han)),plt.xlabel('Normalized frequency'),
plt.ylabel('Magnitude [dB]'),plt.title('Spectrum of Hanning window'),plt.axis([ 0.5, 0.5, -120, 0]),
plt.subplot(2,2,4), plt.plot(freq, 20 * np.log10(W_Hamm)),plt.xlabel('Normalized frequency'),
plt.ylabel('Magnitude [dB]'),plt.title('Spectrum of Hamming window')
plt.axis([-0.5, 0.5, -120, 0]),plt.tight_layout()
```

Fig. 7.29 Python code to plot the window functions in time and frequency domain

Experiment 7.15 Plotting Windows in the Time and Frequency Domain
The python code which plots the window function in the time and frequency domain
is shown in Fig. 7.29, and the corresponding output is shown in Figs. 7.30 and 7.31.
Different window functions considered in this example include (1) rectangular,
(2) triangular or Bartlett, (3) Hanning and (4) Hamming window.

Inference
From Figs. 7.30 and 7.31, the following inferences can be made:

1. Four different types of windows chosen are (a) rectangular, (b) triangular,
   (c) Hamming and (d) Hanning.
2. Fourier transform of different types of window functions results in the sinc
   functions.

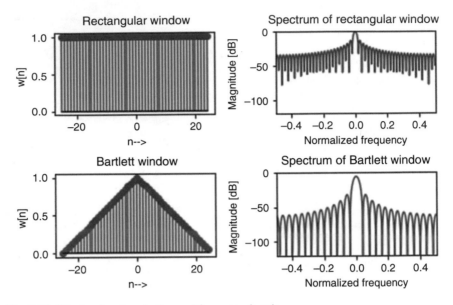

Fig. 7.30 Window functions in time and frequency domain

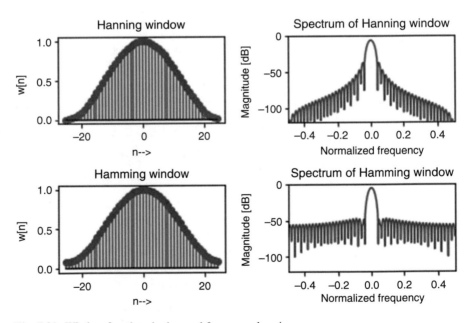

Fig. 7.31 Window functions in time and frequency domain

3. The spectrum of different windows differs with respect to main lobe width and
   side lobe height.

7.4 Frequency Sampling-Based FIR Filter Design

The steps followed in frequency sampling method of FIR filter design are summarized below:

1. The design step starts with a prescribed magnitude response.
2. The prescribed magnitude response is sampled at enough points.
3. Take the inverse Fourier transform of the samples obtained in step (2). This will result in the filter's impulse response.

Experiment 7.16 Frequency Sampling-Based FIR Filter Design

The built-in function '*firwin2*' available in '*scipy*' library is used here to generate FIR filter coefficients. The following python code helps one to obtain the coefficients of Type-I, Type-II, Type-III and Type-IV FIR filters. It is known that Type-I and Type-II FIR filter exhibits even symmetry with odd and even number of coefficients, respectively. Type-III, and Type-IV FIR filter exhibits odd symmetry with odd and even number of coefficients, respectively. The python code, which obtains the response of four types of FIR filter, is shown in Fig. 7.32, and the corresponding output is shown in Fig. 7.33.

Inferences

From Fig. 7.32, the following observations can be made:

1. The built-in function '*firwin2*' is used here to generate the FIR filter coefficients. In the program, the variables '*f1* to *f4*' represent the desired normalized frequency in the range 0 to 1. The variables '*m1* to *m4*' represent the desired magnitude response. '1' represents passband, and '0' represents the stopband. In the program, '*N1* to *N4*' represents the number of coefficients of the filter. For Type-I and Type-III, the number of coefficients has to be odd. For Type-II and Type-IV, the number of coefficients has to be even.
2. The keyword '*asymmetric = false*' implies even symmetry, and '*asymmetric = true*' represents odd symmetry. Type-I and Type-II FIR filters exhibit even symmetry, whereas Type-III and Type-IV FIR filters exhibit odd symmetry.

From Fig. 7.33, the following interpretations can be made:

1. Type-I and Type-II FIR filters act as lowpass filter; Type-III FIR filter act as bandpass filter. Type-IV FIR filter act as a highpass filter.
2. All four types of FIR filters exhibit linear phase characteristics in the passband.
3. It is to be noted that Type-II FIR filter cannot be used as a highpass filter. Type-III FIR filter cannot be used as a lowpass and highpass filters. Type-IV FIR filter cannot be used as a lowpass filter.

```python
#Characteristics of FIR filter using frequency sampling method
import numpy as np
from scipy import signal
import matplotlib.pyplot as plt
#Type-I FIR filter
f1 = [0,0.5,0.5,1]; #desired frequencies(w/pi)
m1 = [1,1,0,0];    #desired magnitude at f
N1 = 51        #samples
h1 = signal.firwin2(N1, f1, m1,antisymmetric=False)
w1, H1 = signal.freqz(h1)
plt.figure(1),plt.subplot(2,2,1),plt.plot(w1,(10 * np.log10(abs(H1)))),
plt.ylabel('$|H(e^{j^\omega})|$ in dB'),plt.xlabel('$\omega/\pi$'),
plt.title('Type-I FIR filter'),plt.subplot(2,2,2),plt.plot(w1,np.unwrap(np.angle(H1)))
plt.ylabel('$\phi(e^{j^\omega})$'),plt.xlabel('$\omega/\pi$'),plt.title('Type-I FIR filter')
#Type-II FIR filter
f2 = [0,0.6,0.6,1]; #desired frequencies(w/pi)
m2 = [1,1,0,0];   #desired magnitude at f
N2 = 50        #samples
h2 = signal.firwin2(N2, f2, m2,antisymmetric=False)
w2, H2 = signal.freqz(h2)
plt.subplot(2,2,3),plt.plot(w2,(10 * np.log10(abs(H2)))),
plt.ylabel('$|H(e^{j^\omega})|$ in dB'),plt.xlabel('$\omega/\pi$'),
plt.title('Type-II FIR filter'),plt.subplot(2,2,4),plt.plot(w2,np.unwrap(np.angle(H2)))
plt.ylabel('$\phi(e^{j^\omega})$'),plt.xlabel('$\omega/\pi$'),plt.title('Type-II FIR
filter'),plt.tight_layout()
#Type-III FIR filter
f3 = [0,0.2,0.4,0.6,0.8,1]; #desired frequencies(w/pi)
m3 = [0,0,1,0,0,0];   #desired magnitude at f
N3= 101        #samples
h3 = signal.firwin2(N3, f3, m3,antisymmetric=True)
w3, H3 = signal.freqz(h3);
plt.figure(2),plt.subplot(2,2,1),plt.plot(w3,(10 * np.log10(abs(H3)))),
plt.ylabel('$|H(e^{j^\omega})|$ in dB'),plt.xlabel('$\omega/\pi$'),
plt.title('Type-III FIR filter'),plt.subplot(2,2,2),plt.plot(w3,np.unwrap(np.angle(H3)))
plt.ylabel('$\phi(e^{j^\omega})$'),plt.xlabel('$\omega/\pi$'),plt.title('Type-III FIR filter')
#Type-IV FIR filter
f4 = [0,0.6,0.6,1]; #desired frequencies(w/pi)
m4 = [0,0,1,1];   #desired magnitude at f
N4 = 150        #samples
h4 = signal.firwin2(N4, f4, m4,antisymmetric=True)
w4, H4 = signal.freqz(h4);
plt.subplot(2,2,3),plt.plot(w4,(10 * np.log10(abs(H4)))),plt.ylabel('$|H(e^{j^\omega})|$ in dB'),
plt.xlabel('$\omega/\pi$'),plt.title('Type-IV FIR filter'),plt.subplot(2,2,4),
plt.plot(w4,np.unwrap(np.angle(H4))),plt.ylabel('$\phi(e^{j^\omega})$'),
plt.xlabel('$\omega/\pi$'),plt.title('Type-IV FIR filter'), plt.tight_layout()
```

Fig. 7.32 Python code to obtain the characteristics of four types of FIR filter using frequency sampling method

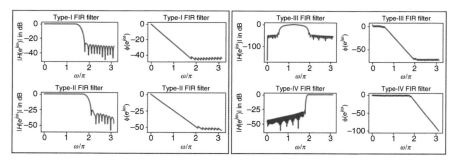

Fig. 7.33 Magnitude and phase responses of four types of FIR filter

7.5 Design of Optimal FIR filter

The optimal equiripple FIR filter design is due to Parks and McClellan. The optimal method provides an FIR filter coefficient representing the best approximation to the desired frequency response in a Chebyshev sense. The term optimal can be defined in various ways. The Parks-McClellan package uses the Remez exchange algorithm to optimize the filter design by selecting the impulse response, which minimizes the peak ripple in the passband and stopband. The filter designed by this approach is termed as '*equiripple*' filter. It is also termed as '*minimax filter*' because the maximum ripple deviation is minimized in the optimization procedure.

Experiment 7.17 Design of Optimal FIR Filter

The built-in functions available in '*scipy*' library like, '*remez*' and '*firls*' can be used to obtain the coefficients of optimal filter using Remez exchange algorithm and the least square approach, respectively.

The aim of this experiment is to design the lowpass, highpass, bandpass and band reject filters using the built-in function 'remez' available in '*scipy*' library. The python code, which generates the filter coefficients and plots the magnitude responses of these four filters, is shown in Fig. 7.34, and the corresponding output is shown in Fig. 7.35.

Inferences

From Fig. 7.34, the following observations can be made:

1. The sampling frequency of the four filters is kept at 1000 Hz, the number of taps of the filter of all four filters is kept as 125 and the transition width of the four filters is kept as 25 Hz.
2. The lowpass filter cut-off frequency is kept at 100 Hz. This means that the filter should pass all frequencies till 100 Hz and block frequency components greater than 100 Hz.
3. The cut-off frequency of a highpass filter is fixed as 200 Hz.
4. For the bandpass and band reject filters, the lower and upper cut-off frequencies are fixed as 100 Hz and 200 Hz, respectively.

```
#Filter design using built-in function remez
import numpy as np
import matplotlib.pyplot as plt
from scipy import signal
fs,N,trans_width=1000,125,25 #Low pass filter design
fc_lp=100 #LPF cut off frequency
h_lp= signal.remez(N, [0, fc_lp, fc_lp + trans_width, 0.5*fs], [1, 0], Hz=fs)
w1, H_lp = signal.freqz(h_lp,1)
plt.subplot(2,2,1),plt.plot(0.5*fs*w1/np.pi, 20*np.log10(np.abs(H_lp)))
plt.xlabel('Frequency (Hz)'),plt.ylabel('Gain (dB)'),plt.title('Magnitude response of LPF')
#High pass filter design
fc_hp = 200.0   # High pass filter cut off frequency
h_hp = signal.remez(N, [0, fc_hp - trans_width, fc_hp, 0.5*fs],[0, 1], Hz=fs)
w2, H_hp = signal.freqz(h_hp, [1])
plt.subplot(2,2,2),plt.plot(0.5*fs*w2/np.pi, 20*np.log10(np.abs(H_hp)))
plt.xlabel('Frequency (Hz)'),plt.ylabel('Gain (dB)'),plt.title('Magnitude response of HPF')
#Band pass filter
band = [100, 200] # Desired pass band, Hz
edges = [0, band[0] - trans_width, band[0], band[1], band[1] + trans_width, 0.5*fs]
h_bpf = signal.remez(N, edges, [0, 1, 0], Hz=fs)
w3, H_bpf = signal.freqz(h_bpf,1)
plt.subplot(2,2,3),plt.plot(0.5*fs*w3/np.pi, 20*np.log10(np.abs(H_bpf)))
plt.xlabel('Frequency (Hz)'),plt.ylabel('Gain (dB)'),plt.title('Magnitude response of BPF')
#Band reject filter
h_brf = signal.remez(N, edges, [1, 0, 1], Hz=fs)
w4, H_brf = signal.freqz(h_brf,1)
plt.subplot(2,2,4),plt.plot(0.5*fs*w4/np.pi, 20*np.log10(np.abs(H_brf)))
plt.xlabel('Frequency (Hz)'),plt.ylabel('Gain (dB)')
plt.title('Magnitude response of BRF'),plt.tight_layout()
```

Fig. 7.34 Python code to obtain the filter coefficients using the built-in function 'remez'

From Fig. 7.35, it is possible to observe that the magnitude responses of the four filters are as per the specification.

7.6   Applications of FIR Filter

The coefficients of FIR filters are either symmetric or anti-symmetric. FIR filter exhibits linear phase characteristics, because of which, there is no phase distortion. The group delay of FIR filter is constant. Since the poles of FIR filter occur at the origin, FIR filters are inherently stable. Because of these characteristics, FIR filters are used in many areas of signal processing like multirate signal processing, adaptive signal processing, etc. In multirate signal processing, FIR filters are preferred to design perfect reconstruction filter bank. In adaptive signal processing, FIR filters are preferred in system identification, adaptive notch filter, inverse system modelling, echo cancellation and variety of such applications. In this section, two simple

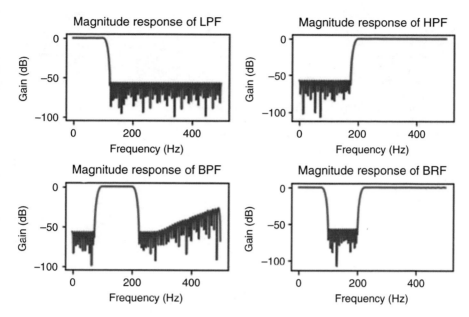

Fig. 7.35 Magnitude responses of the filters

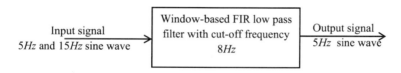

Fig. 7.36 Block diagram of problem statement

applications are discussed. One is signal separation, and the other is signal denoising.

Experiment 7.18 Separation of Signals Using FIR Filter

The signal $x(t)$ is an addition of two signals $x_1(t)$ and $x_2(t)$. The frequencies of the two signals $x_1(t)$ and $x_2(t)$ are 5 Hz and 15 Hz, respectively. The signal $x(t)$ is now passed through a lowpass filter whose cut-off frequency is 8 Hz, the order of the filter is 20 and the window chosen is Hanning.

The problem statement is depicted in the form of block diagram, which is shown in Fig. 7.36.

The python code, which performs lowpass filtering of the input sine wave with 5 and 15 Hz frequency components, is shown in Fig. 7.37, and the corresponding output is in Fig. 7.38.

Inferences

From Fig. 7.37, the following observations can be made:

```
#Low pass filtering of sine wave
import numpy as np
import matplotlib.pyplot as plt
from scipy import signal
from scipy.fftpack import fft,fftfreq
#Step 1: Defining the parameters of sine wave
f1,f2,ph=5,15,0  #Frequency of signal 1 and 2 and phase
#Step 2: Defining the sampling frequency and number of points in FFT
fs, N=100, 256  #Sampling frequency
T=1/fs  #Sampling period
#Step 3: Generation of sine wave
t=np.linspace(0,N*T,N)
x1=np.sin(2*np.pi*f1*t+ph)
x2=np.sin(2*np.pi*f2*t+ph)
x=x1+x2
#Step 4: Design of LPF
f_cut=8  # Cut-off frequency
Nyquist_freq, numtaps=fs/2, 21
f_cutoff=f_cut/Nyquist_freq
h=signal.firwin(numtaps, f_cutoff,window='hann')
y=signal.lfilter(h,1,x) #Step 5: Filtering of the signal
f_axis=fftfreq(N,T)[0:N//2] #Step 5: Obtaining the spectrum of input and output signal
X=fft(x)
Y=fft(y)
#Step 6: Plotting the results
plt.subplot(2,2,1),plt.plot(t,x),plt.xlabel('Time'),plt.ylabel('Amplitude'),plt.title('Input Signal'),
plt.subplot(2,2,2),plt.plot(t,y),plt.xlabel('Time'),plt.ylabel('Amplitude'),
plt.title('Filtered Signal'),plt.subplot(2,2,3),plt.plot(f_axis,2/N*np.abs(X[0:N//2]))
plt.xlabel('Frequency'),plt.ylabel('Magnitude'),plt.title('Spectrum of input Signal')
plt.subplot(2,2,4),plt.plot(f_axis,2/N*np.abs(Y[0:N//2]))
plt.xlabel('Frequency'),plt.ylabel('Magnitude'),plt.title('Spectrum of filtered Signal')
plt.tight_layout()
```

Fig. 7.37 Python code to perform lowpass filtering of the input signal

1. The input signal is the sum of two sine wave frequencies, 5 and 15 Hz.
2. Window-based FIR filter is designed with a cut-off frequency of 8 Hz, the number of taps is 21 and the window chosen is Hanning. The sampling frequency chosen is 100 Hz.
3. 'The built-in function '*firwin*', which is available in '*scipy*' package, is used to design the filter.

From Fig. 7.38, the following observations can be made:

1. The input signal is a mixture of 5 and 15 Hz sine wave.
2. The output signal is a lowpass filtered signal which retains a 5 Hz sine wave.
3. The spectrum of the input signal shows peaks corresponding to 5 and 15 Hz frequency components.

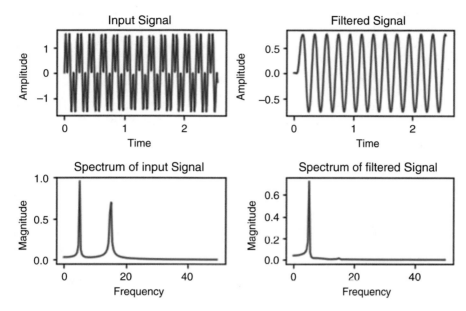

Fig. 7.38 Lowpass filtering using window-based FIR filter

4. The spectrum of the output signal shows peak at 5 Hz, which implies that the filter allows 5 Hz frequency component of the input signal, and it blocks the 15 Hz frequency component of the input signal.

Experiment 7.19 Denoising of the Signal Using FIR Filter

The signal $x(t)$ is a 5 Hz sine wave. This signal $x(t)$ is corrupted by white noise, which follows uniform distribution in the range [0, 1]. The noisy signal is then passed through FIR lowpass filter. The FIR filter coefficients are generated using the windowing technique. Plot results of the clean, noisy and filtered signals.

The python code which performs the above-mentioned task is shown in Fig. 7.39, and the corresponding output is shown in Fig. 7.40.

Inferences

The following observations can be made from this experiment:

1. From Fig. 7.39, it is possible to observe that the built-in function '*random. uniform*' available in *numpy* library is used here to generate uniformly distributed random noise in the interval 0 to 1.
2. The random noise is added to pure sine wave to create noisy sine wave. The noisy sine wave is then filtered using FIR filter, obtained using the built-in function '*firwin*' available in *scipy* library.
3. From Fig. 7.40, it is possible to observe that the clean sine wave has a frequency of 5 Hz. It is then corrupted by random noise to create noisy sine wave. From the filtered signal, it is possible to observe that the impact of noise is minimized.

```
import numpy as np
import matplotlib.pyplot as plt
from scipy import signal
#Step 1: Generation of clean signal
t=np.linspace(0,1,100)
x=np.sin(2*np.pi*5*t)
n=np.random.uniform(0,1,100) #Step 2: Uniform random noise
x1=x+n #Step 3: Noisy sine wave
h=signal.firwin(21,.2,pass_zero=True)#FIR filter coefficients
y=signal.filtfilt(h,1,x1) #Step 4: Filtered signal
#Step 5: Plotting the result
plt.subplot(3,1,1),plt.plot(t,x),plt.xlabel('Time'),plt.ylabel('Amplitude'),
plt.title('Clean signal'),plt.subplot(3,1,2),plt.plot(t,x1)
plt.xlabel('Time'),plt.ylabel('Amplitude'),plt.title('Noisy signal')
plt.subplot(3,1,3),plt.plot(t,y),plt.xlabel('Time'),plt.ylabel('Amplitude'),
plt.title('Filtered signal'),plt.tight_layout()
```

Fig. 7.39 Python code which performs filtering of noisy signal

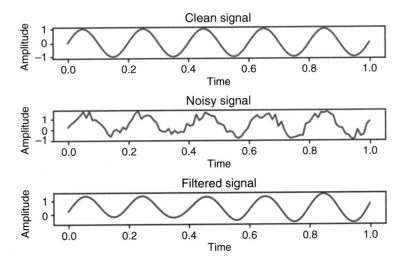

Fig. 7.40 Filtering of signal corrupted by uniform random noise

Exercises

1. Design a linear phase lowpass filter that satisfies the following magnitude response $H(e^{j\omega}) = \begin{cases} 1, & |\omega| < \dfrac{\pi}{4} \\ 0, & \text{otherwise} \end{cases}$. For $N = 5$ and $N = 7$. Assume the window to be rectangular window. Plot the magnitude response of the filter.

Table 7.4 Comparison of different window functions

S.I No.	Name of the window	Time domain expression $-\frac{(N-1)}{2} \leq n \leq \frac{(N-1)}{2}$
1	Rectangular window	$w[n] = 1$
2	Bartlett window or Triangular window	$w[n] = 1 - \frac{2|n|}{N-1}$
3	Hamming window	$w[n] = 0.54 + 0.46\cos\left(\frac{2\pi n}{N-1}\right)$
4	Hanning window	$w[n] = 0.5 + 0.5\cos\left(\frac{2\pi n}{N-1}\right)$
5	Blackman window	$w[n] = 0.42 + 0.5\cos\left(\frac{2\pi n}{N-1}\right) + 0.08\cos\left(\frac{4\pi n}{N-1}\right)$

2. Design a length 7 linear phase highpass filter using windowing method with the cut-off frequency $\omega_c = \frac{\pi}{3}$ radians/sample. Assume the window to be Bartlett window. Plot the magnitude and phase response of the filter.

3. Design a length 5 linear phase bandpass filter with lower cut-off frequency $\omega_{c1} = 0.25\pi$ radians/sample and upper cut-off frequency $\omega_{c2} = 0.75\pi$ radians/sample. Assume the window to be rectangular window. Plot the magnitude and phase response of the filter.

4. Design a length 7 linear phase band reject filter with lower cut-off frequency $\omega_{c1} = 0.15\pi$ radians/sample and upper cut-off frequency $\omega_{c2} = 0.45\pi$ radians/sample. Assume the window to be Hamming window. Plot the magnitude and phase response of the filter.

5. The signal $x(t)$ is an addition of two signals $x_1(t)$ and $x_2(t)$. The frequencies of the two signals $x_1(t)$ and $x_2(t)$ are 5 Hz and 10 Hz, respectively. The signal $x(t)$ is now passed through a highpass filter whose cut-off frequency is 8 Hz, the order of the filter is 10 and the window chosen is Hamming window. Plot the input signal and the filtered signal and comment on the observed result.

6. The time domain expression for different window functions is given in the following Table 7.4.

   Write a python code to plot the above window functions and comment on the observed result. Assume the value of N as 31.

7. Write a python code to design a length 9 linear phase highpass filter using windowing method with the cut-off frequency $\omega_c = 4$ radians/sample. Assume the window to be Bartlett window. Plot the pole-zero plot of the filter, and observe that the zeros of the filter occur in conjugate pair.

8. The impulse response of 5-tap linear phase lowpass filter is given by $h_1[n] = \{0.159, 0.225, 0.25, 0.225, 0.159\}$. Derive another filter from this lowpass filter, whose impulse response is given by $h_2[n]$. The relationship between the impulse responses is given by $h_2[n] = (-1)^n h_1[n]$. Plot the magnitude responses of the two filters and comment on the observed result.

9. Design a lowpass FIR filter using a frequency sampling technique having cut-off frequency of $\pi/2$ radians/sample. The length of the filter is 21. Plot the magnitude response of the filter.

10. Design a digital FIR lowpass filter with the following specifications: (a) Passband cut-off frequency: $f_p = 1$ kHz. (b) Stopband cut-off frequency:

$f_s = 4$ kHz. (c) Passband ripple: $R_p = 0.25$ dB. (d) Stopband attenuation: $R_s = 0.25$ dB. (e) Sampling frequency: $f_s = 20$ kHz. Use subplot to plot the magnitude response, phase response, impulse response and pole-zero plot of the filter.

Objective Questions

1. Assertion: FIR filter exhibits linear phase characteristics.
   Reason: The coefficients of FIR filter are either symmetric or anti-symmetric:

   A. Both assertion and reason are true.
   B. Assertion is true; reason is false.
   C. Assertion is false; reason may be true.
   D. Both assertion and reason are false.

2. If 'N' represents the number of coefficients of the FIR filter, then the group delay of the filter is expressed as

   A. $\tau_g = \frac{N}{2}$
   B. $\tau_g = \frac{N}{2} - 1$
   C. $\tau_g = \frac{N-1}{2}$
   D. $\tau_g = N$

3. Identify the statement that is FALSE with respect to FIR filter

   A. FIR filter is all-zero filter.
   B. FIR filter is all-pole filter.
   C. FIR filter is inherently stable filter.
   D. Group delay of FIR filter is constant.

4. The filter which exhibits even symmetry with odd number of coefficient is

   A. Type-I FIR filter
   B. Type-II FIR filter
   C. Type-III FIR filter
   D. Type-IV FIR filter

5. The built-in function available in *scipy* library to design window-based FIR filter is

   A. signal.firwin()
   B. signal.firwin2()
   C. signal.remez()
   D. signal.firls()

6. The built-in function available in scipy library to design frequency sampling-based FIR filter is

   A. signal.firwin(),
   B. signal.firwin2()
   C. signal.remez()
   D. signal.firls()

7. The following python command $h = signal.firwin(5,0.5)$ generates

   A. Five coefficients of lowpass filter
   B. Four coefficients of lowpass filter
   C. Five coefficients of highpass filter
   D. Four coefficients of highpass filter

8. The type of FIR filter that has zero at $\omega = 0$ and at $\omega = \pi$ is

   A. Type-I FIR filter
   B. Type-II FIR filter
   C. Type-III FIR filter
   D. Type-IV FIR filter

9. Type-II FIR filter cannot be used as

   A. Lowpass filter
   B. Highpass filter
   C. Band pass filter
   D. Band reject filter

10. The frequency response of a linear phase filter is given by $H(e^{j\omega}) = e^{-j4\omega}R(\omega)$, where $R(\omega)$ represents the magnitude response. The group delay of the filter is

    A. 1
    B. 2
    C. 3
    D. 4

11. If a zero occurs at z_0 of a real-valued linear phase filter than the other zeros are at

    A. $\frac{1}{z_0}$ only
    B. z_0^* only
    C. $\frac{1}{z_0}$ and at $\frac{1}{z_0^*}$ only
    D. $\frac{1}{z_0}$, z_0^* and at $\frac{1}{z_0^*}$.

12. Let $h[n]$ represent the impulse response of lowpass filter and then the impulse response $(-1)^n h[n]$ represent

    A. Lowpass filter
    B. Highpass filter
    C. Band pass filter
    D. Band reject filter

13. Match the following

Transfer function of the filter	Type of FIR filter
(P) $H_1(z) = 1 + 2z^{-1} + z^{-2}$	(i) Type-I FIR filter
(Q) $H_2(z) = 1 - z^{-1}$	(ii) Type-II FIR filter

(continued)

Transfer function of the filter	Type of FIR filter
(R) $H_3(z) = 1 + z^{-1}$	(iii) Type-III FIR filter
(S) $H_4(z) = 1 - z^{-2}$	(iv) Type-IV FIR filter

    A. P-(i), Q-(ii), R-(iii), S-(iv)
    B. P-(i), Q-(iv), R-(ii), S-(iii)
    C. P-(iv), Q-(iii), R-(ii), S-(i)
    D. P-(iii), Q-(iv), R-(ii), S-(i)

14. Upon executing the following python code, what will be the impulse response (h) of the filter?

```
import numpy as np
a=[1,2]
a.insert(2,0)
h=np.append([a],-np.flip(a[0:2]))
```

    A. $h = \{1,2,0,2,1\}$
    B. $h = \{1,2,0,-2,-1\}$
    C. $h = \{1,2,0,1,2\}$
    D. $h = \{1,2,0,-1,-2\}$

15. The impulse response of a filter is obtained using the following python code. The filter is

```
import numpy as np
a=[1,2]
h=np.append([a],-np.flip(a))
```

    A. Type-I FIR filter
    B. Type-II FIR filter
    C. Type-III FIR filter
    D. Type-IV FIR filter

Bibliography

1. Vijay Madisetti, "The Digital Signal Processing Handbook", CRC Press, 1997.
2. Paulo S. R. Diniz, Eduardo A.B. da Silva and Sergio L. Netto, "Digital Signal Processing: System Analysis and Design", Cambridge University Press, 2010.
3. T.W. Parks, and C.S. Burrus, "Digital Filter Design", John Wiley and Sons, 1987.
4. Wai-Kai Chen, "Passive, Active and Digital Filters", CRC Press, 2006.
5. Robert J. Schilling and Sandra L. Harris, "Fundamentals of Digital Signal Processing using MATLAB", Cengage Learning, 2012.

Chapter 8
Infinite Impulse Response Filter

Learning Objectives

After reading this chapter, the reader is expected to:

- Design and analyse Butterworth filter.
- Design and analyse Chebyshev and inverse Chebyshev filters.
- Design and analyse elliptic filter.
- Implement different mapping techniques to convert analogue filter into an equivalent digital filter.

Roadmap of the Chapter

This chapter starts with the types of infinite impulse response (IIR) filter and discusses the different mapping methods for converting analogue filters into digital filters. Finally, the design of IIR filters is discussed in this chapter. Roadmap of this chapter is illustrated below.

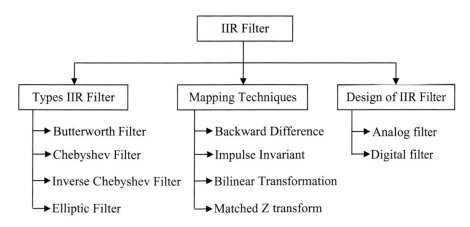

PreLab Questions

1. What is a recursive filter?
2. Examine whether IIR filter is recursive or not. Justify your answer.
3. What is a ripple in the filter's frequency response?
4. List the different types of IIR filters based on the ripples.
5. Why the Butterworth filter is termed as 'maximally flat frequency response' filter?
6. Mention the techniques to convert the analogue filter transfer function into an equivalent digital filter transfer function.
7. Which criterion is important in mapping an analogue filter into an equivalent digital filter?
8. List the steps involved in obtaining the digital filter transfer function from the analogue filter transfer function using the impulse invariant technique.
9. What are the drawbacks of impulse invariant technique?
10. How does the bilinear transformation technique avoid aliasing while performing the mapping process?
11. What is frequency warping with respect to bilinear transformation technique? Suggest a solution to overcome the frequency warping problem in bilinear transformation technique.
12. Tabulate the difference between Butterworth filter, Chebyshev filter, inverse Chebyshev and elliptic filter with respect to (a) ripples in passband and stopband, (b) transition width and (c) order of the filter required to meet the filter specification.
13. Elaborate on the steps involved in the design of a digital IIR filter.

8.1 IIR Filter

In IIR filter, the current output is a function of the current and previous inputs and past outputs. The relationship between the input and output of an IIR filter is given by

$$y[n] = b_0 x[n] + b_1 x[n-1] + \cdots + b_M x[n-M] - a_1 y[n-1] + a_2 y[n-2] \\ + \cdots + a_N y[n-N]$$

$$(8.1)$$

From Eq. (8.1), it is possible to observe that the current output is a function of current and previous inputs and past outputs. Thus, IIR filters are 'recursive filters'. In a recursive filter, the current output depends on both the input and previously calculated outputs. The word 'recursive' literally means 'running back' and refers to previously calculated output values that go back into calculating the current output along with input values.

Upon taking Z-transform of the input-output relation given in Eq. (8.1), the transfer function expression for IIR filter is given by

$$H(z) = \frac{\sum_{k=0}^{M} b_k z^{-k}}{1 + \sum_{k=1}^{N} a_k z^{-k}} \qquad (8.2)$$

The impulse response of the IIR filter can be obtained upon taking inverse Z-transform of the transfer function. As the name suggests, the impulse response is not of finite duration. The impulse response is not guaranteed to be either symmetric or anti-symmetric; hence, it is not possible to obtain linear phase characteristics in the IIR filter. The group delay is not constant in IIR filter. If the pole of the IIR filter lies outside the unit circle, the filter is unstable. This implies that stability is not guaranteed in IIR filter. The main advantage of IIR filter is that it is possible to meet the filter specification with the minimum number of coefficients.

Experiment 8.1 Computation of Impulse Response $h[n]$ of the Recursive Filter
The filter's impulse response $h[n]$ can be obtained from the input and output relation. Let us consider the linear constant coefficient difference equation

$$y[n] + \frac{1}{2}y[n-1] = x[n] \qquad (8.3)$$

where $y[n]$ denotes the output and $x[n]$ represents the input of the equation. From Eq. (8.3), it is possible to observe that the current output ($y[n]$) is a function of current input ($x[n]$) and past output ($y[n-1]$). The impulse response of this filter can be computed by replacing $y[n]$ as $h[n]$ and $x[n]$ as $\delta[n]$. Hence, Eq. (8.3) can be rewritten as

$$h[n] + \frac{1}{2}h[n-1] = \delta[n]$$

The above equation can be rewritten as

$$h[n] = \delta[n] - \frac{1}{2}h[n-1]$$

Assume that the system is initially at rest (i.e. $y[n] = 0$ for $n < 0$) and substituting $n = 0, 1, 2, 3, \ldots$, in the above equation, we get

$$h[0] = \delta[0] = \frac{1}{2}h[0-1] = 1 - 0 = 1$$

$$h[1] = \delta[1] - \frac{1}{2}h[1-1] = 0 - \frac{1}{2}h[0] = -\frac{1}{2} \times 1 = -\frac{1}{2}$$

$$h[2] = \delta[2] - \frac{1}{2}h[2-1] = 0 - \frac{1}{2}h[1] = -\frac{1}{2} \times -\frac{1}{2} = \frac{1}{4}$$

```
#Python code to obtain the impulse response
import numpy as np
from scipy import signal
import matplotlib.pyplot as plt
n=np.arange(-5,11,1)
xn=(n==0) # Define impulse sequence
b=np.array([[1]])# Input coefficients
a=np.array([[1,0.5]])#Output coefficients
y=signal.lfilter(b, a, xn)
plt.stem(n,y),plt.xlabel('n-->'),plt.ylabel('h[n]'),
plt.title('Impulse response')
plt.tight_layout()
```

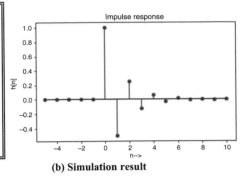

(a) Python code   (b) Simulation result

Fig. 8.1 Python code and its result of impulse response computation. (**a**) Python code. (**b**) Simulation result

$$h[3] = \delta[3] - \frac{1}{2}h[3-1] = 0 - \frac{1}{2}h[2] = -\frac{1}{2} \times \frac{1}{4} = -\frac{1}{8}$$

$$h[n] = \left(-\frac{1}{2}\right)^n u[n]$$

The python code to obtain the impulse response from the input and output coefficients mentioned in the difference equation is shown in Fig. 8.1a, and the corresponding output is shown in Fig. 8.1b.

Inference
1. From Fig. 8.1a, it is possible to observe that the 'scipy' library is used for the filtering, and 'signal.lfilter' is used for the computation of impulse response from the input and output coefficients of the filter equation.
2. From Fig. 8.1b, it is possible to infer that the simulation result is in agreement with the theoretical result.

Task
1. Write a python code to obtain the impulse response of the filter whose difference equation is given by $y[n] - \frac{1}{2}y[n-1] = x[n]$ and comment on the observed output.

8.2 Mapping Techniques in the Design of IIR Filter

Two common approaches in the design of IIR filters are:

Approach 1: Design an analogue IIR filter to meet the given design requirement and convert the analogue filter into an equivalent digital filter using mapping techniques like backward difference method, impulse invariant technique (IIT), bilinear transformation technique (BLT), matched Z-transform technique and so on.

Approach 2: IIR filter is designed using an algorithmic design procedure by solving a set of linear and non-linear equations using a computer or dedicated hardware.

This section focuses on the first approach in which analogue filters are converted into an equivalent digital filter using mapping techniques. While performing the mapping techniques, care must be taken to map stable analogue filter into a stable digital filter.

8.2.1   Backward Difference Method

The relationship between 's' domain and 'z' domain using the backward difference method is given by

$$s = \frac{1 - z^{-1}}{T} \tag{8.4}$$

and

$$z = \frac{1}{1 - sT} \tag{8.5}$$

This section displays the mapping of analogue filter into equivalent digital filter using backward difference method.

Experiment 8.2 Mapping of S-Plane to Z-Plane Using Backward Difference Method

The main objective of this experiment is to prove that stable analogue filter will be mapped to a stable digital filter using backward difference method. The python code, which performs the mapping from S-plane to Z-plane, is given in Fig. 8.2.

Inference

1. From Fig. 8.3a, it is possible to infer that points in the left half of S-plane (i.e. $\sigma \leq 0$) are mapped into inside and on the unit circle in Z-plane.
2. This confirms that the stable analogue filter can be mapped into a stable digital filter using the backward difference mapping technique.
3. On the other hand, from Fig. 8.3b, it is evident that the points on the right side of the S-plane (i.e. $\sigma > 0$) are mapped into outside the unit circle of the Z-plane.

Experiment 8.3 Conversion of Analogue Filter to Digital Filter Using Backward Difference Method

This experiment discusses the conversion of analogue filter to digital filter using backward difference approach. Let us consider the transfer function of the analogue filter is $H(s) = \frac{1}{(s^2+3s+2)}$. Assume a sampling period is 0.1 s. The relationship between 'S-domain' and 'Z-domain' in the backward difference method is given by

```
#Mapping between S to Z plane
import numpy as np
import matplotlib.pyplot as plt
omega1=np.linspace(-15, 15, 20)
omega2 = np.array([1 in range(len(omega1))])
omega2=omega1
sigma1=np.linspace(-10, 0, 20)
sigma2 = np.array([1 in range(len(sigma1))])
sigma2 = sigma1
#np.linspace(-10, 0, 100)
S=[[0 for i in range(len(sigma2))] for j in range(len(omega2))]
Z=S
T=0.1
theta = np.linspace(0, np.pi*2, 500)
circle = np.exp(1j*theta)
plt.subplot(2,1,2),plt.plot(circle.real, circle.imag, 'k'),
for j in range(len(sigma2)):
  for i in range(len(omega2)):
    S[i][j]=complex(sigma1[i],omega1[j])
    Z[i][j]=1/(1-(S[i][j]*(T)))
    plt.subplot(2,1,1),plt.plot(sigma1[i],omega1[j],'bx')
    plt.title('s-plane'), plt.xlabel('$\sigma$'),plt.ylabel('$j\u03A9})$')
    plt.subplot(2,1,2),plt.plot(np.real(Z[i][j]),np.imag(Z[i][j]),'rx')
    plt.title('z-plane'), plt.xlabel('$\sigma$'),plt.ylabel('$j\omega})$')
    plt.tight_layout()
```

Fig. 8.2 Python code for mapping S to Z plane using backward difference method. (a) $\sigma \leq 0$. (b) $\sigma > 0$

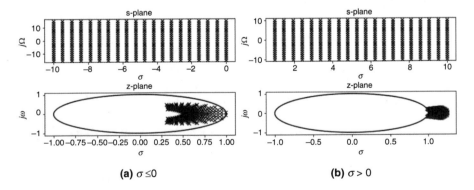

Fig. 8.3 Result of python code given in Fig. 8.2

```
import control as ss
s1 = ss.tf(1, [1,3,2])
print(s1)
yd = s1.sample(0.1, method='backward_diff')
print(yd)
```

```
          1
    -------------
    s^2 + 3 s + 2

0.007576 z^2 - 2.22e-16 z + 2.22e-16
------------------------------------
     z^2 - 1.742 z + 0.7576

dt = 0.1
```

| (a) Python program | (b) Simulation result |

Fig. 8.4 Python code and simulation result for backward difference mapping method. (**a**) Python program. (**b**) Simulation result

$$s = \frac{1 - z^{-1}}{T}$$

Substituting the value of $T = 0.1$ in the above expression, we get

$$s = \frac{1 - z^{-1}}{0.1}$$

The above expression can be simplified as

$$s = 10 - 10z^{-1}$$

Substituting the above relation in the transfer function of the analogue filter, we get

$$H(z) = \frac{1}{\left((10 - 10z^{-1})^2 + 3(10 - 10z^{-1}) + 2\right)}$$

Simplifying the above expression, we get

$$H(z) = \frac{0.007576z^2}{z^2 - 1.742z + 0.7576}$$

The frequency responses of both the analogue and digital filters are computed using python code, shown in Fig. 8.4a, and its corresponding simulation result is shown in Fig. 8.4b. Before execution of the python code, the '*python-control*' package must be installed using the *pip* command, which is given by '*pip install control*'. The new python commands used in this python program are (1) *xx.tf* and (2) *yy.sample*. The result of the python code given in Fig. 8.4a is shown in Fig. 8.4b.

Inference

The following inferences can be made from this experiment:

1. From the simulation result shown in Fig. 8.4b, it is evident that the denominator polynomial function is exactly matched with the theoretical result.
2. The denominator polynomial corresponds to the poles of the system. Thus, analogue filter is mapped into an equivalent digital filter, which is in agreement with the theoretical result.

Experiment 8.4 Mapping Stable Analogue Filter to a Stable Digital Filter Using Backward Difference Method
This experiment deals with mapping a stable analogue filter to a stable digital filter using the backward difference method. The transfer function of a stable second-order filter considered in this example is given by

$$H(s) = \frac{4}{s^2 + 2.82s + 4}$$

This filter is converted into an equivalent digital filter $H(z)$ using backward difference method. The python code to verify the conversion of a second-order stable analogue filter into a stable digital filter is given in Fig. 8.5, and the simulation result is depicted in Fig. 8.6. This python code will work for the second-order filter only. The simulation result of this experiment is shown in Fig. 8.6.

Inference
From Fig. 8.6b, it is possible to observe that the poles of $H(s)$ lie left half of the S-plane, confirming that the analogue filter is stable. Similarly, from this figure, it is possible to know that the poles of $H(z)$ lie inside the unit circle. Hence, the backward difference mapping method preserves the stability criterion during mapping.

8.2.2 Impulse Invariant Technique

In impulse invariant technique, the digital filter is designed by sampling the impulse response of the analogue filter. The pole at $s = s_\mathrm{p}$ is mapped to a pole at $z = e^{s_\mathrm{p}T}$ in the digital filter. Impulse invariant technique performs many-to-one mapping; hence, it suffers from an aliasing problem. Thus, impulse invariant technique is useful if the analogue filter is band-limited. The step followed in impulse invariant technique is given in Fig. 8.7.

The impulse response of the analogue filter is represented by $h(t)$. It is sampled to get $h[nT]$. Upon taking Z-transform of the sampled impulse response, the transfer function of the digital filter is obtained, which is represented as $H(z)$.

Experiment 8.5 Mapping of S-Plane to Z-Plane Using Impulse Invariant Technique
This experiment deals with mapping the S-plane to Z-plane using impulse invariant technique.

Case 1: Mapping the points on the $j\Omega$ axis of the S-plane

```
import numpy as np
import control as ss
from scipy import signal
import matplotlib.pyplot as plt
num=[4]
den=[1,2*np.sqrt(2),4]
T=0.1
fsam=1/T
s1=ss.tf(num,den)
print('H(s) =', s1)
a1=1
b1=-(2+den[1]*T)/(1+(den[1]*T)+den[2]*(T**2))
c1=1/(1+(den[1]*T)+den[2]*(T**2))
num1=[num[0]*c1*(T**2),0,0]
den1=[a1,b1,c1]
yd = s1.sample(T, method='backward_diff')
print('Using Built in function: H(z) =',yd)
s2=ss.tf(num1,den1,T)
print('H(z) =', s2)
ps,zs=ss.pzmap(s1)
# Pole-zero plot
plt.subplot(2,1,1),plt.plot(ps.real, ps.imag, 'kx', ms=10),plt.xlabel('$\sigma$'),
plt.ylabel('$j\Omega$'),plt.title('Pole-zero plot of H(s)'),plt.grid()
z, p, k = signal.tf2zpk(num1,den1)
theta = np.linspace(0, np.pi*2, 500)
circle = np.exp(1j*theta)
plt.subplot(2,1,2),plt.plot(circle.real, circle.imag, 'k--')
plt.plot(p.real, p.imag, 'rx', ms=7.5),plt.xlabel('$\sigma$'),
plt.ylabel('$j\omega$'),plt.title('Pole-zero plot of H(z)'),plt.grid()
plt.tight_layout(),plt.show()
```

Fig. 8.5 Python code for mapping analogue filter to digital filter

The python code for mapping the points on the $j\Omega$ axis onto unit circle is given in Fig. 8.8, and its corresponding output is shown in Fig. 8.9. Figure 8.9 confirms that the points on the $j\Omega$ axis in the S-plane are mapped onto a unit circle in the Z-plane.

Case 2: Mapping the left half of the S-plane

The python code for mapping the points in left half of S-plane are mapped into inside the unit circle is given in Fig. 8.10, and its corresponding output is shown in Fig. 8.11. Figure 8.11 confirms that the points in the left half of S-plane are mapped into within the unit circle of the Z-plane.

Case 3: Mapping right half of S-plane

The points in the right half of S-plane are mapped into outside the unit circle in the Z-plane. The python code for mapping the points in right half of S-plane are mapped into outside the unit circle is given in Fig. 8.12, and its corresponding output is

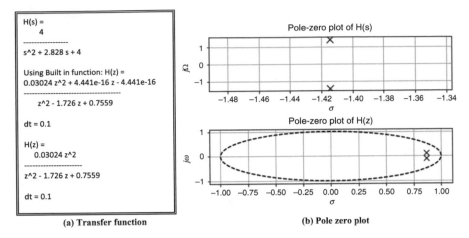

```
H(s) =
       4
-----------------
s^2 + 2.828 s + 4

Using Built in function: H(z) =
0.03024 z^2 + 4.441e-16 z - 4.441e-16
------------------------------------
     z^2 - 1.726 z + 0.7559

dt = 0.1

H(z) =
     0.03024 z^2
----------------------
z^2 - 1.726 z + 0.7559

dt = 0.1
```

(a) Transfer function   **(b) Pole zero plot**

Fig. 8.6 Simulation result. (**a**) Transfer function. (**b**) Pole-zero plot

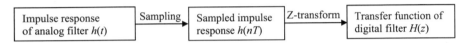

Impulse response of analog filter $h(t)$	Sampling →	Sampled impulse response $h(nT)$	Z-transform →	Transfer function of digital filter $H(z)$

Fig. 8.7 Steps in impulse invariant technique

```
#Mapping between S to Z plane using IIT
import numpy as np
import matplotlib.pyplot as plt
omega1=np.linspace(-15, 15, 50)
sigma1=np.zeros(len(omega1))
S=sigma1+1j*omega1
T=1
theta = np.linspace(0, np.pi*2, 500)
circle = np.exp(1j*theta)
plt.subplot(2,1,2),plt.plot(circle.real, circle.imag, 'k'),
plt.subplot(2,1,1),plt.plot(sigma1,omega1,'bx')
plt.title('S-plane'),plt.xlabel('$\sigma$'),plt.ylabel('$j\Omega$')
z=np.exp(S*T)
plt.subplot(2,1,2),plt.plot(np.real(z),np.imag(z),'rx')
plt.title('Z-plane'),plt.xlabel('$\sigma$'),plt.ylabel('$j\omega})$')
plt.tight_layout()
```

Fig. 8.8 Python code for mapping points on the $j\Omega$ axis onto the unit circle

shown in Fig. 8.13. From Fig. 8.13, it is possible to infer that the points in the right half of the S-plane are mapped outside the unit circle in the Z-plane.

Thus, a stable analogue filter can be mapped to an equivalent stable digital filter using the impulse invariant technique.

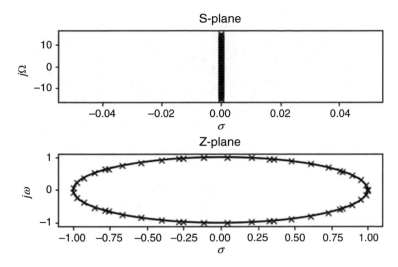

Fig. 8.9 Result of the python code given in Fig. 8.8

```
import numpy as np
import matplotlib.pyplot as plt
omega1=np.linspace(-15, 15, 30)
omega2 = np.array([1 in range(len(omega1))])
omega2=omega1
sigma1=np.linspace(-10, -0.1, 30)
sigma2 = np.array([1 in range(len(sigma1))])
sigma2 = sigma1
S=[[0 for i in range(len(sigma2))] for j in range(len(omega2))]
Z=S
T=0.9
theta = np.linspace(0, np.pi*2, 500)
circle = np.exp(1j*theta)
plt.subplot(2,1,2),plt.plot(circle.real, circle.imag, 'k'),
for j in range(len(sigma2)):
    for i in range(len(omega2)):
        S[i][j]=complex(sigma1[i],omega1[j])
        Z[i][j]=np.exp(S[i][j]*(T))
        plt.subplot(2,1,1),plt.plot(sigma1[i],omega1[j],'bx')
        plt.title('S-plane'), plt.xlabel('$\sigma$'),plt.ylabel('$j\Omega$')
        plt.subplot(2,1,2),plt.plot(np.real(Z[i][j]),np.imag(Z[i][j]),'rx')
        plt.title('Z-plane'), plt.xlabel('$\sigma$'),plt.ylabel('$j\omega})$')
        plt.tight_layout()
```

Fig. 8.10 Python code of mapping left half of S-plane

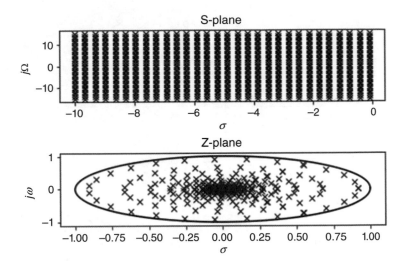

Fig. 8.11 Result of python code given in Fig. 8.10

```
import numpy as np
import matplotlib.pyplot as plt
omega1=np.linspace(-2, 2, 30)
omega2 = np.array([1 in range(len(omega1))])
omega2=omega1
sigma1=np.linspace(0.1, 2, 30)
sigma2 = np.array([1 in range(len(sigma1))])
sigma2 = sigma1
S=[[0 for i in range(len(sigma2))] for j in range(len(omega2))]
Z=S
T=1
theta = np.linspace(0, np.pi*2, 500)
circle = np.exp(1j*theta)
plt.subplot(2,1,2),plt.plot(circle.real, circle.imag, 'k'),
for j in range(len(sigma2)):
    for i in range(len(omega2)):
        S[i][j]=complex(sigma1[i],omega1[j])
        Z[i][j]=np.exp(S[i][j]*(T))
        plt.subplot(2,1,1),plt.plot(sigma1[i],omega1[j],'bx')
        plt.title('S-plane'),plt.xlabel('$\sigma$'),plt.ylabel('$j\Omega$')
        plt.subplot(2,1,2),plt.plot(np.real(Z[i][j]),np.imag(Z[i][j]),'rx')
        plt.title('Z-plane'),plt.xlabel('$\sigma$'),plt.ylabel('$j\omega})$')
        plt.tight_layout()
```

Fig. 8.12 Python code of mapping right half of S-plane

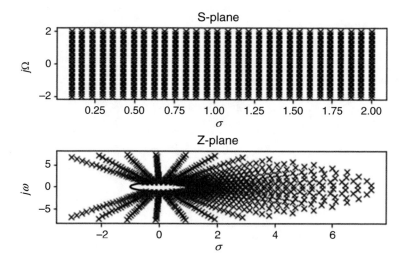

Fig. 8.13 Result of python code given in Fig. 8.12

Experiment 8.6 Many-to-One Mapping in the Impulse Invariant Mapping Technique

This experiment tries to prove that many to one mapping will have happened in the impulse invariant mapping technique. The python code to verify the many-to-one mapping for the impulse invariant technique is given in Fig. 8.14, and its corresponding output is shown in Fig. 8.15. In Fig. 8.15, the symbol '◊' represents points that are mapped between the analogue frequency range from $-\frac{\pi}{T}$ to $\frac{\pi}{T}$ and the digital frequency, whereas the symbol 'x' represents the points that are mapped between analogue frequency in the range $-\frac{3\pi}{T}$ to $\frac{3\pi}{T}$ the digital frequency.

Inferences

The following inferences can be made from this experiment:

1. Overlapping of the symbols in Fig. 8.15 implies that impulse invariant technique is basically a many-to-one mapping.
2. Many-to-one mapping leads to an 'aliasing problem' in impulse invariant technique.
3. Hence, impulse invariant technique is suitable for the design of lowpass and bandpass filters.
4. It is not advisable to use it in highpass and bandstop filters design.

Experiment 8.7 Conversion of Analogue Filter into a Digital Filter Using IIT

This experiment discusses the conversion of an analogue filter to a digital filter using impulse invariant technique. Let us consider the transfer function of the analogue filter $H(s) = \frac{0.5(s+3)}{(s+1)(s+4)}$ to be converted into the transfer function of the digital filter $H(z)$ using the impulse invariant technique. Assume the sampling frequency to be 20 Hz.

```
import numpy as np
import matplotlib.pyplot as plt
T=0.5
omega1=np.linspace(-np.pi/T, np.pi/T, 15)
omega2 = np.array([1 in range(len(omega1))])
omega2=omega1
omega11=np.linspace(-3*(np.pi/T), 3*(np.pi/T), 15)
omega21 = np.array([1 in range(len(omega11))])
omega21=omega11
sigma1=np.linspace(0, 0, 15)
sigma2 = np.array([1 in range(len(sigma1))])
sigma2 = sigma1
S=[[0 for i in range(len(sigma2))] for j in range(len(omega2))]
S1=[[0 for i in range(len(sigma2))] for j in range(len(omega21))]
Z,Z1=S,S1
theta = np.linspace(0, np.pi*2, 500)
circle = np.exp(1j*theta)
plt.subplot(2,1,2),plt.plot(circle.real, circle.imag, 'k'),
for j in range(len(sigma2)):
  for i in range(len(omega2)):
    S[i][j]=complex(sigma1[i],omega1[j])
    Z[i][j]=np.exp(S[i][j]*(T))
    S1[i][j]=complex(sigma1[i],omega11[j])
    Z1[i][j]=np.exp(S1[i][j]*(T))
    plt.subplot(2,1,1),plt.plot(sigma1[i],omega1[j],'bx'),plt.title('S-plane')
    plt.xlabel('$\sigma$'),plt.ylabel('$j\Omega$')
    plt.subplot(2,1,2),plt.plot(np.real(Z[i][j]),np.imag(Z[i][j]),'rd',markersize=12)
    plt.subplot(2,1,2),plt.plot(np.real(Z1[i][j]),np.imag(Z1[i][j]),'kx',markersize=6)
    plt.title('Z-plane'),plt.xlabel('$\sigma$'),plt.ylabel('$j\omega})$')
    plt.tight_layout()
```

Fig. 8.14 Python code for many-to-one mapping of impulse invariance technique

Step 1: By using partial fraction expansion, the given analogue transfer function $H(s)$ can be written as

$$H(s) = \frac{A}{(s+1)} + \frac{B}{(s+4)}$$

From the above expression, it is possible to write as

$$A(s+4) + B(s+1) = 0.5(s+3) \tag{8.6}$$

Substituting $s = -4$ in the above equation, we get $B(-3) = -0.5, B = \frac{1}{6}$.
Substituting $s = -1$ in Eq. (8.6), we get $3A = 1, A = \frac{1}{3}$.
Substituting the values of 'A' and 'B' in the expression of $H(s)$, we get

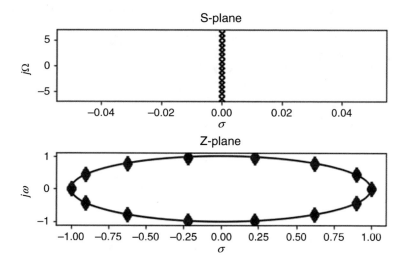

Fig. 8.15 Result of python code given in Fig. 8.14

$$H(s) = \frac{1/3}{(s+1)} + \frac{1/6}{(s+4)}$$

Step 2: Transfer function $H(z)$

The digital transfer function $H(z)$ using impulse invariant technique is given by

$$H(z) = \sum_{i=1}^{N} \frac{A_i}{1 - e^{s_i T} z^{-1}}$$

In this case, $N = 2$, the expression for $H(z)$ is given by

$$H(z) = \sum_{i=1}^{2} \frac{A_i}{1 - e^{s_i T} z^{-1}} = \frac{A_1}{1 - e^{s_1 T} z^{-1}} + \frac{A_2}{1 - e^{s_2 T} z^{-1}}$$

Substituting $A_1 = 1/3$, $A_2 = 1/6$, $s_1 = -1$, $s_2 = -4$ and $T = 1/20 = 0.05$ in the above expression, we get

$$H(z) = \frac{1/3}{(1 - e^{-0.05} z^{-1})} + \frac{1/6}{(1 - e^{-0.2} z^{-1})}$$

Python code to verify the theoretical result with the simulation result is given in Fig. 8.16, and its corresponding output is shown in Fig. 8.17. From Fig. 8.16, the following new tool imported for the simulation, they are (1) *sympy*, (2) *control* and (3) *scipy*. In addition, the following new python commands are used for this simulation are (1) *symbols*—used to define the symbol (z^(−1) and '+'),

Fig. 8.16 Python code for analogue filter to digital filter conversion using impulse invariance method

```
import control as ss
import numpy as np
from sympy import symbols
from scipy import signal
z=symbols('z^-1')
z1=symbols(' + ')
s1 = ss.tf([0.5,1.5], [1,5,4])
print('H(s) ='),print(s1)
T=1/20
y=['+']
yy=signal.residue([0.5,1.5],[1,5,4])
A=[0 for i in range(len(yy[0]))]
S=[0 for i in range(len(yy[1]))]
A[0]=yy[0][0]
S[0]=yy[1][0]
y1=np.append((A[0]/(1-np.exp(S[0]*T)*z)),y)
for i in range(len(yy[0])-1):
    A[i+1]=yy[0][i+1]
    S[i+1]=yy[1][i+1]
    y1=np.append(y1,(A[i+1]/(1-np.exp(S[i+1]*T)*z)))
print('H(z) = '),print(y1)
```

```
H(s) =

 0.5 s + 1.5
-------------
s^2 + 5 s + 4

H(z) =
[0.333333333333333/(1 - 0.951229424500714*z^-1) '+'
 0.166666666666667/(1 - 0.818730753077982*z^-1)]
```

Fig. 8.17 Simulation result of python code given in Fig. 8.16

(2) *residue*—used to obtain the residues and poles. The result of transfer function in S domain and Z domain is displayed in Fig. 8.17.

Note: This python code converts the analogue filter into the digital filter, which has distinct poles.

Inference

From Fig. 8.17, it is possible to observe that the simulation result is on par with the theoretical result.

```
import control as ss
import numpy as np
import matplotlib.pyplot as plt
from sympy import symbols
from scipy import signal
z=symbols('z^-1')
num,den=[1,2],[1,6,5]
s1 = ss.tf(num, den)
print('H(s) =',s1)
T=1/20
zs,ps,ks=signal.tf2zpk(num,den)
r,p,k=signal.residue(num,den)
theta = np.linspace(0, np.pi*2, 500)
zz=np.zeros(len(theta))
plt.subplot(2,1,1),plt.plot(zz, theta, 'k--'),plt.plot(ps.real,ps.imag, 'bx',ms=10)
plt.plot(zs.real,zs.imag, 'go',ms=10),plt.xlabel('$\sigma$'),
plt.ylabel('$j\Omega$'),plt.title('Pole-zero plot H(s)')
y1=np.append((r[0]/(1-np.exp(p[0]*T)*z)),(r[1]/(1-np.exp(p[1]*T)*z)))
print('H(z) = ',y1)
z=np.exp(p*T)
theta = np.linspace(0, np.pi*2, 500)
circle = np.exp(1j*theta)
plt.subplot(2,1,2),plt.plot(circle.real, circle.imag, 'k--'),
plt.plot(z.real,z.imag, 'rx',ms=10),
plt.xlabel('$\sigma$'),plt.ylabel('$j\omega$'),plt.title('Pole-zero plot H(z)')
```

Fig. 8.18 Python code to map poles in *S*-domain to *Z*-domain using IIT

Experiment 8.8 Conversion of a Stable Analogue Filter to a Stable Digital Filter Using the Impulse Invariance Technique (IIT)

This experiment is to verify the stability of the analogue filter to digital filter during conversion using impulse invariant technique. The python code to verify the stability of the analogue filter to digital conversion using IIT is given in Fig. 8.18. Here, we have considered the second-order stable analogue filter with two poles at -5 and -1 and a zero at -2. This stable analogue filter is converted into a digital filter using the impulse invariance method. The simulation result of the python code, which is given in Fig. 8.18, is shown in Fig. 8.19.

Inference

1. From Fig. 8.19, it is possible to observe that the poles of the analogue filter lay left half of *S*-plane; it is evident that the analogue filter is stable.
2. Similarly, the pole-zero plot of $H(z)$ is shown in Fig. 8.19. From the figure, it is possible to observe that all poles are lying inside the unit circle.
3. This implies that the digital filter is also stable. Therefore, the stability is retained while mapping the analogue filter into the digital filter using the impulse invariance method.

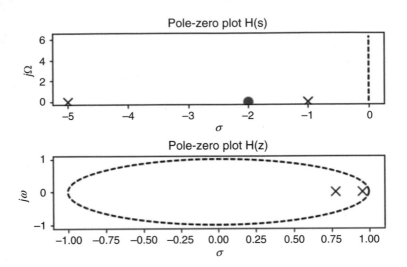

Fig. 8.19 Pole-zero plot

8.2.3 Bilinear Transformation Technique (BLT)

In bilinear transformation technique, the mapping from S-plane to Z-plane is done using the relation

$$s = \frac{2}{T} \frac{1 - z^{-1}}{1 + z^{-1}} \qquad (8.7)$$

Unlike impulse invariant technique (IIT), there is no aliasing problem in bilinear transformation technique. The relationship between the analogue frequency (Ω) and digital frequency (ω) in BLT is given in Eq. (8.5)

$$\Omega = \frac{2}{T} \tan\left(\frac{\omega}{2}\right) \qquad (8.8)$$

The above equation can also be expressed as

$$\omega = 2 \tan^{-1}\left(\frac{\Omega T}{2}\right) \qquad (8.9)$$

From Eqs. (8.8) and (8.9), it is possible to infer that there exists a non-linear relationship between analogue and digital frequency in bilinear transformation technique, which is termed as *warping*.

```
import numpy as np
import matplotlib.pyplot as plt
T=[1,0.5,0.25,0.1]
omega=np.linspace(-15,15,20)
for i in range(len(T)):
    domega=2*np.arctan(omega*(T[i]/2))
    plt.plot(omega,domega,'-*'),plt.title('BLT Mapping')
    plt.xlabel('$\Omega$ rad/sec'),plt.ylabel('$\omega$ rad/sample')
plt.legend(['$T={}$ Sec'.format(T[0]),'$T={}$ Sec'.format(T[1]),'$T={}$
Sec'.format(T[2]),'$T={}$ Sec'.format(T[3])],loc=0)
plt.tight_layout()
```

Fig. 8.20 Python code to display the relationship between analogue and digital frequency

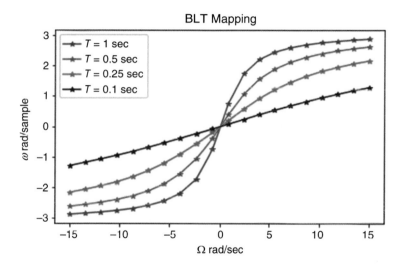

Fig. 8.21 Relationship between analogue and digital frequencies

Experiment 8.9 Display the Relationship Between Analogue Frequency and Digital Frequency Using Bilinear Transformation (BLT) Technique

This experiment displays the relationship between analogue frequency and digital frequency using BLT. The python code given in Fig. 8.20 gives the relationship between analogue and digital frequency using BLT. The simulation result of the python code given in Fig. 8.20 is shown in Fig. 8.21. The sampling intervals are considered as [1, 0.5, 0.25, 0.1]. The relationship between analogue and digital frequency for different sampling intervals is shown in Fig. 8.21.

Inferences

1. From this figure, it is possible to observe that with the value of $T > 0.1$, the mapping between analogue and digital frequency is linear at the low-frequency range and non-linear at the high-frequency range.

```
import numpy as np
import matplotlib.pyplot as plt
omega1=np.linspace(-15, 15, 20)
omega2 = np.array([1 in range(len(omega1))])
omega2=omega1
#sigma1=np.linspace(0, 0, 20)# for sigma=0
#sigma1=np.linspace(-15, -0.5, 20)# for sigma < 0
sigma1=np.linspace(0.5, 5, 20)# for sigma > 0
sigma2 = np.array([1 in range(len(sigma1))])
sigma2 = sigma1
S=[[0 for i in range(len(sigma2))] for j in range(len(omega2))]
Z=S
T=0.9
theta = np.linspace(0, np.pi*2, 500)
circle = np.exp(1j*theta)
plt.subplot(2,1,2),plt.plot(circle.real, circle.imag, 'k'),
for j in range(len(sigma2)):
    for i in range(len(omega2)):
        S[i][j]=complex(sigma1[i],omega1[j])
        Z[i][j]=(1+(S[i][j]*(T/2)))/(1-(S[i][j]*(T/2)))
        plt.subplot(2,1,1),plt.plot(sigma1[i],omega1[j],'bx')
        plt.title('S-plane'), plt.xlabel('$\sigma$'),plt.ylabel('$j\Omega$')
        plt.subplot(2,1,2), plt.plot(np.real(Z[i][j]),np.imag(Z[i][j]),'rx')
        plt.title('Z-plane'), plt.xlabel('$\sigma$'),plt.ylabel('$j\omega})$')
        plt.tight_layout()
```

Fig. 8.22 Python code for frequency mapping using BLT

2. This non-linear relationship is termed as 'warping'.
3. In order to overcome this warping effect, prewarping is necessary for bilinear transformation technique.
4. When $T = 0.1$, the mapping is almost linear for both low- and high-frequency ranges.

Experiment 8.10 Illustration of BLT Preserves Stability Criterion

The objective of this experiment is to prove that stable analogue filter will be mapped to a stable digital filter using BLT. The stability conditions are verified with the python code. The python code maps S-plane into an equivalent Z-plane using BLT. The S-plane is represented as $s = \sigma + j\Omega$, whereas the Z-plane is represented as $z = re^{j\omega}$. Three different cases considered in this example are $\sigma = 0$, $\sigma < 0$ and $\sigma > 0$. The python code to verify the stability preservation of BLT is given in Fig. 8.22. Varying the values of $\sigma = 0$, $\sigma < 0$ and $\sigma > 0$ in the python code, the mapping results are shown in Fig. 8.23.

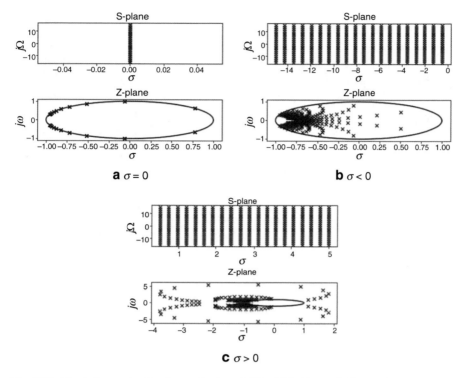

Fig. 8.23 Simulation result of python code given in Fig. 8.22. (**a**) $\sigma = 0$. (**b**) $\sigma < 0$ (**c**) $\sigma > 0$

Inferences

1. When $\sigma = 0$ indicate $j\Omega$ in the S-plane. From Fig. 8.23a, it is possible to observe that the entire $j\Omega$ axis is mapped to points on the unit circle in the Z-plane.
2. When $\sigma < 0$ represent the left half of S-plane. All the points in the LHS of S-plane are mapped into within the unit circle of the Z-plane.
3. When $\sigma > 0$ are the points in the RHS of the S-plane. These points are mapped into the outside of the unit circle in the Z-plane.
4. An analogue filter is stable if the poles lie in the left half of S-plane. A digital filter is stable if the poles lie within the unit circle. The BLT ensures that the stable analogue filter will be mapped as a stable digital filter.

8.2.4 Matched Z-Transform Technique

In matched Z-transform technique, the poles and zeros of the analogue filter are mapped to Z-plane using the relation $z = e^{sT}$. The matched Z-transform gives the same pole location as impulse invariant technique but a different zero location.

Experiment 8.11 Conversion of Analogue Filter to Equivalent Digital Filter Using Matched Z-Transform Method

This experiment discusses the conversion of an analogue filter into its equivalent digital filter using matched Z transform. Let us consider a second-order analogue filter transfer function is $H(s) = \frac{(s+3)}{(s+1)(s+4)}$ to be converted into digital filter using matched Z-transform. Assume $T = 0.1$ s. The matched Z transform (MZT) method directly maps the poles and zeros of an analogue filter into the poles and zeros of the digital filter. The transfer function of the analogue filter is given as

$$H(s) = \frac{\prod_{k=1}^{M}(s - z_k)}{\prod_{k=1}^{N}(s - p_k)} \tag{8.10}$$

The transfer function of the equivalent digital filter is computed by replacing the term $(s - z_k)$ with $(1 - e^{z_k T} z^{-1})$. The transfer function of the equivalent digital filter can be written as

$$H(z) = \frac{\prod_{k=1}^{M}(1 - e^{z_k T} z^{-1})}{\prod_{k=1}^{N}(1 - e^{p_k T} z^{-1})} \tag{8.11}$$

where T is the sampling period. The transfer function of the analogue filter considered in this example is given by

$$H(s) = \frac{(s + 3)}{(s + 1)(s + 4)}$$

From the above equation, the zeros and poles are computed as $z_1 = -3$, and $p_1 = -1, p_2 = -4$, respectively.

Substituting the values of zeros, poles and $T = 0.1$ in Eq. (8.11), we get

$$H(z) = \frac{(1 - e^{-0.3} z^{-1})}{(1 - e^{-0.1} z^{-1})(1 - e^{-0.4} z^{-1})}$$

Simplifying the above equation, we get

$$H(z) = \frac{(1 - 0.7408 z^{-1})}{(1 - 0.9048 z^{-1})(1 - 0.6703 z^{-1})}$$

Further simplifying the above equation, we get

$$H(z) = \frac{z(z - 0.7408)}{(z^2 - 1.5751 z + 0.6065)}$$

```
import numpy as np
from scipy import signal
import control as ss
T=0.1
z=[-3]
p =[-1, -4]  # analog poles
b_s=signal.convolve([1],[1,-z[0]])
a_s=signal.convolve([1,-p[0]],[1,-p[1]])
Hs=ss.tf(b_s,a_s)
print('Transfer function H(s) =',Hs)
b_z=signal.convolve([1,0], [1, -(np.exp(z[0]*T))])
a_z = signal.convolve([1, -(np.exp(p[0]*T))], [1, -(np.exp(p[1]*T))])
Hz=ss.tf(b_z,a_z,T)
print('Transfer function H(z) =',Hz)
```

Fig. 8.24 Python code for Experiment 8.11

Fig. 8.25 Result of the python code is given in Fig. 8.24

```
Transfer function H(s) =
   s + 3
 -------------
s^2 + 5 s + 4

Transfer function H(z) =
   z^2 - 0.7408 z
 ----------------------
z^2 - 1.575 z + 0.6065

dt = 0.1
```

Python code to verify this experiment is given in Fig. 8.24, and the corresponding simulation result is shown in Fig. 8.25.

Inferences

1. Figure 8.24 shows that the *signal.convole* command is used here to compute the product of two zeros or two poles term.
2. This python code indicates that direct mapping has existed between analogue filter poles to digital filter poles and analogue filter zeros to digital filter zeros.
3. From Fig. 8.25, it is possible to observe that the transfer function of the digital filter is on par with the theoretical result.

8.3   Analog Frequency Transformation

A normalized lowpass filter can be transformed into a desired lowpass, highpass, bandpass or bandstop filter by frequency transformation technique. Table 8.1 summarizes different analogue frequency transformations.

Experiment 8.12 Analogue Frequency Transformation
The objective of this experiment is to convert a normalized first-order lowpass Butterworth filter into an equivalent bandpass and band reject filter with the lower and upper cut-off frequencies of 3 and 5 rad/s, respectively, using the frequency transformation method. A normalized first-order lowpass Butterworth filter transfer function is given by $H(s) = \frac{1}{(s+1)}$. Convert this filter into an equivalent (1) bandpass filter and (2) band reject filter with the lower and upper cut-off frequencies of 3 and 5 rad/s, respectively, using the frequency transformation method.

Step 1: Converting the prototype filter into its equivalent bandpass filter

The prototype filter can be converted into its equivalent band pass filter by using the frequency transformation $s \rightarrow \frac{s^2 + \Omega_u \Omega_l}{s(\Omega_u - \Omega_l)}$. In this experiment, the value of $\Omega_u = 5$ rad/s and $\Omega_l = 3$ rad/s. Hence, the s can be computed as

$$s \rightarrow \frac{s^2 + 15}{s(2)} = \frac{s^2 + 15}{2s}$$

Now replacing s in the normalized lowpass filter transfer function by $s \rightarrow \frac{s^2 + 15}{2s}$, we get

$$H_1(s) = \frac{1}{\left(\frac{s^2 + 15}{2s} + 1\right)}$$

Simplifying the above expression, we get

$$H_1(s) = \frac{2s}{s^2 + 2s + 15}$$

Step 2: Obtaining the transfer function of the band reject filter

Table 8.1 Analogue frequency transformation

S. No.	Type of transformations	Transformation
1	Lowpass to lowpass	$s \rightarrow \frac{s}{\Omega_c}$
2	Lowpass to highpass	$s \rightarrow \frac{\Omega_c}{s}$
3	Lowpass to bandpass	$s \rightarrow \frac{s^2 + \Omega_u \Omega_l}{s(\Omega_u - \Omega_l)}$
4	Lowpass to bandstop	$s \rightarrow \frac{s(\Omega_u - \Omega_l)}{s^2 + \Omega_u \Omega_l}$

```
import numpy as np
import matplotlib.pyplot as plt
import control as ss
num,den=[1],[1,1]
num1,den1=[2, 0],[1,2,15]
num2,den2=[1,0,15],[1,2,15]
s = ss.tf(num, den)
print(s)
s1 = ss.tf(num1, den1)
print(s1)
s2 = ss.tf(num2, den2)
print(s2)
omega1=np.linspace(0, 15, 100)
mag, phase, omega1=ss.freqresp(s, omega1)
mag1, phase1, omega1=ss.freqresp(s1, omega1)
mag2, phase2, omega1=ss.freqresp(s2, omega1)
plt.subplot(2,1,1),plt.plot(omega1,mag,'-.',omega1,mag1,'--',omega1,mag2,'-')
plt.title('Magnitude response'),plt.legend(['Prototype','BPF','BRF'],loc=0)
plt.xlabel('$\Omega$ in rad/sec'),plt.ylabel('|$H(j\Omega)$|')
plt.subplot(2,1,2),plt.plot(omega1,phase,'-.',omega1,phase1,'--',omega1,phase2,'-')
plt.title('Phase response'),plt.legend(['Prototype','BPF','BRF'],loc=0)
plt.xlabel('$\Omega$ in rad/sec'),plt.ylabel('∠$H(j\Omega)$)')
plt.tight_layout()
```

Fig. 8.26 Python code to obtain the magnitude and phase response

The given prototype filter can be converted into a band reject filter using the frequency transformation $s \rightarrow \frac{s(\Omega_u - \Omega_l)}{s^2 + \Omega_u \Omega_l}$. In this problem, the value of $\Omega_u = 5$ rad/s and $\Omega_l = 3$ rad/s. Hence, the equivalent s is calculated as

$$s \rightarrow \frac{s(\Omega_u - \Omega_l)}{s^2 + \Omega_u \Omega_l} = \frac{2s}{s^2 + 15}$$

Now replacing s in the normalized lowpass filter transfer function by $s \rightarrow \frac{2s}{s^2 + 15}$, we get

$$H_2(s) = \frac{1}{\left(\frac{2s}{s^2 + 15} + 1\right)}$$

Simplifying the above expression, we get

$$H_2(s) = \frac{s^2 + 15}{s^2 + 2s + 15}$$

The following python code helps us to understand that the prototype filter $H(s)$ is converted into bandpass filter $H_1(s)$ and band reject filter $H_2(s)$. The python code is given in Fig. 8.26, and the corresponding simulation result is shown in Fig. 8.27.

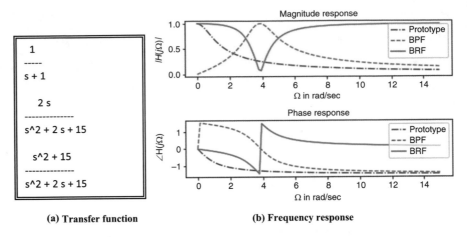

(a) Transfer function   (b) Frequency response

Fig. 8.27 Simulation result of the python code given in Fig. 8.26. (a) Transfer function. (b) Frequency response

Inferences

1. The transfer function of prototype normalized lowpass filter, desired bandpass filter and band reject filter are shown in Fig. 8.27a, which is in agreement with the theoretical result.
2. From the magnitude response shown in Fig. 8.27b, it is possible to observe that the filter response is in agreement with the desired result.
3. From the phase response, it is possible to observe that the phase response is non-linear.

8.4   Butterworth Filter

The squared magnitude response of Nth order Butterworth lowpass filter is given by

$$\left|H_N\left(e^{j\Omega}\right)\right|^2 = \frac{1}{1 + \left(\frac{\Omega}{\Omega_c}\right)^{2N}} \qquad (8.12)$$

where Ω_c is cut-off frequency and N denotes the order of the filter. Butterworth filter exhibits maximally flat response in both passband and stopband. Therefore, these filters are called maximally flat filters or flat-to-flat filters. The salient features of lowpass Butterworth filter are:

1. The magnitude response is a monotonically decreasing function of frequency.
2. The maximum gain occurs at $\Omega = 0$.

```
import numpy as np
import matplotlib.pyplot as plt
omega=np.linspace(0, 5, 100)
omegac=np.array([2])
N=[1,2,3,4,5]
for i in range(len(N)):
    H=1/(1+(omega/omegac)**(2*N[i]))
    plt.plot(omega,np.abs(H)),plt.title('Squared Magnitude Response')
    plt.xlabel('$\Omega$ in rad/sec'),
    plt.ylabel('|$H(j\Omega)|^2$')
plt.legend(['N = 1','N =2','N = 3','N = 4','N = 5'])
plt.tight_layout()
```

Fig. 8.28 Python code for squared magnitude response of Butterworth lowpass filter

3. The first $(2N - 1)$ derivatives of an Nth order lowpass Butterworth filter are zero at $\Omega = 0$. Hence, Butterworth filters are termed as maximally flat magnitude filters.
4. The high-frequency roll off of a Nth order Butterworth filter is $20N$ dB/decade.

Experiment 8.13 Magnitude Response of Butterworth Filter for Different Filter Order

The objective of this experiment is to obtain the squared magnitude response of the Butterworth lowpass filter for different orders. The python code to generate a squared magnitude response of different orders of Butterworth lowpass filter is given in Fig. 8.28, and the corresponding result is shown in Fig. 8.29.

Inferences

1. The squared magnitude response of Butterworth lowpass filter with different orders is shown in Fig. 8.29. Here, the cut-off frequency is chosen as 2 rad/s, and the orders are varied from 1 to 5.
2. From Fig. 8.29, it is possible to observe that the transition width decreases when the order of the filter (N) increases. Also, there is no ripple in the passband and stopband.
3. The squared magnitude response is a monotonically decreasing function of frequency.

Experiment 8.14 Computing the Order of Butterworth Filter

The objective of this experiment is to compute the order of Butterworth filter for the given filter specifications. The given specifications are as follows: (1) The passband gain at 2 rad/s is 0 dB. (2) The stopband attenuation at 5 rad/s is 30 dB. (3) Passband cut-off frequency $\Omega_p = 2$ rad/s. (4) Stopband cut-off frequency is $\Omega_s = 5$ rad/s.

The order of Butterworth filter can be computed using the following formula

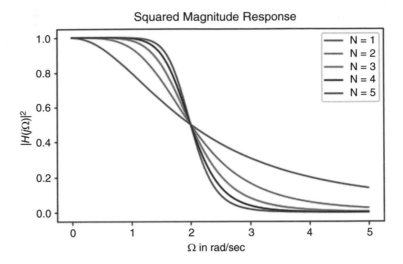

Fig. 8.29 Squared magnitude response of Butterworth lowpass filter

$$N = \left\lceil \frac{\log\left(\frac{10^{-\frac{A_p}{10}}-1}{10^{-\frac{A_s}{10}}-1}\right)}{2\log\left(\frac{\Omega_p}{\Omega_s}\right)} \right\rceil \tag{8.13}$$

The symbol $\lceil . \rceil$ denotes ceiling operator. Substituting the values given in the specifications in the above equation, we get

$$N = \left\lceil \frac{\log\left(\frac{10^{-\frac{3}{10}}-1}{10^{-\frac{30}{10}}-1}\right)}{2\log\left(\frac{2}{5}\right)} \right\rceil$$

The order of the filter is calculated as

$$N = \left\lceil \frac{\log(0.001)}{2\times(-0.3979)} \right\rceil = \left\lceil \frac{-3}{-0.7958} \right\rceil = \lceil 3.77 \rceil = 4$$

The python code to compute the order of the Butterworth filter is given in Fig. 8.30.

Inference
After executing the python code is given in Fig. 8.30. The result is *Order of the filter (N) = 4.0*. It is in agreement with the theoretical result.

Fig. 8.30 Python code for Butterworth filter order calculation

```
import numpy as np
Ap, As=-3, -30
omega_p,omega_s=2, 5
num1=10**(-Ap/10)-1
num2=10**(-As/10)-1
num=np.log(num1/num2)
den=2*np.log(omega_p/omega_s)
N=np.ceil(num/den)
print('Order of the filter (N) =')
```

Fig. 8.31 Python code to generate transfer function of Butterworth filter

```
import control as ss
import numpy as np
from scipy import signal
N=np.array([1,2,3,4,5,6,7,8],dtype=int)
for i in range(len(N)):
    b,a=signal.butter(N[i],1,'low', analog=True)
    s1=ss.tf(b,a)
    print('H(s) for N ={} '.format(N[i]))
    print(s1)
```

Task

1. Change the value of A_p and A_s in the python code, given in Fig. 8.30, execute the program and observe the result of the order of the filter.

Experiment 8.15 Transfer Function of Normalized Butterworth Filter of Different Orders

The objective of this experiment is to obtain the transfer function of normalized Butterworth lowpass filter with different orders of the filter. The python code to obtain the transfer function of Butterworth filter is given in Fig. 8.31. The simulation result is given in Fig. 8.32.

Inference

1. Using the python code given in Fig. 8.31, Nth order normalized Butterworth lowpass filter transfer function can be obtained.
2. The simulation result of the python code given in Fig. 8.31 is shown in Fig. 8.32, and the maximum value of the order N is chosen as 8.
3. From Fig. 8.32, it possible to see that the transfer function of the normalized Butterworth lowpass filter from first order to eighth order.

Experiment 8.16 Design of Butterworth Filter for a Given Specifications

The aim of this experiment is to design a Butterworth filter using bilinear transformation technique (BLT) that has a passband gain of 0 to -3 dB cut-off frequency of 2 kHz, and attenuation of at least 20 dB for frequencies greater than 5 kHz. Assume the sampling frequency to be 20 kHz and sampling period $T = 1$ s.

```
H(s) for N =1
  1
-----
s + 1
H(s) for N =2
      1
-----------------
s^2 + 1.414 s + 1
H(s) for N =3
      1
--------------------
s^3 + 2 s^2 + 2 s + 1
H(s) for N =4
          1
-----------------------------------------
s^4 + 2.613 s^3 + 3.414 s^2 + 2.613 s + 1
H(s) for N =5
           1
-------------------------------------------------------
s^5 + 3.236 s^4 + 5.236 s^3 + 5.236 s^2 + 3.236 s + 1
H(s) for N =6
             1
-----------------------------------------------------------------
s^6 + 3.864 s^5 + 7.464 s^4 + 9.142 s^3 + 7.464 s^2 + 3.864 s + 1
H(s) for N =7
               1
---------------------------------------------------------------------------
s^7 + 4.494 s^6 + 10.1 s^5 + 14.59 s^4 + 14.59 s^3 + 10.1 s^2 + 4.494 s + 1
H(s) for N =8
                 1
-------------------------------------------------------------------------------------------
s^8 + 5.126 s^7 + 13.14 s^6 + 21.85 s^5 + 25.69 s^4 + 21.85 s^3 + 13.14 s^2 + 5.126 s + 1
```

Fig. 8.32 Simulation result of the python code given in Fig. 8.31

The specifications given in this experiment as follows:

1. Sampling frequency $f_{samp} = 20$ kHz
2. Pass band gain $A_p = -3$ dB
3. Stop band attenuation $A_s = -20$ dB
4. Pass band frequency $f_p = 2$ kHz
5. Stop band frequency $f_s = 5$ kHz

Converting the pass band and stop band frequencies from Hz to radians per sample, it is necessary to compute the ω_p and ω_s from the frequency f_p and f_s, which are given below

$$\omega_{\mathrm{p}} = 2\pi \left(\frac{f_{\mathrm{p}}}{f_{\mathrm{samp}}} \right)$$

Substituting the values of f_{p} and f_{samp} in the above expression, we get

$$\omega_{\mathrm{p}} = 2\pi \left(\frac{2}{20} \right) = 0.2\pi$$

Now the expression for ω_{s} is given by

$$\omega_{\mathrm{s}} = 2\pi \left(\frac{f_{\mathrm{s}}}{f_{\mathrm{samp}}} \right)$$

Substituting the values of f_{s} and f_{samp} in the above expression, we get

$$\omega_{\mathrm{s}} = 2\pi \left(\frac{5}{20} \right) = 0.5\pi$$

Step 1: Prewarping

The prewarping process must be done for the bilinear transformation technique to preserve one-to-one mapping in the frequency transformation from digital to analogue.

The corresponding analogue frequencies Ω_{p} and Ω_{s} are obtained using prewarping technique which is given by

$$\Omega_{\mathrm{p}} = \frac{2}{T} \tan \left(\frac{\omega_{\mathrm{p}}}{2} \right)$$

Substituting the value of $\omega_{\mathrm{p}} = 0.2\pi$ and $T - 1$ in the above expression, we get

$$\Omega_{\mathrm{p}} = 2 \tan \left(\frac{0.2\pi}{2} \right) = 0.650$$

Similarly, the value of Ω_{s} is computed as

$$\Omega_{\mathrm{s}} = 2 \tan \left(\frac{0.5\pi}{2} \right) = 2$$

Step 2: To determine the order of the filter

The expression for the order of Butterworth filter is given by

$$N = \left\lceil \frac{\log\left(\frac{10^{-\frac{A_p}{10}} - 1}{10^{-\frac{A_s}{10}} - 1}\right)}{2\log\left(\frac{\Omega_p}{\Omega_s}\right)} \right\rceil$$

Substituting the value of $A_p = -3$ dB, $A_s = -20$ dB, $\Omega_p = 0.650$ and $\Omega_s = 2$ in the above expression, we get

$$N = \left\lceil \frac{\log\left(\frac{10^{\left(\frac{3}{10}\right)} - 1}{10^{\left(\frac{20}{10}\right)} - 1}\right)}{2\log\left(\frac{0.650}{2}\right)} \right\rceil$$

Simplifying the above expression, we get

$$N = \lceil 2.0463 \rceil \approx 3$$

The order of the filter is calculated as 3.

Step 3: To determine the cut-off frequency

The expression for cut-off frequency is given by

$$\Omega_c = \frac{\Omega_p}{\left\{10^{-\frac{A_p}{10}} - 1\right\}^{\frac{1}{2N}}}$$

Substituting $A_p = -3$ dB, $\Omega_p = 0.650$ and $N = 3$ in the above expression, we get

$$\Omega_c = 0.650\left(10^{\frac{3}{10}} - 1\right)^{-\frac{1}{6}}$$

Simplifying the above expression, the value of Ω_c is computed as

$$\Omega_c = 0.650$$

Step 4: Transfer function of normalized lowpass filter

The transfer function of normalized lowpass filter for $N = 3$ is given using the Butterworth polynomial as

$$H_{lp}(s) = \frac{1}{B(s)}$$

The Butterworth polynomial $B(s)$ for $N = 3$ is $(s + 1)(s^2 + s + 1)$. Substituting this in the above expression, we get

$$H_{\text{lp}}(s) = \frac{1}{(s+1)(s^2+s+1)}$$

Step 5: Converting normalized lowpass filter to the desired lowpass filter using frequency transformation

$$H(s) = H_{\text{lp}}(s)\big|_{s=\frac{s}{\Omega_c}}$$

$$H(s) = \frac{1}{(s+1)(s^2+s+1)}\bigg|_{s=\frac{s}{\Omega_c}}$$

Substituting the value of $\Omega_c = 0.650$ (from Step 3), the above expression is given by

$$H(s) = \frac{1}{\left(\frac{s}{0.650}+1\right)\left(\left(\frac{s}{0.650}\right)^2+\frac{s}{0.650}+1\right)}$$

Simplifying the above expression, we get

$$H(s) = \frac{0.2751}{s^3+1.301s^2+0.8459s+0.2751}$$

Step 6: Converting the analogue filter into an equivalent digital filter
The digital equivalent of the analogue filter using BLT Technique is given by

$$H(z) = H(s)\big|_{s=\frac{2z-1}{Tz+1}}$$

Substituting the expression for $H(s)$ from Step 5, we get

$$H(z) = \frac{0.2751}{s^3+1.301s^2+0.8459s+0.2751}\bigg|_{s=\frac{2z-1}{Tz+1}}$$

Substituting $T = 1$ s, the above expression can be written as

$$H(z) = \frac{0.2751}{\left(2\frac{z-1}{z+1}\right)^3+1.301\left(2\frac{z-1}{z+1}\right)^2+0.8459\left(2\frac{z-1}{z+1}\right)+0.2751}$$

Simplifying the above expression, we get

$$H(z) = \frac{0.0181z^3+0.0544z^2+0.0544z+0.0181}{z^3-1.759z^2+1.182z-0.2778}$$

```
import numpy as np
#import matplotlib.pyplot as plt
from scipy import signal
import control as ss
# Specifications of Filter
fsam=20000 # Sampling frequency
fp=2000 # Pass band frequency
fs=5000 # Stop abnd frequency
Ap, As, Td=3,20, 1
wp=2*np.pi*(fp/fsam) # pass band freq in radian per sample
ws=2*np.pi*(fs/fsam) # Stop band freq in radian per sample
# prewarping process
omega_p=(2/Td)*np.tan(wp/2)
omega_s=(2/Td)*np.tan(ws/2)
#Computation of order and normalized cut-off frequency
N, omega_c=signal.buttord(omega_p,omega_s,Ap,As,analog=True)
print('Order of the Filter N =', N)
print('Cut-off frequency= {:.4f} rad/s '. format(omega_c))
# Computation of H(s)
b, a=signal.butter(N,omega_c,'low', analog=True)
s1 = ss.tf(b, a)
print('Transfer function H(s)=',s1)
bz, az=signal.bilinear(b, a, Td)
z1 = ss.tf(bz,az,Td)
print('Transfer function H(z)=',z1)
```

Fig. 8.33 Python code to verify the result of Experiment 8.16

Inference
1. The theoretical result is verified with the python code, which is shown in Fig. 8.33. The built-in function *signal.buttord* helps to obtain the order and cut-off frequency of the analogue filter.
2. The numerator and denominator coefficients of the analogue filter are computed using *signal.butter* built-in function.
3. The analogue filter coefficients are converted into digital filter coefficients using bilinear transformation, the built-in function used for bilinear transformation is *signal.bilinear*.
4. After executing the python code, which is given in Fig. 8.33, the simulation result is shown in Fig. 8.34. From this figure, it is possible to observe that the simulation result is on par with the theoretical result.

Task
1. Reduce the gap between passband and stopband cut-off frequencies and observe the order of the filter.

```
Order of the Filter N = 3

Cut-off frequency= 0.6504 rad/s

Transfer function H(s)=

       0.2751

-----------------------------------

s^3 + 1.301 s^2 + 0.8459 s + 0.2751

Transfer function H(z)=

0.01813 z^3 + 0.0544 z^2 + 0.0544 z + 0.01813

---------------------------------------------

   z^3 - 1.759 z^2 + 1.182 z - 0.2778
```

Fig. 8.34 Simulation result of the python code is given in Fig. 8.33

Experiment 8.17 Design of Butterworth Filter for Given Specifications

The objective of this experiment is to obtain the Butterworth lowpass filter coefficients, transfer function and frequency response for the following filter specifications: (1) order of the filter $(N) = 2$, sampling frequency $(F_s) - 8$ kHz and (2) cut-off frequency $(f_c) = 2$ kHz. Use BLT method for transformation. The python code for this experiment is given in Fig. 8.35, and its corresponding output is shown in Figs. 8.36 and 8.37. Figure 8.37 gives the magnitude and phase responses of the Butterworth filter.

Inference

1. From the magnitude response, it is possible to observe that the filter behaves like a lowpass filter.
2. The gain drops beyond 2000 Hz, which is in agreement with the filter specification.
3. It is also possible to observe that the phase response is not linear.

Experiment 8.18 Frequency Transformation

The objective of this experiment is to convert the normalized Butterworth analogue filter into desired Butterworth analogue filter using frequency transformation. Using frequency transformation technique, the normalized lowpass filter is converted into desired lowpass, highpass, bandpass and band reject filter. The filter considered for frequency transformation is a second-order normalized lowpass filter. This filter is converted to a desired lowpass and highpass filter for a cut-off frequency of 2 rad/s. For the bandpass and band reject filter, the lower and upper cut-off frequency is considered as 3 and 5 rad/s, respectively. The python code for this experiment is given in Fig. 8.38. The corresponding result is shown in Figs. 8.39 and 8.40.

```
import numpy as np
import matplotlib.pyplot as plt
from scipy import signal
import control as ss
# Specifications of Filter
fsam=8000 # Sampling frequency in Hz
fc=2000 # cut off frequency in Hz
N, T = 2, 1/fsam
wc1=2*np.pi*fc # Cut off frequency in rad/sec
print('Cut-off frequency (in rad/sec)=', wc1)
wc = (2/T)*np.tan(wc1*T/2) # Prewarp the analog frequency
# Design analog Butterworth filter using signal.butter function
b, a = signal.butter(N, wc, 'low', analog='True')
s1 = ss.tf(b,a)
print('Transfer function H(s)=',s1)
# Perform bilinear Transformation
bz, az = signal.bilinear(b, a, fs=fsam)
# Print numerator and denomerator coefficients of the filter
print('Numerator Coefficients:', bz),print('Denominator Coefficients:', az)
z1 = ss.tf(bz,az,T)
print('Transfer function H(z)=',z1)
# Compute frequency response of the filter using signal.freqz function
wz, hz = signal.freqz(bz, az, 512)
fig = plt.figure(figsize=(10, 8))
Mag = 20*np.log10(abs(hz)) # Calculate Magnitude in dB
Freq = wz*fsam/(2*np.pi) # Calculate frequency in Hz
# Plot Magnitude response
sub1 = plt.subplot(2, 1, 1),sub1.plot(Freq, Mag, 'r', linewidth=2),sub1.axis([1, fsam/2, -60, 5])
sub1.set_title('Magnitude Response', fontsize=15),
sub1.set_xlabel('Frequency [Hz]', fontsize=15),sub1.set_ylabel('Magnitude [dB]', fontsize=15)
sub1.grid()
# Plot phase angle
sub2 = plt.subplot(2, 1, 2)
Phase = np.unwrap(np.angle(hz))*180/np.pi # Calculate phase angle in degree from hz
sub2.plot(Freq, Phase, 'g', linewidth=2),sub2.set_ylabel('Phase (degree)', fontsize=15)
sub2.set_xlabel(r'Frequency (Hz)', fontsize=15),sub2.set_title(r'Phase response', fontsize=15)
sub2.grid(),plt.subplots_adjust(hspace=0.5),fig.tight_layout(),plt.show()
```

Fig. 8.35 Python code for Experiment 8.17

From Fig. 8.39, it is possible to observe that the transfer function of normalized Butterworth lowpass filter is on par with the theoretical transfer function, which is given by

$$H(s) = \frac{1}{s^2 + 1.414s + 1}$$

Here, we have considered the order of the filter is ($N = 2$), the cut-off frequency of lowpass and highpass filter is chosen as 2. The highest degree of the denominator polynomial function of lowpass and highpass filters is 2, whereas the highest degree of the denominator polynomial function of bandpass and band reject filters is 4. The

Cut-off frequency (in rad/sec)= 12566.370614359172
Transfer function H(s)=
          2.56e+08
----
s^2 + 2.263e+04 s + 2.56e+08
Numerator Coefficients: [0.29289322 0.58578644 0.29289322]
Denominator Coefficients: [1.00000000e+00 -2.04583550e-16 1.71572875e-01]
Transfer function H(z)=
0.2929 z^2 + 0.5858 z + 0.2929
----
  z^2 - 2.046e-16 z + 0.1716
dt = 0.000125

Fig. 8.36 Simulation result of python code is given in Fig. 8.35

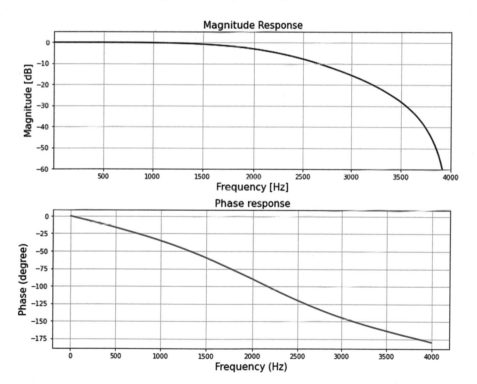

Fig. 8.37 Magnitude and phase responses

```
import numpy as np
import matplotlib.pyplot as plt
from scipy import signal
import control as ss
N, omega_c, wn = 2, 2, [3,5] # Order, cutoff freq. of (LPF and HPF),(BPF and BSF),
omega1=np.linspace(0, 10, 100)
b, a = signal.butter(N, 1, 'low', analog=True)
b1, a1 = signal.butter(N, omega_c, 'low', analog=True)
b2, a2 = signal.butter(N, omega_c, 'high', analog=True)
b3, a3 = signal.butter(N, wn, 'bandpass', analog=True)
b4, a4 = signal.butter(N, wn, 'bandstop', analog=True)
s1 = ss.tf(b, a)
print('Normalized Butterworth filter H(s)=', s1)
mag, phase, omega1=ss.freqresp(s1, omega1)
plt.figure(1),plt.plot(omega1,np.abs(mag)),plt.xlabel('$\Omega$-->'),
plt.ylabel('$|H(j\Omega)|$'),plt.title('Magnitude response of Normalized LPF')
s2 = ss.tf(b1, a1)
print('Desired Butterworth LPF H1(s)=', s2)
mag1, phase1, omega1=ss.freqresp(s2, omega1)
plt.figure(2),plt.subplot(2,2,1),plt.plot(omega1,np.abs(mag1))
plt.xlabel('$\Omega$-->'),plt.ylabel('$|H_1(j\Omega)|$'),plt.title('Desired LPF')
s3 = ss.tf(b2, a2)
print('Desired Butterworth HPF H2(s)=', s3)
mag2, phase2, omega1=ss.freqresp(s3, omega1)
plt.subplot(2,2,2),plt.plot(omega1,np.abs(mag2)),plt.xlabel('$\Omega$-->'),
plt.ylabel('$|H_2(j\Omega)|$'),plt.title('Desired HPF')
s4 = ss.tf(b3, a3)
print('Desired Butterworth BPF H3(s)=', s4)
mag3, phase3, omega1=ss.freqresp(s4, omega1)
plt.subplot(2,2,3),plt.plot(omega1,np.abs(mag3)),plt.xlabel('$\Omega$-->'),
plt.ylabel('$|H_3(j\Omega)|$'),plt.title('Desired BPF')
s5 = ss.tf(b4, a4)
print('Desired Butterworth BSF H4(s)=', s5)
mag4, phase4, omega1=ss.freqresp(s5, omega1)
plt.subplot(2,2,4),plt.plot(omega1,np.abs(mag4)),plt.xlabel('$\Omega$-->'),
plt.ylabel('$|H_4(j\Omega)|$'),plt.title('Desired BSF')
plt.tight_layout()
```

Fig. 8.38 Python code to convert normalized Butterworth filter to desired filters

bandpass and band reject filter contains rising and falling transition widths, each transition width takes second-order roll-off rate; hence, the order of the filter is 4.

Inferences

1. The magnitude responses of the normalized lowpass and desired lowpass, highpass, bandpass and bandstop filters are shown in Fig. 8.40.
2. From Fig. 8.40a, it is clearly understood that the cut-off frequency is 2 Hz.
3. From Fig. 8.40b, the cut-off frequency of lowpass and highpass filters is 2 Hz, while bandpass and band reject filters are 3 and 5 Hz.

Normalized Butterworth filter H(s)=

$$\frac{1}{s^2 + 1.414\,s + 1}$$

Desired Butterworth LPF H1(s)=

$$\frac{4}{s^2 + 2.828\,s + 4}$$

Desired Butterworth HPF H2(s)=

$$\frac{s^2}{s^2 + 2.828\,s + 4}$$

Desired Butterworth BPF H3(s)=

$$\frac{4\,s^2}{s^4 + 2.828\,s^3 + 34\,s^2 + 42.43\,s + 225}$$

Desired Butterworth BSF H4(s)=

$$\frac{s^4 + 30\,s^2 + 225}{s^4 + 2.828\,s^3 + 34\,s^2 + 42.43\,s + 225}$$

Fig. 8.39   Transfer functions of normalized and desired Butterworth filters

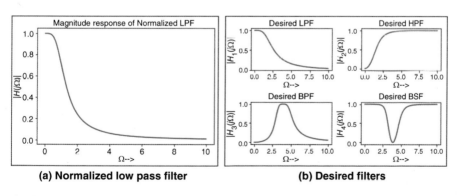

(a) Normalized low pass filter   **(b) Desired filters**

Fig. 8.40   Magnitude responses of normalized and desired Butterworth filters. (**a**) Normalized lowpass filter. (**b**) Desired filters

4. The conclusion that can be made from this experiment is that the desired filter can be obtained from the normalized lowpass filter using analogue frequency transformation.

```
import numpy as np
import matplotlib.pyplot as plt
from scipy import signal
import control as ss
# Specifications of Filter
fsam=8000 # Sampling frequency in Hz
fc1,fc2=[1500,2500],[1000,3000] # cut off frequency in Hz
T = 1/fsam
wcp1=2*np.pi*fc1[0] # Pass band Cut off frequency in rad/sec
wcp2=2*np.pi*fc1[1] # Pass band Cut off frequency in rad/sec
wcs1=2*np.pi*fc2[0] # Stop band Cut off frequency in rad/sec
wcs2=2*np.pi*fc2[1] # Stop band Cut off frequency in rad/sec
pwcp1 = (2/T)*np.tan(wcp1*T/2) # Prewarp the analog frequency
pwcp2 = (2/T)*np.tan(wcp2*T/2) # Prewarp the analog frequency
pwcs1 = (2/T)*np.tan(wcs1*T/2) # Prewarp the analog frequency
pwcs2 = (2/T)*np.tan(wcs2*T/2) # Prewarp the analog frequency
N,wn=signal.buttord([pwcp1,pwcp2],[pwcs1,pwcs2],1,30,True)
print('Order of the filter (N) = ',N)
# Design analog Butterworth filter using signal.butter function
b, a = signal.butter(N, wn, 'bandpass', analog='True')
s1 = ss.tf(b,a)
print('Transfer function H(s)=',s1)
# Perform bilinear Transformation
bz, az = signal.bilinear(b, a, fs=fsam)
z1 = ss.tf(bz,az,T)
print('Transfer function H(z)=',z1)
# Compute frequency response of the filter using signal.freqz function
wz, hz = signal.freqz(bz, az, 512)
fig = plt.figure(figsize=(10, 8))
Mag = 10*np.log10(abs(hz)) # Calculate Magnitude in dB
Freq = wz*fsam/(2*np.pi) # Calculate frequency in Hz
# Plot Magnitude response
sub1 = plt.subplot(2, 1, 1)
sub1.plot(Freq, Mag, 'r', linewidth=2),sub1.axis([1, fsam/2, -60, 5])
sub1.set_title('Magnitude Response', fontsize=15),
sub1.set_xlabel('Frequency [Hz]', fontsize=15),sub1.set_ylabel('Magnitude [dB]', fontsize=15)
sub1.grid()
# Plot phase angle
sub2 = plt.subplot(2, 1, 2)
Phase = np.unwrap(np.angle(hz))*180/np.pi # Calculate phase angle in degree from hz
sub2.plot(Freq, Phase, 'g', linewidth=2),sub2.set_ylabel('Phase (degree)', fontsize=15)
sub2.set_xlabel(r'Frequency (Hz)', fontsize=15),sub2.set_title(r'Phase response', fontsize=15)
sub2.grid(),plt.subplots_adjust(hspace=0.5),fig.tight_layout(),plt.show()
```

Fig. 8.41   Python code for Butterworth bandpass filter design

Experiment 8.19 Design of Butterworth Bandpass Filter

The objective of this experiment is to write a python code to design a Butterworth digital bandpass filter for the following specifications: (1) Passband frequencies are 1500 Hz and 2500 Hz. (2) Stopband frequencies are 1000 Hz and 3000 Hz. (3) Sampling frequency (F_s) = 8 kHz. (4) Passband ripple is 1 dB and stopband attenuation is 30 dB. Use BLT method for transformation.

Order of the filter (N) = 5
Transfer function H(s)=

$$8.041e{+}20\ s^5$$

----

s^{10} + 4.91e+04 s^9 + 2.485e+09 s^8 + 6.857e+13 s^7 + 1.753e+18 s^6 + 2.947e+22
s^5 + 4.487e+26 s^4 + 4.494e+30 s^3 + 4.17e+34 s^2 + 2.109e+38 s + 1.1e+42

Transfer function H(z)=
     0.005376 z^{10} - 0.02688 z^8 - 7.894e-18 z^7 + 0.05376 z^6 - 3.552e-17 z^5 -
0.05376 z^4 - 7.894e-18 z^3 + 0.02688 z^2 - 0.005376

----

z^{10} + 5.999e-16 z^9 + 2.156 z^8 + 3.663e-15 z^7 + 2.281 z^6 + 1.516e-15 z^5 + 1.297
z^4 + 1.516e-15 z^3 + 0.3952 z^2 + 2.684e-16 z + 0.05031
dt = 0.000125

Fig. 8.42   Simulation result of python code is given in Fig. 8.41

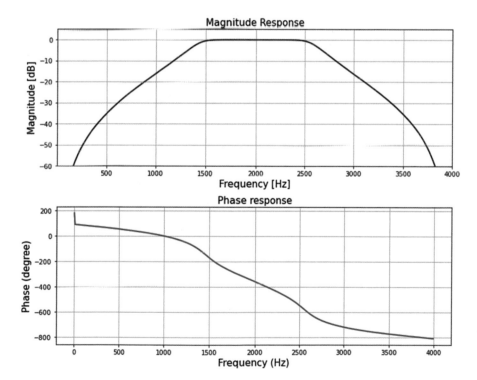

Fig. 8.43   Magnitude and phase responses

Table 8.2 Built-in functions used in the program

S. No.	Built-in function	Library	Use
1	tan	Numpy	To obtain the tan value
2	buttord	Scipy	To obtain the order and cut-off frequency of the Butterworth filter
3	butter	Scipy	To obtain the coefficients of Butterworth filter
4	bilinear	Scipy	To convert analogue transfer function into an equivalent digital transfer function using the bilinear transformation mapping
5	freqz	Scipy	To obtain the frequency response of the digital filter

The python code for this experiment is given in Fig. 8.41, and its corresponding output is shown in Figs. 8.42 and 8.43. Figure 8.43 gives the magnitude and phase response of the Butterworth digital bandpass filter. The libraries used in this program are (1) *numpy*, (2) matplotlib, (3) *scipy* and (4) *control*. The built-in functions used in this program are summarized in Table 8.2.

Inference
From the magnitude response shown in Fig. 8.43, it is possible to observe that the passband of the digital filter is 1500–2500 Hz. This is in agreement with the specification of the filter. The phase response of the filter is non-linear.

Task
1. Change the passband cut-off frequencies of the bandpass filter and see the changes in the magnitude and phase responses.

8.5   Chebyshev Filter

The squared magnitude response of Chebyshev filter is given by

$$\left|H\left(e^{j\Omega}\right)\right|^2 = \frac{1}{1 + \varepsilon^2 T_N^2(\Omega)} \tag{8.14}$$

The parameter ε sets the ripple amplitude. Chebyshev filters can be classified into two types, namely, Type I Chebyshev filters and Type II Chebyshev filters. Type I Chebyshev filter exhibits ripple in passband, whereas Type II Chebyshev filter exhibits ripple in stopband. Chebyshev polynomial $T_N(x)$ for different order is given in Table 8.3.

Experiment 8.20 Plotting Chebyshev Polynomial for Different Order
This experiment tries to plot the Chebyshev polynomial functions of different order using python. The python code is used here to plot the Chebyshev polynomial functions of different order and the corresponding output is shown in Figs. 8.44

Table 8.3 Chebyshev polynomial function

Order (N)	Polynomial function
0	$T_0(x) = 1$
1	$T_1(x) = x$
2	$T_2(x) = 2x^2 - 1$
3	$T_3(x) = 4x^3 - 3x$
4	$T_4(x) = 8x^4 - 8x^2 + 1$
5	$T_5(x) = 16x^5 - 20x^3 + 5x$
6	$T_6(x) = 32x^6 - 48x^4 + 18x^2 - 1$
7	$T_7(x) = 64x^7 - 112x^5 + 56x^3 - 7x$
8	$T_8(x) = 128x^8 - 256x^6 + 160x^4 - 32x^2 + 1$

```
import numpy as np
import matplotlib.pyplot as plt
x=np.linspace(-1,1,50)
T0=np.ones(len(x)) # zeroth degree polynomial
T1=x # First degree polynomial
T2=2*x*T1-T0 # Second degree polynomial
T3=2*x*T2-T1 # Third degree polynomial
T4=2*x*T3-T2 # Fourth degree polynomial
T5=2*x*T4-T3 # Fifth degree polynomial
plt.subplot(3,2,1),plt.plot(x,T0,'k--',linewidth=3.5), plt.xlabel('x'), plt.ylabel('T_0(x)'),
plt.grid(),plt.subplot(3,2,2),plt.plot(x,T1,'k--',linewidth=3.5), plt.xlabel('x'),
plt.ylabel('T_1(x)'),plt.grid(),plt.subplot(3,2,3),plt.plot(x,T2,'k--',linewidth=3.5), plt.xlabel('x'),
plt.ylabel('T_2(x)'),plt.grid(),plt.subplot(3,2,4),plt.plot(x,T3,'k--',linewidth=3.5),
plt.xlabel('x'),plt.ylabel('T_3(x)'),plt.grid(), plt.subplot(3,2,5),plt.plot(x,T4,'k--',linewidth=3.5),
plt.xlabel('x'),plt.ylabel('T_4(x)'),plt.grid(),plt.subplot(3,2,6),plt.plot(x,T5,'k--',linewidth=3.5),
plt.xlabel('x'),plt.ylabel('T_5(x)'),plt.grid(),plt.tight_layout()
```

Fig. 8.44 Python code to plot the Chebyshev polynomial

and 8.45, respectively. From Fig. 8.45, it is possible to infer that the Chebyshev polynomial function is created based on the recursive formula, which is given by

$$T_N(x) = 2xT_{N-1}(x) - T_{N-2}(x) \qquad (8.15)$$

Inference

Figure 8.45 shows that whenever $N = 1$, 3, 5, etc., the graph passes through the origin. For $N = 0$, 2, 4, etc., the graph does not pass through the origin.

Task

1. Write a python code to plot the Chebyshev polynomial function of order 8. Mention the number of zero crossings that exist in it.

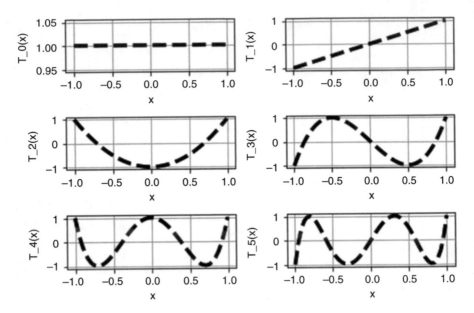

Fig. 8.45 Chebyshev polynomials of different order

Experiment 8.21 Design of Type I Chebyshev Filter

This experiment discusses the design of a Type I Chebyshev filter using bilinear transformation technique (BLT) that has a passband of 0 to −2 dB cut-off frequency of 3 kHz and attenuation of at least 20 dB for frequencies greater than 5 kHz. Assume the sampling frequency to be 15 kHz. Assume $T = 1$ s.

Given data

1. Sampling frequency $f_{samp} = 15$ kHz.
2. Gain in pass band $A_p = -2$ dB.
3. Stop band attenuation $A_s = -20$ dB.
4. Pass band cut off frequency $f_p = 3$ kHz.
5. Stop band cut off frequency $f_s = 5$ kHz.

Step 1: Prewarping

First, it is necessary to compute the ω_p and ω_s from the frequency f_p and f_s, which are given below

$$\omega_p = 2\pi\left(\frac{f_p}{f_{samp}}\right)$$

Substituting the values of f_p and f_{samp} in the above expression, we get

$$\omega_p = 2\pi \left(\frac{3}{15}\right) = 0.4\pi$$

Now the expression for ω_s is given by

$$\omega_s = 2\pi \left(\frac{f_s}{f_{samp}}\right)$$

Substituting the values of f_s and f_{samp} in the above expression, we get

$$\omega_s = 2\pi \left(\frac{5}{15}\right) = 0.667\pi$$

The corresponding analogue frequencies Ω_p and Ω_s are obtained using prewarping by using BLT relation as

$$\Omega_p = \frac{2}{T} \tan\left(\frac{\omega_p}{2}\right)$$

Substituting the value of $\omega_p = 0.4\pi$ and $T = 1$ in the above expression, we get

$$\Omega_p = \frac{2}{1} \tan\left(\frac{0.4\pi}{2}\right) = 1.4531$$

Similarly, the value of Ω_s is computed as

$$\Omega_s = \frac{2}{1} \tan\left(\frac{0.667\pi}{2}\right) = 3.4633$$

Step 2: To determine the passband ripple factor (ε)
The passband ripple factor is calculated as

$$\varepsilon = \sqrt{10^{\left(\frac{A_p}{10}\right)} - 1}$$

$$\varepsilon = \sqrt{10^{\left(\frac{2}{10}\right)} - 1} = \sqrt{10^{(0.2)} - 1}$$

$$\varepsilon = \sqrt{1.5849 - 1} = \sqrt{0.5849} = 0.7648$$

Step 3: To determine the order of the filter
The expression for the order of Chebyshev filter is given by

$$N \geq \frac{\cosh^{-1}\left(\sqrt{\frac{1}{\varepsilon^2}\left(10^{-\frac{A_s}{10}} - 1\right)}\right)}{\cosh^{-1}\left(\frac{\Omega_s}{\Omega_p}\right)}$$

Substituting the value of $\varepsilon = 0.7648$, $A_s = -20$ dB, $\Omega_p = 1.4531$ and $\Omega_s = 3.4633$, in the above expression, we get

$$N \geq \frac{\cosh^{-1}\left(\sqrt{\frac{1}{(0.7648)^2}\left(10^{-\frac{-20}{10}} - 1\right)}\right)}{\cosh^{-1}\left(\frac{3.4633}{1.4531}\right)}$$

The above expression is simplified as

$$N \geq \frac{\cosh^{-1}\left(\sqrt{\frac{1}{(0.7648)^2}\left(10^2 - 1\right)}\right)}{\cosh^{-1}(2.3834)}$$

The above expression is further simplified as

$$N \geq \frac{\cosh^{-1}\left(\sqrt{\frac{1}{(0.7648)^2}\left(100 - 1\right)}\right)}{1.5144}$$

The further simplification of the above expression, we get

$$N \geq \frac{\cosh^{-1}\left(\sqrt{\frac{99}{0.5849}}\right)}{1.5144}$$

The above expression is simplified as

$$N \geq \frac{\cosh^{-1}\left(\sqrt{169.2597}\right)}{1.5144}$$

The simplified version of the above equation, we get

$$N \geq \frac{3.2574}{1.5144} = 2.1510 = 3$$

From the above result, the order of the filter is computed as three ($N = 3$).

Step 4: Computation of the left half of the S plane poles

From the order of filter 3, 6 poles will be there in this filter. Only 3 poles will be calculated for further process. That is, those poles must be lying left half of S-plane.

The computation of the poles are given by

$$s_k = \sigma_k + j\Omega_k = \quad -\sin\left(\frac{(2k-1)\pi}{2N}\right)\sinh\left(\frac{1}{N}\sinh^{-1}\left(\frac{1}{\varepsilon}\right)\right)$$
$$+ j\cos\left(\frac{(2k-1)\pi}{2N}\right)\cosh\left(\frac{1}{N}\sinh^{-1}\left(\frac{1}{\varepsilon}\right)\right) \tag{8.16}$$

In the above equation substituting $k = 1, 2, 3$, we may get 3 numbers of poles. For $k = 1$, the above equation can be written as

$$s_1 = \sigma_1 + j\Omega_1 = \quad -\sin\left(\frac{(2.1-1)\pi}{2\times 3}\right)\sinh\left(\frac{1}{3}\sinh^{-1}\left(\frac{1}{0.7648}\right)\right)$$
$$+ j\cos\left(\frac{(2.1-1)\pi}{2\times 3}\right)\cosh\left(\frac{1}{3}\sinh^{-1}\left(\frac{1}{0.7648}\right)\right)$$

The above equation can be simplified as

$$s_1 = -\sin\left(\frac{(2.1-1)\pi}{2\times 3}\right)\sinh\left(\frac{1}{3}(1.0831)\right) + j\cos\left(\frac{(2.1-1)\pi}{2\times 3}\right)\cosh\left(\frac{1}{3}(1.0831)\right)$$

The above equation further simplified as

$$s_1 = -\sin\left(\frac{\pi}{2\times 3}\right)\sinh(0.3610) + j\cos\left(\frac{\pi}{2\times 3}\right)\cosh(0.3610)$$

Simplifying the above equation, we get

$$s_1 = -\sin\left(\frac{\pi}{6}\right)(0.3689) + j\cos\left(\frac{\pi}{6}\right)(1.0659)$$

We know that $\sin\left(\frac{\pi}{6}\right) = 0.5$ and $\cos\left(\frac{\pi}{2}\right) = 0.8660$, substituting this results in the above equation, we get

$$s_1 = -(0.5)(0.3689) + j(0.8660)(1.0659)$$
$$s_1 = -0.1845 + j0.9231$$

Substituting $k = 2$, in Eq. (8.16), we get

$$s_2 = \sigma_2 + j\Omega_2 = \quad -\sin\left(\frac{(2.2-1)\pi}{2\times 3}\right)\sinh\left(\frac{1}{3}\sinh^{-1}\left(\frac{1}{0.7648}\right)\right)$$
$$+ j\cos\left(\frac{(2.2-1)\pi}{2\times 3}\right)\cosh\left(\frac{1}{3}\sinh^{-1}\left(\frac{1}{0.7648}\right)\right)$$

The above equation can be simplified as

$$s_2 = - \sin\left(\frac{(4-1)\pi}{2\times 3}\right) \sinh\left(\frac{1}{3}(1.0831)\right) + j\cos\left(\frac{(4-1)\pi}{2\times 3}\right) \cosh\left(\frac{1}{3}(1.0831)\right)$$

The above equation further simplified as

$$s_2 = - \sin\left(\frac{3\pi}{6}\right) \sinh(0.3610) + j\cos\left(\frac{3\pi}{6}\right) \cosh(0.3610)$$

Simplifying the above equation, we get

$$s_2 = - \sin\left(\frac{\pi}{2}\right)(0.3689) + j\cos\left(\frac{\pi}{2}\right)(1.0659)$$

We know that $\sin\left(\frac{\pi}{2}\right) = 1$ and $\cos\left(\frac{\pi}{2}\right) = 0$, substituting this results in the above equation, we get

$$s_2 = -0.3689$$

Substituting $k = 3$, in Eq. (8.16), we get

$$\begin{aligned} s_3 = \sigma_3 + j\Omega_3 = \quad & - \sin\left(\frac{(2.3-1)\pi}{2\times 3}\right) \sinh\left(\frac{1}{3}\sinh^{-1}\left(\frac{1}{0.7648}\right)\right) \\ & + j\cos\left(\frac{(2.3-1)\pi}{2\times 3}\right) \cosh\left(\frac{1}{3}\sinh^{-1}\left(\frac{1}{0.7648}\right)\right) \end{aligned}$$

The above equation can be simplified as

$$s_3 = - \sin\left(\frac{(6-1)\pi}{2\times 3}\right) \sinh\left(\frac{1}{3}(1.0831)\right) + j\cos\left(\frac{(6-1)\pi}{2\times 3}\right) \cosh\left(\frac{1}{3}(1.0831)\right)$$

The above equation further simplified as

$$s_3 = - \sin\left(\frac{5\pi}{6}\right) \sinh(0.3610) + j\cos\left(\frac{5\pi}{6}\right) \cosh(0.3610)$$

Simplifying the above equation, we get

$$s_3 = - \sin\left(\frac{5\pi}{6}\right)(0.3689) + j\cos\left(\frac{5\pi}{6}\right)(1.0659)$$

The simplification of the above result, we get

$$s_3 = -0.1845 - j0.9231$$

The results of this step are $s_1 = -0.1845 + j0.9231$, $s_2 = -0.3689$ and $s_3 = -0.1845 - j0.9231$.

Step 5: Calculate the normalized frequency transfer function (i.e. $\Omega_p = 1$)
The normalized frequency transfer function is obtained by

$$H(s) = K \frac{(-1)^N s_1 \times s_2 \times \cdots \times s_N}{(s - s_1)(s - s_2)\cdots(s - s_N)}$$

where

$$K = \begin{cases} \dfrac{1}{\sqrt{1 + \varepsilon^2}} & \text{for } N \text{ is Even} \\ 1 & \text{for } N \text{ is odd} \end{cases}$$

In this example, the order of the filter is 3; it shows that odd, hence the gain $K = 1$.
The transfer function is formed as

$$H(s) = K \frac{(-1)^3 \times s_1 \times s_2 \times s_3}{(s - s_1)(s - s_2)(s - s_3)}$$

Substituting the values of $s_1 = -0.1845 + j0.9231$, $s_2 = -0.3689$ and $s_3 = -0.1845 - j0.9231$ in the above equation, we get

$$H(s) = K \frac{(-1) \times (-0.1845 + j0.9231) \times (-0.3689) \times (-0.1845 - j0.9231)}{(s - (-0.1845 + j0.9231))(s - (-0.3689))(s - (-0.1845 - j0.9231))}$$

The above equation can be simplified as

$$H(s) = \frac{0.3269}{(s + 0.1845 - j0.9231))(s + 0.3689))(s + 0.1845 + j0.9231))}$$

The above equation can be further simplified as

$$H(s) = \frac{0.3269}{(s^3 + 0.7378s^2 + 1.0222s + 0.3269)}$$

Step 6: Calculate the transfer function of the desired frequency
The desired passband cut-off frequency of the filter is $\Omega_p = 1.4531$. The transfer function is given by

$$H_a\left(\frac{s}{1.4531}\right) = \frac{0.3269}{\left(\left(\frac{s}{1.4531}\right)^3 + 0.7378\left(\frac{s}{1.4531}\right)^2 + 1.0222\left(\frac{s}{1.4531}\right) + 0.3269\right)}$$

The above result can be further simplified as

$$H_a\left(\frac{s}{1.4531}\right) = \frac{1.003}{(s^3 + 1.072s^2 + 2.158s + 1.003)}$$

Step 7: Converting the analogue filter into an equivalent digital filter
The digital equivalent of the analogue filter using bilinear transformation technique is given by

$$H(z) = H_a(s)\big|_{s = \frac{2}{T}\frac{1-z^{-1}}{1+z^{-1}}}$$

Substituting the expression for $H(s)$ from Step 5, we get

$$H(z) = \frac{1.003}{(s^3 + 1.072s^2 + 2.158s + 1.003)}\Bigg|_{s = 2\frac{1-z^{-1}}{1+z^{-1}}}$$

The above expression can be written as

$$H(z) = \frac{1.003}{\left(\left(2\frac{1-z^{-1}}{1+z^{-1}}\right)^3 + 1.072\left(2\frac{1-z^{-1}}{1+z^{-1}}\right)^2 + 2.158\left(2\frac{1-z^{-1}}{1+z^{-1}}\right) + 1.003\right)}$$

Simplifying the above expression, we get

$$H(z) = \frac{0.05696z^3 + 0.1709z^2 + 0.1709z + 0.05696}{z^3 - 1.191z^2 + 1.045z - 0.399}$$

The python code for this experiment is given in Fig. 8.45. The built-in function *signal.cheb1ord* is used here to compute the order and cut-off frequency of the filter. The built-in function *signal.cheby1* is used here to obtain the numerator and denominator coefficients of filter, *ss.tf* is used for the computation of transfer function and *signal.bilinear* helps to convert analogue filter coefficients into digital filter coefficients using BLT approach.

Inferences
1. After executing the python code given in Fig. 8.46, the simulation results are shown in Figs. 8.47 and 8.48.
2. The order of the filter is calculated as ($N = 3$); the cut-off frequency is obtained as 1.4531 rad/s, which is equivalent to passband cut-off frequency, and the analogue

```
import numpy as np
import matplotlib.pyplot as plt
from scipy import signal
import control as ss
# Specifications of Filter
fsam=15000 # Sampling frequency
fp=3000 # Passband frequency
fs=5000 # Stopband frequency
Ap, As, Td=2,20, 1
wp=2*np.pi*(fp/fsam) # passband freq in radian per sample
ws=2*np.pi*(fs/fsam) # Stopband freq in radian per sample
# prewarping process
omega_p=(2/Td)*np.tan(wp/2)
omega_s=(2/Td)*np.tan(ws/2)
N, omega_c=signal.cheb1ord(omega_p,omega_s,Ap,As,analog=True)
print('Order of the Filter N =', N)
print('Cut-off frequency= {:.4f} rad/s'. format(omega_c))
# Computation of H(s)
b_s, a_s=signal.cheby1(N,Ap,omega_c,'low', analog=True)
s1 – ss.tf(b_s, a_s)
print('Transfer function H(s)=',s1)
bz, az=signal.bilinear(b_s, a_s, Td)
z1 = ss.tf(bz,az,Td)
print('Transfer function H(z)=',z1)
ws, hs = signal.freqz(bz, az) # Calculate Magnitude from hz in dB
Mag = 20*np.log10(abs(hs)) # Calculate phase angle in degree from hz
Phase = np.unwrap(np.arctan2(np.imag(hs), np.real(hs)))*(180/np.pi)
# Calculate frequency in Hz from wz
Freq = ws*fsam/(2*np.pi)
# Plot filter magnitude and phase responses using subplot.
fig = plt.figure(figsize=(10, 6))# Plot Magnitude response
sub1 = plt.subplot(2, 1, 1)
sub1.plot(Freq, Mag, 'r', linewidth=2)
sub1.axis([1, fsam/2, -100, 5])
sub1.set_title('Magnitude Response', fontsize=15)
sub1.set_xlabel('Frequency [Hz]', fontsize=15)
sub1.set_ylabel('Magnitude [dB]', fontsize=15)
sub1.grid()
 # Plot phase angle
 sub2 = plt.subplot(2, 1, 2)
 sub2.plot(Freq, Phase, 'g', linewidth=2),sub2.set_ylabel('Phase (degree)', fontsize=15)
 sub2.set_xlabel(r'Frequency (Hz)', fontsize=15)
 sub2.set_title(r'Phase response', fontsize=15)
 sub2.grid(),plt.subplots_adjust(hspace=0.5),fig.tight_layout(),plt.show()
```

Fig. 8.46 Python code to get transfer function of Chebyshev lowpass filter

Order of the Filter N = 3

Cut-off frequency= 1.4531 rad/s

Transfer function H(s)=

$$\qquad \frac{1.003}{s^3 + 1.072\,s^2 + 2.158\,s + 1.003}$$

Transfer function H(z)=

$$\frac{0.05696\,z^3 + 0.1709\,z^2 + 0.1709\,z + 0.05696}{z^3 - 1.191\,z^2 + 1.045\,z - 0.399}$$

dt = 1

Fig. 8.47   Simulation result

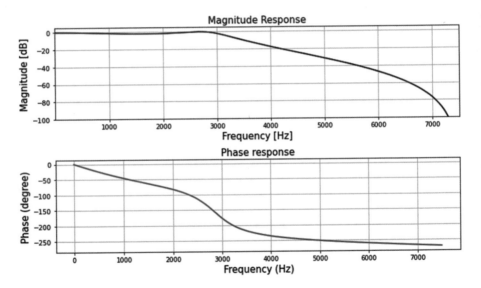

Fig. 8.48   Magnitude and phase response

and digital filters transfer functions are the same as the theoretical result, which is shown in Fig. 8.47.

3. The frequency response of the Chebyshev Type I filter is shown in Fig. 8.48. This figure clearly shows that the passband frequency is up to 3 kHz.

Experiment 8.22 Design of Chebyshev Type I Digital Highpass Filter
This experiment deals with the design of a Chebyshev Type I digital highpass filter for the following specifications: (1) Passband frequency is 2500 Hz. (2) Stopband frequency is 1500 Hz. (2) Sampling frequency (F_s) = 8 kHz. (4) Passband ripple is

```
import numpy as np
import matplotlib.pyplot as plt
from scipy import signal
import control as ss
# Specifications of Filter
fsam=8000 # Sampling frequency in Hz
fc1=1500 # Stop band cut-off frequency in Hz
fc2=2500#  Pass band cut-off frequency in Hz
Ap,As,T =3,40, 1/fsam
wc1=2*np.pi*fc1 # Stopband Cut off frequency in rad/sec
wc2=2*np.pi*fc2 # Passband Cut off frequency in rad/sec
#print('Cut-off frequency (in rad/sec)=', wc1)
pwc1 = (2/T)*np.tan(wc1*T/2) # Prewarp the analog frequency
pwc2 = (2/T)*np.tan(wc2*T/2) # Prewarp the analog frequency
# Design analog Butterworth filter using signal.butter function
N, Wn = signal.cheb1ord(pwc2, pwc1, Ap, As,analog=True)
print('Order of the filter (N) = ',N)
b, a = signal.cheby1(N, Ap, Wn, 'high',analog=True)
s1 = ss.tf(b,a)
print('Transfer function H(s)=',s1)
# Perform bilinear transformation
bz, az = signal.bilinear(b, a, fs=fsam)
# Print numerator and denominator coefficients of the filter
print('Numerator Coefficients:', bz)
print('Denominator Coefficients:', az)
z1 = ss.tf(bz,az,T)
print('Transfer function H(z)=',z1)
# Compute frequency response of the filter using signal.freqz function
wz, hz = signal.freqz(bz, az, 512)
fig = plt.figure(figsize=(10, 8))
Mag = 20*np.log10(abs(hz)) # Calculate Magnitude in dB
Freq = wz*fsam/(2*np.pi) # Calculate frequency in Hz
# Plot Magnitude response
sub1 = plt.subplot(2, 1, 1)
sub1.plot(Freq, Mag, 'r', linewidth=2),sub1.axis([1, fsam/2, -60, 5])
sub1.set_title('Magnitude Response', fontsize=15),sub1.set_xlabel('Frequency [Hz]', fontsize=15)
sub1.set_ylabel('Magnitude [dB]', fontsize=15),sub1.grid()
# Plot phase angle
sub2 = plt.subplot(2, 1, 2)
# Calculate phase angle in degree from hz
Phase = np.unwrap(np.angle(hz))*180/np.pi
sub2.plot(Freq, Phase, 'g', linewidth=2),sub2.set_ylabel('Phase (degree)', fontsize=15)
sub2.set_xlabel(r'Frequency (Hz)', fontsize=15),sub2.set_title(r'Phase response', fontsize=15)
sub2.grid(),plt.subplots_adjust(hspace=0.5),fig.tight_layout(),plt.show()
```

Fig. 8.49 Python code for Butterworth bandpass filter design

3 dB and stop attenuation is 40 dB. Use BLT method for transformation. The python code for this experiment is given in Fig. 8.49, and its corresponding output is shown in Figs. 8.50 and 8.51. Figure 8.50 gives the magnitude and phase responses of the Chebyshev Type I digital highpass filter.

Inference

From this experiment, the following inferences can be made:

Order of the filter (N) = 4
Transfer function H(s)=
                   0.7079 s^4

----

s^4 + 5.476e+04 s^3 + 3.788e+09 s^2 + 4.512e+13 s + 1.858e+18
Numerator Coefficients: [0.01208527 -0.04834108 0.07251162 -0.04834108
0.01208527]
Denominator Coefficients: [1.   2.12648692 2.50060541 1.60804914 0.50706501]
Transfer function H(z)=
0.01209 z^4 - 0.04834 z^3 + 0.07251 z^2 - 0.04834 z + 0.01209

----

    z^4 + 2.126 z^3 + 2.501 z^2 + 1.608 z + 0.5071
dt = 0.000125

Fig. 8.50 Simulation result of python code is given in Fig. 8.49

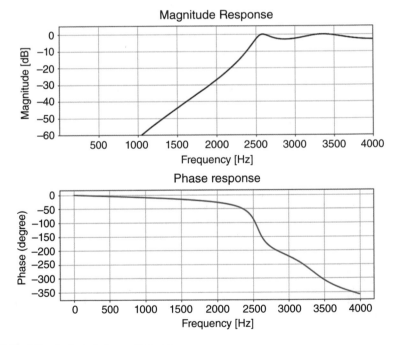

Fig. 8.51 Magnitude and phase responses

1. From the magnitude response, it is possible to observe that passband gain is high after the frequency of 2000 Hz, and there is a ripple in the passband.
2. From the phase response, it is possible to confirm that the curve is not linear; hence, it cannot provide the linear phase in the output.

Tasks

1. Write a python code to obtain the magnitude and phase response of Chebyshev Type I bandpass filter with the passband frequencies of 1500–3000 Hz. Choose the order of the filter that is 3.
2. Write a python code to obtain the magnitude and phase response of Chebyshev Type I band reject filter with the stopband frequencies of 1500–3000 Hz. Choose the order of the filter that is 3.

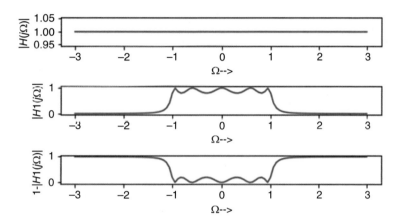

Fig. 8.52 First step procedure

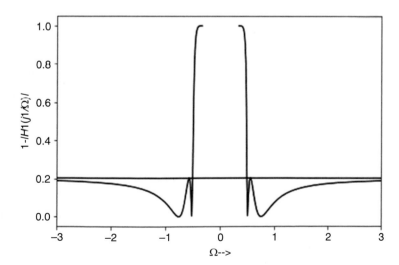

Fig. 8.53 Second step procedure

```
import numpy as np
import matplotlib.pyplot as plt
from scipy import signal
import control as ss
N, rs, omega_c=3, 30, [5]
omega1=np.linspace(0, 10, 100)
b_s, a_s=signal.cheby2(N,rs,omega_c,'low', analog=True)
s1 = ss.tf(b_s, a_s)
print('H(s) = ', s1)
z1=np.roots(b_s)
print('Zeros : ', z1)
mag, phase, omega1=ss.freqresp(s1, omega1)
plt.figure,plt.plot(omega1,np.abs(mag))
plt.xlabel('$\Omega$-->'),plt.ylabel('$|H(j\Omega)|$')
plt.title('Magnitude response of Chebyshev Type II LPF')
plt.tight_layout()
```

Fig. 8.54 Python code for Chebyshev Type II LPF

8.6   Chebyshev Type II IIR Filter

The Chebyshev Type II IIR filter has an equiripple in stopband and monotonic response in passband, which is inverse of Chebyshev Type I IIR filter; hence, it is also called as 'inverse Chebyshev filter'. Two-step procedures can obtain the frequency response of this filter. In the first step, subtract the frequency response of Chebyshev Type I filter ($H(\omega)$) from 1, which is illustrated in Fig. 8.52. In the second step, convert the ω by $\frac{1}{\omega}$. This result will give the frequency response of Chebyshev Type II IIR filter, which is shown in Fig. 8.53.

From Fig. 8.53, it is possible to observe that the frequency response has monotonic in the passband and an equiripple in the stop band.

Using the two step procedures, the mathematical expression for the frequency response of the Chebyshev Type II filter can be written as

$$H(\Omega) = 1 - \frac{1}{1 + \varepsilon^2 T_N^2 (1/\Omega)} \tag{8.17}$$

Simplifying the above expression, we get

$$H(\Omega) = \varepsilon^2 T_N^2 \left(\frac{1/\Omega}{1 + \varepsilon^2 T_N^2 (1/\Omega)} \right) \tag{8.18}$$

From the above expression, it is possible to understand that the Chebyshev Type II filter has zeros as the numerator, and all the zeros lie on the imaginary axis.

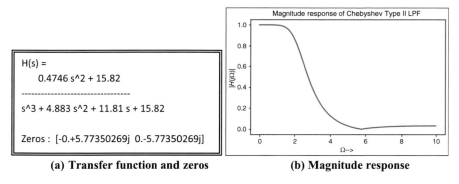

| (a) Transfer function and zeros | (b) Magnitude response |

Fig. 8.55 Simulation result of the python code is given in Fig. 8.54. (**a**) Transfer function and zeros. (**b**) Magnitude response

Experiment 8.23 Design of Chebyshev Type II Lowpass Filter

This experiment discusses the design of Chebyshev Type II lowpass filter using python with a stopband frequency of 5 Hz. Assume the order of the filter is 3, and stopband attenuation is 30 dB.

The python code to design a Chebyshev Type II lowpass filter is shown in Fig. 8.54, and its corresponding simulation result is shown in Fig. 8.55. From Fig. 8.54, it is possible to infer that the *signal.cheby2* built-in command can be used to obtain the numerator and denominator coefficients. After executing the python code in Fig. 8.54, the transfer function of Chebyshev Type II lowpass filter is shown in Fig. 8.55a. This figure confirms that the Chebyshev Type II filter has zeros, which occur on the imaginary axis. The magnitude response of the Chebyshev Type II lowpass filter is shown in Fig. 8.55b. This figure confirms the monotonic response in the passband and equiripple in the stopband.

Inference

From Fig. 8.55b, it is possible to confirm that the ripple exists in the stopband and monotonic response in the passband.

Task

1. Write a python code to plot the magnitude response of Chebyshev Type II highpass filter with the lower cut-off frequency of 5 Hz. Assume the order of the filter is 3 and stopband attenuation is 30 dB.

Experiment 8.24 Design of Chebyshev Type II Digital Bandstop Filter

This experiment deals with the design of a Chebyshev Type II digital bandstop filter for the following specifications: (1) Stopband frequencies are from 1500 to 2500 Hz with an attenuation of 40 dB. (2) Passband frequencies are below 1000 Hz and above 3000 Hz with the passband ripple of 3 dB. Sampling frequency $(F_s) = 8$ kHz. Use BLT method for transformation. The python code for this experiment is given in Fig. 8.56, and its corresponding output is shown in Figs. 8.57 and 8.58. Figure 8.58 gives the magnitude and phase responses of the Chebyshev Type II digital bandstop filter.

```
import numpy as np
import matplotlib.pyplot as plt
from scipy import signal
import control as ss
# Specifications of Filter
fsam=8000 # Sampling frequency in Hz
fc1,fc2=[1500,2500],[1000,3000] # cut off frequency in Hz
Ap, As, T = 2, 40, 1/fsam
wcp1=2*np.pi*fc1[0] # Pass band Cut off frequency in rad/sec
wcp2=2*np.pi*fc1[1] # Pass band Cut off frequency in rad/sec
wcs1=2*np.pi*fc2[0] # Stop band Cut off frequency in rad/sec
wcs2=2*np.pi*fc2[1] # Stop band Cut off frequency in rad/sec
pwcs1 = (2/T)*np.tan(wcp1*T/2) # Prewarp the analog frequency
pwcs2 = (2/T)*np.tan(wcp2*T/2) # Prewarp the analog frequency
pwcp1 = (2/T)*np.tan(wcs1*T/2) # Prewarp the analog frequency
pwcp2 = (2/T)*np.tan(wcs2*T/2) # Prewarp the analog frequency
N,wn=signal.cheb2ord([pwcp1,pwcp2],[pwcs1,pwcs2],Ap,As,analog=True)
print('Order of the filter (N) = ',N)
# Design analog Chebyshev Type 2 filter using signal.cheby2 function
b, a = signal.cheby2(N, As, wn, 'bandstop', analog='True')
s1 = ss.tf(b,a)
print('Transfer function H(s)=',s1)
# Perform bilinear transformation
bz, az = signal.bilinear(b, a, fs=fsam)
z1 = ss.tf(bz,az,T)
print('Transfer function H(z)=',z1)
# Compute frequency response of the filter using signal.freqz function
wz, hz = signal.freqz(bz, az, 512)
fig = plt.figure(figsize=(10, 8))
Mag = 10*np.log10(abs(hz)) # Calculate Magnitude in dB
Freq = wz*fsam/(2*np.pi) # Calculate frequency in Hz
# Plot Magnitude response
sub1 = plt.subplot(2, 1, 1)
sub1.plot(Freq, Mag, 'r', linewidth=2),sub1.axis([1, fsam/2, -60, 5])
sub1.set_title('Magnitude Response', fontsize=15),
sub1.set_xlabel('Frequency [Hz]', fontsize=15),sub1.set_ylabel('Magnitude [dB]', fontsize=15)
sub1.grid()
# Plot phase angle
sub2 = plt.subplot(2, 1, 2)
Phase = np.unwrap(np.angle(hz))*180/np.pi # Calculate phase angle in degree from hz
sub2.plot(Freq, Phase, 'g', linewidth=2),sub2.set_ylabel('Phase (degree)', fontsize=15)
sub2.set_xlabel(r'Frequency (Hz)', fontsize=15),sub2.set_title(r'Phase response', fontsize=15)
sub2.grid(),plt.subplots_adjust(hspace=0.5),fig.tight_layout(),plt.show()
```

Fig. 8.56 Python code for Chebyshev Type II bandstop filter design

Inferences

1. From Fig. 8.58, the magnitude response shows no ripple in the passbands and a ripple in the stop band.
2. From the phase response, it is possible to confirm that IIR filters do not have linear phase characteristics; hence, the phase response is not linear.

Order of the filter (N) = 4

Transfer function H(s)=

$$s^8 + 1.249e+09\ s^6 + 5.145e+17\ s^4 + 8.183e+25\ s^2 + 4.295e+33$$

$s^8 + 6.844e+04\ s^7 + 3.59e+09\ s^6 + 1.094e+14\ s^5 + 2.338e+18\ s^4 + 2.802e+22\ s^3 +$
$2.353e+26\ s^2 + 1.148e+30\ s + 4.295e+33$

Transfer function H(z)=

$0.1535\ z^8 - 5.028e-11\ z^7 + 0.4981\ z^6 - 1.309e-10\ z^5 + 0.7012\ z^4 - 1.309e-10\ z^3 +$
$0.4981\ z^2 - 5.028e-11\ z + 0.1535$

$z^8 - 1.944e-10\ z^7 + 0.3006\ z^6 - 9.283e-11\ z^5 + 0.5759\ z^4 - 6.545e-11\ z^3 + 0.0987$
$z^2 - 9.75e-12\ z + 0.02925$

dt = 0.000125

Fig. 8.57 Simulation result of python code is given in Fig. 8.56

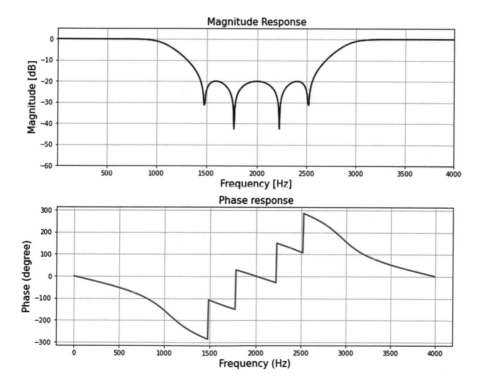

Fig. 8.58 Magnitude and phase responses

Task

1. Write a python code to design a Chebyshev Type II digital bandpass filter for the following specifications: (a) Passband frequencies are from 1500 to 2500 Hz with an attenuation of 3 dB. (b) Stopband frequencies are below 1000 Hz and above 3000 Hz with the passband ripple of 40 dB. Sampling frequency $(F_s) = 8$ kHz. Use BLT method for transformation.

8.7   Elliptic Filter

The elliptic filter has an equiripple in both the passband and stopband. It is also called as 'Cauer filter'. The elliptic filter gives minimal error between the desired and ideal filter response for a given set of error tolerances. The elliptical filter maximizes the rate of transition between the frequency response passband and stopband, at the expense of equiripple in both bands. Also, it increases ringing in the step response.

The square magnitude response of 'nth' order elliptic filter is given by

$$|H(j\Omega)|^2 = \frac{1}{1 + \varepsilon^2 R_n^2 \left(\xi, \Omega/\Omega_c \right)} \tag{8.19}$$

In the above expression, R_n is the nth order elliptic rational function, 'ε' is the ripple factor and 'ξ' is the selectivity factor. The ripple factor specifies the passband ripple, and a combination of the ripple factor and the selectivity factor specifies the stopband ripple. As the ripple in the stopband approaches zero, the filter tends to become Chebyshev Type I filter. As the ripple in the passband approaches zero, the filter tends to become Chebyshev Type II filter. If both passband and stopband ripples approach zero, then the filter tends to become a Butterworth filter. Elliptic filter meets a standard magnitude specification with lower filter order than other filter approximations.

Experiment 8.25 Design of an Elliptic Lowpass Filter

This experiment tries to design an elliptic lowpass filter with a passband frequency of 5 Hz. Assume the order of the filter is 3, passband ripple is 2 dB and stopband attenuation is 30 dB. The python code to design an elliptic lowpass filter is shown in Fig. 8.59, and its corresponding simulation result is shown in Fig. 8.60. From Fig. 8.59, it is possible to infer that the *signal.ellip* built-in command is used here to obtain the numerator and denominator coefficients. After executing the python code given in Fig. 8.59, the transfer function of elliptic lowpass filter is shown in Fig. 8.60a. This figure confirms that the elliptic filter has zeros, and those zeros occur on the imaginary axis. The magnitude response of the elliptic lowpass filter is shown in Fig. 8.60b. From this figure, it is possible to observe the equiripples in both the passband and stopband.

```
import numpy as np
import matplotlib.pyplot as plt
from scipy import signal
import control as ss
N, rp,rs, omega_c=3, 2, 20, [5]
omega1=np.linspace(0, 10, 100)
b_s, a_s=signal.ellip(N,rp,rs,omega_c,'low', analog=True)
s1 = ss.tf(b_s, a_s)
print('H(s) = ', s1)
z1=np.roots(b_s)
print('Zeros : ', z1)
mag, phase, omega1=ss.freqresp(s1, omega1)
plt.figure,plt.plot(omega1,np.abs(mag))
plt.xlabel('$\u03A9$-->'),plt.ylabel('$|H({j\u03A9})|$')
plt.title('Magnitude response of Elliptic LPF')
plt.tight_layout()
```

Fig. 8.59 Python code for elliptic LPF

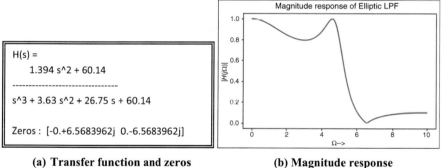

H(s) =

   1.394 s^2 + 60.14

----

s^3 + 3.63 s^2 + 26.75 s + 60.14

Zeros : [-0.+6.5683962j 0.-6.5683962j]

(a) **Transfer function and zeros**   (b) **Magnitude response**

Fig. 8.60 Simulation result of the python code is given in Fig. 8.59. (a) Transfer function and zeros. (b) Magnitude response

Inference

From the simulation result of this experiment, it is possible to observe that ripples present in both the passband and stopband, and also it infers that the width of the transition band is reduced.

Task

1. Write a python code to design an elliptic highpass filter with a lower cut-off frequency of 5 Hz. Assume the order of the filter is 3, passband ripple is 2 dB and stopband attenuation is 30 dB.

```
import numpy as np
import matplotlib.pyplot as plt
from scipy import signal
import control as ss
# Specifications of Filter
fsam=8000 # Sampling frequency in Hz
fc1,fc2=[1500,2500],[1000,3000] # cut off frequency in Hz
Ap, As, T = 2, 40, 1/fsam
wcp1=2*np.pi*fc1[0] # Pass band Cut off frequency in rad/sec
wcp2=2*np.pi*fc1[1] # Pass band Cut off frequency in rad/sec
wcs1=2*np.pi*fc2[0] # Stop band Cut off frequency in rad/sec
wcs2=2*np.pi*fc2[1] # Stop band Cut off frequency in rad/sec
pwcs1 = (2/T)*np.tan(wcp1*T/2) # Prewarp the analog frequency
pwcs2 = (2/T)*np.tan(wcp2*T/2) # Prewarp the analog frequency
pwcp1 = (2/T)*np.tan(wcs1*T/2) # Prewarp the analog frequency
pwcp2 = (2/T)*np.tan(wcs2*T/2) # Prewarp the analog frequency
N,wn=signal.ellipord([pwcp1,pwcp2],[pwcs1,pwcs2],Ap,As,analog=True)
print('Order of the filter (N) = ',N)
# Design analog Elliptic filter using signal.ellip function
b, a = signal.ellip(N, Ap, As, wn, 'bandstop', analog='True')
s1 = ss.tf(b,a)
print('Transfer function H(s)=',s1)
# Perform bilinear transformation
bz, az = signal.bilinear(b, a, fs=fsam)
z1 = ss.tf(bz,az,T)
print('Transfer function H(z)=',z1)
# Compute frequency response of the filter using signal.freqz function
wz, hz = signal.freqz(bz, az, 512)
fig = plt.figure(figsize=(10, 8))
Mag = 10*np.log10(abs(hz)) # Calculate Magnitude in dB
Freq = wz*fsam/(2*np.pi) # Calculate frequency in Hz
# Plot Magnitude response
sub1 = plt.subplot(2, 1, 1)
sub1.plot(Freq, Mag, 'r', linewidth=2),sub1.axis([1, fsam/2, -60, 5])
sub1.set_title('Magnitude Response', fontsize=15),
sub1.set_xlabel('Frequency [Hz]', fontsize=15),sub1.set_ylabel('Magnitude [dB]', fontsize=15)
sub1.grid()
# Plot phase angle
sub2 = plt.subplot(2, 1, 2)
Phase = np.unwrap(np.angle(hz))*180/np.pi # Calculate phase angle in degree from hz
sub2.plot(Freq, Phase, 'g', linewidth=2),sub2.set_ylabel('Phase (degree)', fontsize=15)
sub2.set_xlabel(r'Frequency (Hz)', fontsize=15),sub2.set_title(r'Phase response', fontsize=15)
sub2.grid(),plt.subplots_adjust(hspace=0.5),fig.tight_layout(),plt.show()
```

Fig. 8.61 Python code for elliptic bandstop filter design

Experiment 8.26 Design of Elliptic Digital Bandstop Filter

The objective of this experiment is to design an elliptic digital bandstop filter for the following specifications: (1) Stopband frequencies are from 1500 to 2500 Hz with an attenuation of 40 dB. (2) Passband frequencies are below 1000 Hz and above 3000 Hz with the passband ripple of 3 dB. Sampling frequency (F_s) = 8 kHz. Use BLT method for transformation. The python code for this experiment is given in Fig. 8.61, and its corresponding output is shown in Figs. 8.62 and 8.63. Figure 8.63

Order of the filter (N) = 3
Transfer function H(s)=

$$s^6 + 9.409e+08\ s^4 + 2.409e+17\ s^2 + 1.678e+25$$

$$s^6 + 9.278e+04\ s^5 + 2.858e+09\ s^4 + 1.391e+14\ s^3 + 7.317e+17\ s^2 + 6.081e+21\ s + 1.678e+25$$

Transfer function H(z)=

$$0.1338\ z^6 - 3.28e{-}11\ z^5 + 0.3241\ z^4 - 5.86e{-}11\ z^3 + 0.3241\ z^2 - 3.28e{-}11\ z + 0.1338$$

$$z^6 - 1.122e{-}10\ z^5 - 0.5182\ z^4 - 1.985e{-}11\ z^3 + 0.7377\ z^2 + 7.843e{-}12\ z - 0.3037$$

dt = 0.000125

Fig. 8.62 Simulation result of python code is given in Fig. 8.61

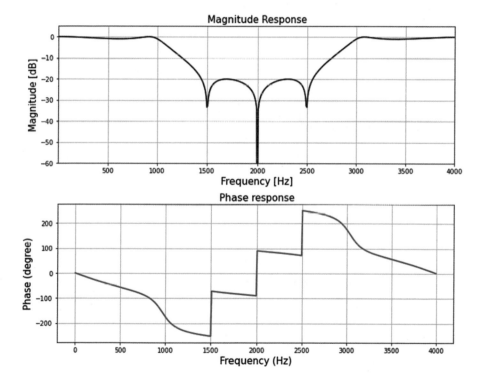

Fig. 8.63 Magnitude and phase responses

shows the magnitude and phase response of the elliptic (Cauer) digital bandstop filter.

Inference

From this experiment, the following observations can be made:

```
#Magnitude response of analog IIR filter
import numpy as np
import matplotlib.pyplot as plt
from scipy import signal
#Step 1: Specification of the filter
N=10 #Order of the filter
wn=100 #Cut-off frequency
rp=5  #Pass band ripple
rs=40  #Stop band ripple
#Step 2: Magnitude response of Butterworth filter
b1,a1=signal.butter(N,wn,'low',analog='true')
w1,H1=signal.freqs(b1,a1)
#Magnitude response of Type-1 Chebyshev filter
b2,a2=signal.cheby1(N,rp,wn,'low',analog='true')
w2,H2=signal.freqs(b2,a2)
#Magnitude response of Type-II Chebyshev filter
b3,a3=signal.cheby2(N,rs,wn,'low',analog='true')
w3,H3=signal.freqs(b3,a3)
#Magnitude response of Elliptic filter
b4,a4=signal.ellip(N,rp,rs,wn,'low',analog='true')
w4,H4=signal.freqs(b4,a4)
#Step 3: Plotting the magnitude responses
plt.subplot(2,2,1),plt.semilogx(w1,20*np.log10(abs(H1))),'k-')
plt.xlabel('Frequency'),plt.ylabel('Magnitude[dB]'),plt.title('Butterworth filter')
plt.subplot(2,2,2),plt.semilogx(w2,20*np.log10(abs(H2))),'k-')
plt.xlabel('Frequency'),plt.ylabel('Magnitude[dB]'),plt.title('Chebyshev (type 1) filter')
plt.subplot(2,2,3),plt.semilogx(w3,20*np.log10(abs(H3))),'k-')
plt.xlabel('Frequency'),plt.ylabel('Magnitude[dB]'),plt.title('Chebyshev (type 2) filter')
plt.subplot(2,2,4),plt.semilogx(w4,20*np.log10(abs(H4))),'k-')
plt.xlabel('Frequency'),plt.ylabel('Magnitude[dB]'),plt.title('Elliptic filter')
plt.tight_layout()
```

Fig. 8.64 Python code to obtain the magnitude response of four analogue IIR filters

1. The magnitude response is having ripples in the passband as well as stopband.
2. The phase response is nonlinear; hence, IIR filters do not have linear phase characteristics.

Experiment 8.27 Comparing the Magnitude Response of Butterworth, Chebyshev (Type I), Chebyshev (Type II) and Elliptic Filters

The objective of this experiment is to compare the magnitude response of analogue IIR filters, namely, Butterworth, Chebyshev (Type I), Chebyshev (Type II) and elliptic filters for the following lowpass filter specifications: (1) order of the filter = 10, (2) cut-off frequency = 100 Hz, (3) passband ripple = 5 dB and (4) stopband ripple = 40 dB. The python code, which obtains the magnitude response of the four filters, is shown in Fig. 8.64, and the corresponding output is shown in Fig. 8.65.

The built-in functions used in the program are summarized in Table 8.4.

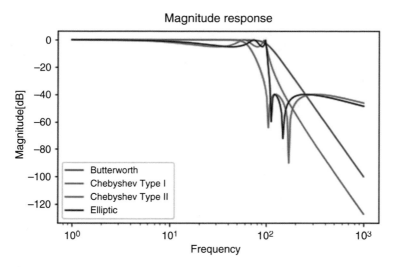

Fig. 8.65 Result of python code shown in Fig. 8.64

Table 8.4 Built-in function used

S. No.	Built-in function	Library	Purpose of built-in function
1	butter	Scipy	To obtain the coefficients of analogue/digital Butterworth filter
2	cheby1	Scipy	To obtain the coefficients of analogue/digital Type I Chebyshev filter
3	cheby2	Scipy	To obtain the coefficients of analogue/digital Type II Chebyshev filter
4	ellip	Scipy	To obtain the coefficients of analogue/digital elliptic filter

Inferences

From Fig. 8.65, the following inferences can be made:

1. Butterworth filter exhibits maximally flat frequency response without ripples in passband and stopband.
2. Type I Chebyshev filter exhibits ripples in only passband.
3. Type II Chebyshev filter exhibits ripples in stopband.
4. Elliptic filter exhibits ripples in both passband and stopband.
5. The roll-off rate of elliptic filter is better than Butterworth and Chebyshev filters.

Experiment 8.28 Comparing the Order of Butterworth, Chebyshev (Type I and Type II) and Elliptic Filters for the Same Filter Specification

The objective of this experiment is to compute the order of Butterworth, Type I Chebyshev, Type II Chebyshev and elliptic filter for the following specifications using bilinear transformation technique (BLT). −3 dB cut-off frequency at 5 Hz, and an attenuation of 40 dB at 10 Hz, use bilinear transformation technique. Assume the

```
#Order of different IIR filters
import numpy as np
import matplotlib.pyplot as plt
from scipy import signal
#Step 1: Filter specification
fsample=1000
f_pass, f_stop, g_pass, g_stop, Td =5, 10, 3, 40, 1
wp=f_pass/(fsample/2)
ws=f_stop/(fsample/2)
#Step 2: Pre-warping
omega_pass=(2/Td)*np.tan(wp/2)
omega_stop=(2/Td)*np.tan(ws/2)
#Step 3: Computing the order of different filters
N1,w1=signal.buttord(omega_pass,omega_stop,g_pass,g_stop)
N2,w2=signal.cheb1ord(omega_pass,omega_stop,g_pass,g_stop)
N3,w3=signal.cheb2ord(omega_pass,omega_stop,g_pass,g_stop)
N4,w4=signal.ellipord(omega_pass,omega_stop,g_pass,g_stop)
#Step 4: Plotting the order of different filters
filter_name=['Butterworth','Cheby1','Cheby2','Elliptic']
order1=[N1,N2,N3,N4]
plt.bar(filter_name,order1),plt.title('Order of different filters')
for i, v in enumerate(order1):
    plt.text(i, v+0.1, str(v),color = 'blue', fontweight = 'bold')
plt.show()
```

Fig. 8.66 Comparing the order of different IIR filters

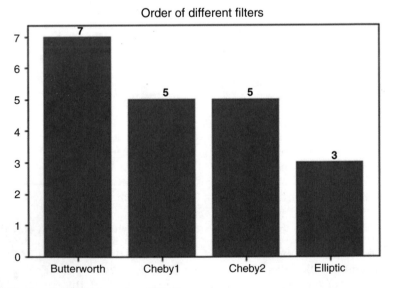

Fig. 8.67 Comparison of the order of IIR filters

```
# Low pass Butterworth filter
import numpy as np
import matplotlib.pyplot as plt
from scipy import signal
#Step 1: Signal generation
t1=np.linspace(0,1,100)
f1, f2, f3=5, 0, 10
x1=np.sin(2*np.pi*f1*t1)
x2=np.sin(2*np.pi*f2*t1)
x3=np.sin(2*np.pi*f3*t1)
x=np.concatenate([x1,x2,x3])
t=np.linspace(0,1,len(x))
plt.figure(1),plt.subplot(2,3,1),plt.plot(t,x),plt.xlabel('Time'),plt.ylabel('Amplitude')
plt.title('Input signal')
#Step 2: Design of filter
N=[2,5,10,20,25] #Order of the filter
fsamp=100 #Sampling frequency
f_cut=8 #Cut-off frequency
fn=fsamp/2
wc=f_cut/fn
for i in range(len(N)):
    b, a = signal.butter(N[i],wc,'low')
#Step 3: Obtaining the output
    plt.figure(1), plt.subplot(2,3,i+2)
    y=signal.lfilter(b,a,x)
    plt.plot(t,y),plt.xlabel('Time'),plt.ylabel('Amplitude')
    plt.title('Filtered signal (N={})'.format(N[i]))
    plt.tight_layout()
```

Fig. 8.68 Lowpass Butterworth filter to filter sinusoidal signal

sampling frequency to be 1000 Hz. The python code, which computes the order of different IIR filter, is shown in Fig. 8.66, and the corresponding output is shown in Fig. 8.67.

Inferences

From Fig. 8.67, it is possible to observe the following facts:

1. The order of Butterworth filter to meet the given filter specification is 7.
2. The order of Chebyshev filter to meet the given filter specification is 5.
3. The order of elliptic filter to meet the given filter specification is 3.
4. The order of Butterworth filter is higher than the order of Chebyshev and elliptic filters.
5. Elliptic filter meets the given filter specification with a minimum order.

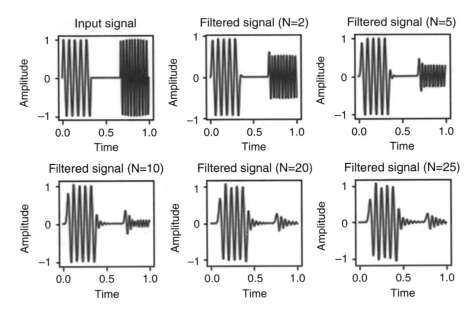

Fig. 8.69 Butterworth lowpass filter output

Experiment 8.29 Filtering of Sinusoidal Signal Using Butterworth Filter of a Different Order

This experiment aims to filter a sinusoidal signal consisting of three frequencies: 5, 0 and 10 Hz. This signal is to be filtered by a Butterworth lowpass filter of different orders, namely, 2, 5, 10, 20 and 25. The cut-off and sampling frequencies chosen are 8 and 100 Hz, respectively. The python code, which implements the above-mentioned task is shown in Fig. 8.68, and the corresponding outputs are shown in Fig. 8.69.

Inferences

The following inferences can be made from Fig. 8.69:

1. The sinusoidal input signal contains three frequencies, namely, 5, 0 and 10 Hz.
2. The input signal is passed through lowpass Butterworth filter of order 2, 5, 10, 20 and 25.
3. The filter retains 5 Hz frequency, and it blocks 10 Hz frequency component.
4. The extent of filtering increases with an increase in the order of the filter. The 10 Hz frequency component is blocked effectively when the filter order is 20 and 25.
5. The group delay increases with an increase in the order of the filter. The delay can be observed in the filtered signal with orders 20 and 25.

Task

1. Use the highpass filter with cut-off frequency of 10 Hz to repeat Experiment 8.29. Compare the result with the result of Experiment 8.29.

```
#Comparing the performances of IIR filters
import numpy as np
import matplotlib.pyplot as plt
from scipy import signal
#Step 1: Signal generation
t1=np.linspace(0,1,100)
f1,f2,f3=5,0,10
x1=np.sin(2*np.pi*f1*t1)
x2=np.sin(2*np.pi*f2*t1)
x3=np.sin(2*np.pi*f3*t1)
x=np.concatenate([x1,x2,x3])
t=np.linspace(0,1,300)
plt.figure(1),plt.plot(t,x),plt.xlabel('Time'),plt.ylabel('Amplitude')
plt.title('Input signal')
#Step 2: Design of filter
N=5 #Order of the filter
fsamp=100 #Sampling frequency
f_cut=8 #Cut-off frequency
fn=fsamp/2
wc=f_cut/fn
b1, a1 = signal.butter(N,wc,'low')
b2, a2 = signal.cheby1(N,3,wc,'low')
b3, a3 = signal.cheby2(N,40,wc,'low')
b4, a4 = signal.ellip(N,3,40,wc,'low')
#Step 3: Obtaining the output of the filter
y1=signal.lfilter(b1,a1,x)
y2=signal.lfilter(b2,a2,x)
y3=signal.lfilter(b3,a3,x)
y4=signal.lfilter(b4,a4,x)
#Step 4: Plotting the output signals
plt.figure(2)
plt.subplot(2,2,1),plt.plot(t,y1),plt.xlabel('Time'),plt.ylabel('Amplitude'),
plt.title('Butterworth filter output')
plt.subplot(2,2,2),plt.plot(t,y2),plt.xlabel('Time'),plt.ylabel('Amplitude'),
plt.title('Chebyshev (Type 1) filter output')
plt.subplot(2,2,3),plt.plot(t,y3),plt.xlabel('Time'),plt.ylabel('Amplitude'),
plt.title('Chebyshev (Type2) filter output')
plt.subplot(2,2,4),plt.plot(t,y4),plt.xlabel('Time'),plt.ylabel('Amplitude'),
plt.title('Elliptic filter output')
plt.tight_layout()
```

Fig. 8.70 Python code to compare the performances of IIR filters

Experiment 8.30 Filtering of Sinusoidal Signal with Different IIR Filters of the Same Order

This experiment aims to filter a sinusoidal signal consisting of three frequencies: 5, 0 and 10 Hz. This signal is to be filtered by a Butterworth, Chebyshev (Type I), Chebyshev (Type II) and elliptic lowpass filters. The order of the filter is fixed as

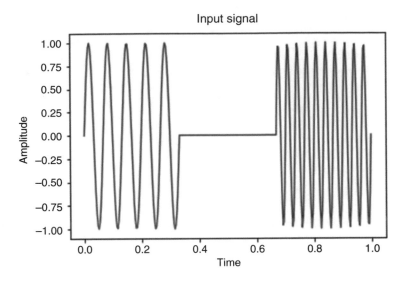

Fig. 8.71 Input signal

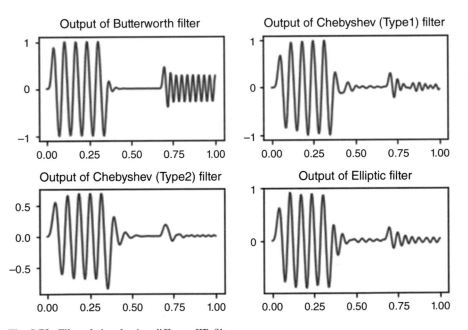

Fig. 8.72 Filtered signal using different IIR filters

5. The cut-off frequency and the sampling frequency chosen are 8 and 100 Hz, respectively.

The python code, which implements the above-mentioned task, is shown in Fig. 8.70, and the corresponding outputs are shown in Figs. 8.71 and 8.72.

Inferences

From Fig. 8.71, it is possible to observe that the input signal has three frequency components, namely 5, 0 and 10 Hz. This

1. Input signal is passed through lowpass Butterworth, Chebyshev (Type I and Type II) and elliptic filters.
2. Butterworth filter performs smoothing action without ripples. But the extent of filtering is poor. Significant high-frequency components appear along with low frequency components.
3. In Chebyshev and elliptic filters, the high-frequency components are filtered effectively; however, the filtering action is not smooth. Ripples appear in the filtered output signal.

Exercises

1. The transfer function of analogue filter is given by $H(s) = \frac{10}{s+10}$. Write a python code to obtain the transfer function of the equivalent digital filter using (a) impulse invariant technique and (b) bilinear transformation technique. Assume the sampling period to be 1 s.
2. A normalized first-order lowpass Butterworth filter transfer function is given by $H(s) = \frac{1}{(s+1)}$. Write a python code to convert this filter into desired highpass filter with a cut-off frequency of 5 rad/s. Using subplot, plot the magnitude response of normalized filter and the desired highpass filter.
3. Write a python code to obtain the order and cut-off frequency of analogue Butterworth filter that has -3 dB cut-off frequency of 20 rad/s and 10 dB of attenuation at 40 rad/s. Plot the magnitude response of the filter.
4. Design an analogue bandpass filter that has the following characteristics:

   (a) -3 dB upper and lower cut-off frequency of 100 Hz and 10 kHz
   (b) Stopband attenuation of 30 dB at 50 Hz and 25 kHz
   (c) Monotonic frequency response

   Plot the magnitude response of the above-mentioned filter.
5. Write a python code to design a digital lowpass Butterworth filter for the following specification using impulse invariant technique (a) -3 dB cut-off frequency at 250 Hz and (b) magnitude of frequency response down at least 10 dB at 500 Hz. Assume the sample to be 1000 samples/s. Plot the magnitude response of the filter.
6. Write a python code to design a digital filter using the bilinear transformation technique for the following specification: (a) maximally flat frequency response with -3 dB cut-off at 10 rad/s, (b) 30 dB of attenuation at all frequencies greater than 20 rad/s and (c) assuming the sampling frequency to be 1000 samples/s. Plot the magnitude response of the filter.
7. Write a python code to design a Type I Chebyshev filter using bilinear transformation technique (BLT) that has a passband of 0 to -3 dB cut-off frequency at 5 kHz and attenuation of at least 30 dB for frequencies greater than 10 kHz.

Assume the sampling frequency to be 20 kHz. Assume $T = 1$ s. Plot the magnitude response of the filter.

8. Write a python code to design a Chebyshev Type II digital bandstop filter for the following specifications: (a) Stopband frequencies are from 1000 to 2000 Hz with an attenuation of 40 dB. (b) Passband frequencies are below 800 Hz and above 2500 Hz with the passband ripple of 3 dB. Sampling frequency $(F_s) = 8$ kHz. Use BLT method for transformation. Plot its frequency response.

9. Write a python code to design an elliptic lowpass filter with the passband frequency of 10 Hz. Assume the order of the filter is 2, passband ripple 3 dB and stopband attenuation is 40 dB.

10. Obtain the order of Butterworth filter, Type I Chebyshev filter, Type II Chebyshev filter and elliptic filter that has -3 dB bandwidth of 10 Hz and an attenuation of 30 dB at 20 Hz using bilinear transformation technique. Assume the sampling frequency to be 1500 Hz.

Objective Questions

1. The filter has maximally flat response at both passband and stopband is called as

   A. Elliptic filter
   B. Butterworth filter
   C. Chebyshev filter
   D. Inverse Chebyshev filter

2. The filter has monotonic response at passband and ripple at stopband

   A. Elliptic filter
   B. Butterworth filter
   C. Chebyshev filter
   D. Inverse Chebyshev filter

3. The filter has ripple at passband and monotonic response at stopband

   A. Elliptic filter
   B. Butterworth filter
   C. Chebyshev filter
   D. Inverse Chebyshev filter

4. The filter has ripple at both passband and stopband

   A. Elliptic filter
   B. Butterworth filter
   C. Chebyshev filter
   D. Inverse Chebyshev filter

5. Mapping between S-plane to Z-plane using approximation derivative method is

   A. $s = \frac{1-z^{-1}}{T}$
   B. $s = \frac{1+z^{-1}}{T}$

C. $s = \frac{T}{1 - z^{-1}}$

D. $s = \frac{T}{1 + z^{-1}}$

6. Identify the wrong statement

    A. Impulse invariant method exists one to one mapping between Ω and ω.
    B. Impulse invariant method does not exist one to one mapping between Ω and ω.
    C. Impulse invariant method retains the stability while converting analogue filter into digital filter.
    D. Impulse invariant method is appropriate for the design of lowpass and bandpass filters only.

7. Mapping between S-plane to Z-plane using bilinear transformation method is

    A. $s = \frac{2}{T} \left\{ \frac{1 + z^{-1}}{1 - z^{-1}} \right\}$

    B. $s = \frac{2}{T} \left\{ \frac{1 - z^{-1}}{1 + z^{-1}} \right\}$

    C. $s = \frac{T}{2} \left\{ \frac{1 + z^{-1}}{1 - z^{-1}} \right\}$

    D. $s = \frac{T}{2} \left\{ \frac{1 - z^{-1}}{1 + z^{-1}} \right\}$

8. Identify the correct statement

    A. The order (N) of the Chebyshev polynomial function $T_N(x)$ is even; the graph will pass through origin.
    B. The order (N) of the Chebyshev polynomial function $T_N(x)$ is even, and the amplitude of the $T_N(x)$ is equal to zero at x equal to zero.
    C. The order (N) of the Chebyshev polynomial function $T_N(x)$ is odd, and the amplitude of the $T_N(x)$ is equal to non-zero at x equal to zero.
    D. The order (N) of the Chebyshev polynomial function $T_N(x)$ is even; the graph will not pass through origin.

9. The poles of Chebyshev Type I analogue filter are located on an S-plane in the form of

    A. Circle
    B. Parabola
    C. Ellipse
    D. Hyperbola

10. The Chebyshev type II analogue filter has

    A. Zeros as the numerator and all the zeros lie on the real axis only.
    B. Zeros as the numerator and all the zeros lie on the imaginary axis only.
    C. Zeros as the numerator and all the zeros lie on both real and imaginary axes.
    D. None of the above.

Bibliography

1. Harry Y.F. Lam, "Analog and Digital Filters: Design and Realization", Prentice Hall, 1979.
2. Andreas Antoniou, "Digital Filters: Analysis, Design and Signal Processing Applications", McGraw Hill, 2018.
3. Dietrich Schlichtharle, "Digital Filters: Basics and Design", Springer, 2011.
4. Leland B. Jackson, "Digital Filters and Signal Processing", Springer, 1988.
5. Thomas Haslwanter, "Hands-on Signal Analysis with Python", Springer, 2022.

Chapter 9
Quantization Effect of Digital Filter Coefficients

Learning Objectives

After completing this chapter, the reader is expected to

- Understand and implement approximation of numbers through flooring, ceiling and rounding operations.
- Analyse the impact of quantizing the finite impulse response filter coefficients.
- Analyse the impact of quantizing the infinite impulse response filter coefficients.
- Demonstrate limit cycle oscillation due to quantization of IIR filter coefficients.

Roadmap of the Chapter

The roadmap of this chapter is shown below. This chapter discusses the effect of fixed-point representations of digital systems and the effect of quantization using rounding, two's complement and magnitude truncation approaches. Also, it discusses the impact of the finite word length effect of FIR and IIR filters.

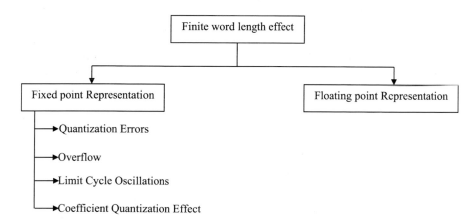

© The Author(s), under exclusive license to Springer Nature Singapore Pte Ltd. 2024   377
S. Esakkirajan et al., *Digital Signal Processing*,
https://doi.org/10.1007/978-981-99-6752-0_9

PreLab Questions
1. What is the finite word length effect in DSP?
2. List out the difference between fixed-point and floating-point number representations.
3. What is the significance of sign-magnitude representation?
4. What are quantization and quantization errors?
5. What is rounding, and how does it the quantization process?
6. What do you mean by two's complement truncation in quantization, and what is the range of quantization error?
7. What is magnitude truncation, and can it suppress the limit cycle oscillation?
8. What is overflow? And mention the types of overflow.
9. What do you mean by limit cycle oscillation, and how it exists in signal processing?
10. What do you mean by coefficient quantization, and how it affects the stability of the result?

9.1   Number Representation

Numeric representation and type of arithmetic profoundly influence the performance of DSP system. The two forms of representation of numbers are fixed-point representation and floating-point representation. Fixed-point arithmetic represents numbers in a fixed range with a finite number of bits of precision. Numbers outside the specified range will either saturate or wrap around. In general, fixed-point representation is preferred for high speed and lower cost. Floating-point arithmetic represents every number in two parts (a) a mantissa and (b) an exponent. Floating-point representation has a higher dynamic range, and there is no need for scaling, which makes it attractive for complex algorithms.

The implementation of digital filters involves the use of finite precision arithmetic. This leads to quantizing the filter coefficients and the results of the arithmetic operations. Such quantization operations are non-linear and cause a filter response substantially different from the response of the underlying infinite precision model.

9.2   Fixed-Point Quantization

Quantization is the process of approximating a quantity X into a quantity $Q(X)$. It is approximately equal to X, but it has some distortion errors. Quantization exists when it represents the real numbers by nearest integers and is to reduce the word length of a binary representation of X by reducing the number of bits after the binary point. The relationship between X and $Q(X)$ is called as 'quantization characteristics'. Quantization commonly comes across in digital signal processing in the form of

1. Rounding
2. Two's complement truncation
3. Magnitude truncation

9.2.1   Fixed-Point Quantization by Rounding

If X is an original value to be quantized using the rounding method, and the quantized output is denoted as $Q_r(X)$. If the quantized value has N bits to the right of the binary point, then the quantization step size is $\Delta = 2^{-N}$. In general, rounding selects the quantized integer nearest to the original value. But the error between the quantized and original value is not more than $\pm \frac{\Delta}{2}$. Hence, the rounding error is denoted as $\varepsilon_r = Q_r(X) - X$. Then, the rounding error is ranging from $-\frac{\Delta}{2} \leq \varepsilon_r \leq \frac{\Delta}{2}$. The error resulting from quantization can be modelled as a random variable uniformly distributed over the appropriate error range. Hence, the round off error can be considered error-free calculations that have been corrupted by additive white noise. The mean value of the rounding error is zero and variance of the rounding error is $\sigma_{\varepsilon_r}^2 = \Delta^2 / 12$.

Experiment 9.1 Perform the Fixed-Point Quantization Using the Rounding Method
This experiment deals with quantizing value using a rounding approach. The first and foremost method of quantization is rounding, which is mathematically written as

$$X_r = \Delta \times \text{round}(x/\Delta) \qquad (9.1)$$

where x denotes the value to be quantized, Δ represents the step size and is computed as $\Delta = 2^{-N}$, N denotes the number of bits and X_r represents the quantized output using rounding. The python code for computing quantization using the rounding approach is given in Fig. 9.1, and its corresponding simulation result is shown in Fig. 9.2.

```
# This program performs quantization using rounding
import numpy as np
h=float(input('Enter the value to be Quantized: '));
B=int(input('Enter the Number of Bits (N): '));
Q = 1/(2**(B))
Qhr=Q*np.round(h/Q)#Rounding
print('The input unquantized value : ',h)
print('The Quantized result using Rounding: ', Qhr)
```

Fig. 9.1 Python code for quantization using rounding

Fig. 9.2 Simulation result
of python code given in
Fig. 9.1

> Enter the value to be Quantized: 0.126
>
> Enter the Number of Bits (N): 2
>
> The input unquantiz ed value : 0.126
>
> The Quantized result using Rounding: 0.25

Inference

1. From Fig. 9.2, it is possible to infer that the value to be quantized is '0.126' and the number of bits that can be used to represent the value is '2'.
2. The two bits can represent the maximum of 4 levels (0–0.25, 0.25–0.5, 0.5–0.75 and 0.75–1.0).
3. Therefore, the input '0.126' is quantized, and the result is 0.25 (0.126 is greater than 0.125).

Task

1. Enter the value to be quantized is 5.15, and 8 bits are used for the quantization. Observe the quantized result.

9.2.2 Fixed-Point Quantization Using Two's Complement Truncation

The two's complement truncation always gives the quantized value less than or equal to the original value. Hence, the truncation error is denoted as $\varepsilon_t = Q_t(X) - X$. Then, the truncation error is ranging from $-\Delta \leq \varepsilon_t \leq 0$. The error resulting from quantization can be modelled as a random variable uniformly distributed over the appropriate error range. Hence, the two's complement truncation error can be considered error-free calculations corrupted by additive white noise. The mean value of the two's complement truncation error can be obtained as $\mu_{\varepsilon_t} = -\frac{\Delta}{2}$. The variance of the two's complement truncation error is calculated as $\sigma_{\varepsilon_t}^2 = \Delta^2/12$.

Experiment 9.2 Perform Fixed-Point Quantization Using Two's Complement
This experiment deals with quantizing the value using two's complement truncation approach. The quantization using two's complement truncation method is denoted as

$$X_t = \Delta \times \lfloor (x/\Delta) \rfloor \tag{9.2}$$

where $\lfloor \; \rfloor$ denotes flooring operation. X_t represents the quantized result using truncation method. The python code, which computes the fixed-point quantization of a number using two's complement truncation method, is shown in Fig. 9.3, and its simulation result is depicted in Fig. 9.4.

```
# This program performs quantization using 2's complement truncation
import numpy as np
h=float(input('Enter the value to be Quantized: '));
B=int(input('Enter the Number of Bits (N): '));
Q = 1/(2**(B))
Qht=Q*np.floor(h/Q)#2's Complement truncation
print('The input unquantized value : ',h)
#print('The Quantized result using Rounding: ', Qhr)
print('The Quantized result using 2s Complement truncation: ', Qht)
```

Fig. 9.3 Python code for quantization using two's complement truncation

```
Enter the value to be Quantized: 0.126
Enter the Number of Bits (N): 2
The input unquantiz ed value :  0.126
The Quantized result using 2s Complement truncation: 0.0
```

Fig. 9.4 Simulation result of python code given in Fig. 9.3

Inference

From Fig. 9.4, it is possible to infer that the value to be quantized is 0.126, and the resultant value is 0 for $N = 2$.

Task

1. Choose the proper value of N and execute the python code given in Fig. 9.3 to get 0.126 as the output.

9.2.3 Fixed-Point Quantization Using Magnitude Truncation

Magnitude truncation gives the quantized value less than the original value for $X > 0$, and the quantized result is greater than original for $X < 0$. Hence, the magnitude truncation error is denoted as $\varepsilon_{mt} = Q_{mt}(X) - X$.

Then, the magnitude truncation error is ranging from $-\Delta \leq \varepsilon_{mt} \leq 0$ for $X > 0$ and $0 \leq \varepsilon_{mt} \leq \Delta$ for $X < 0$. The mean value of the magnitude truncation error is obtained as 0, and variance of the magnitude truncation error is calculated as $\sigma_{\varepsilon_{mt}}^2 = \Delta^2/3$. The specific advantage of magnitude truncation lies in its inherent capability of limit cycle suppression.

```
# This program performs quantization using Magnitude truncation
import numpy as np
h=float(input('Enter the value to be Quantized: '));
B=int(input('Enter the Number of Bits (N): '));
Q = 1/(2**(B))
# Magnitude Truncation
if h > 0:
    Qhmt=Q*np.floor(h/Q)
else:
    Qhmt=Q*np.ceil(h/Q)
print('The input unquantized value : ',h)
print('The Quantized result using Magnitude truncation: ', Qhmt)
```

Fig. 9.5 Python code for quantization using magnitude truncation

```
Enter the value to be Quantized: 0.126
Enter the Number of Bits (N): 2
The input unquantiz ed value :  0.126
The Quantized result using M agnitude truncation: 0.0
```

Fig. 9.6 Simulation result of python code given in Fig. 9.5

Experiment 9.3 Perform Fixed-Point Quantization Using Magnitude Truncation

The quantization using magnitude truncation is represented as

$$X_{mt} = \begin{cases} \Delta \times \lfloor (x/\Delta) \rfloor, & \text{for } x \geq 0 \\ \Delta \times \lceil (x/\Delta) \rceil, & \text{for } x < 0 \end{cases} \tag{9.3}$$

where $\lceil \ \rceil$ denotes ceiling operation. X_{mt} represents the quantized result using magnitude truncation. The python code of this experiment is shown in Fig. 9.5, and the simulation result is displayed in Fig. 9.6.

Inference

The following inferences can be made from this experiment:

1. Figure 9.6 shows that the value to be quantized is 0.126, and the number of bits is chosen as 2. The result of quantization using the magnitude truncation is 0. This simulation result is on par with the theoretical result.
2. The selection of the number of bits is essential for quantization.
3. The quantization method is crucial for the hardware implementation of the digital systems.

Task

1. Execute the python code given in Fig. 9.5, and enter the value to be quantized as '0.126' and obtain the minimum value of 'N', which will give the quantized result that is the same as the input value.

9.3 Coefficient Quantization

The filter coefficients can be obtained using filter design approaches based on the given set of filter specifications. These filter coefficients are represented as a system function of the filter $H(z)$, and they can be used in signal processing applications. These filter coefficients may be integer or non-integer numbers. If the filter coefficient is an integer, then the finite precision format is enough to represent it in digital computation. Otherwise, the coefficient is non-integer, then infinite precision format is necessary to represent it for the accurate result of the applications. However, in real-time scenario the hardware setup in the application may not be able to store the value of filter coefficients as it is due to the limited register size or finite precision processor. The representation of the filter coefficients from infinite precision to finite number precision may give coefficients quantization. This coefficient quantization can change the location of the filter poles and zeros. As a result, after implementing a filter, it may observe that the frequency response of the filter is quite different from that of the original design. The coefficients obtained by design methods are real or complex. These coefficients are often stored in a finite length register for real-world digital signal processing. The coefficients are often rounded to accommodate it in the finite length register. This causes a rounding error, which will influence the filter characteristics. The frequency response of the quantized filter coefficients will differ from the desired frequency response. Sensitivity of the filter response to coefficient quantization is dependent on the type of filter structures. Detail about this will be discussed later in this chapter.

Experiment 9.4 Effect of Quantization Using the Rounding Approach of FIR Filters

This experiment deals with the effect of coefficient quantization of FIR filters. Here, the method to quantize the filter coefficients is considered as rounding method. The FIR filter coefficients are computed by using the built in python command 'a = signal.firwin(n, cutoff = 0.25, window = "hamming")'. Here 'n' denotes the number of the filter coefficients, 'cutoff = 0.25' represents cut-off frequency of the filter is 0.25π rad/sample and the 'window = "hamming"' indicates the window function used for the FIR filter design. The python code for this experiment is shown in Fig. 9.7. In this figure, '$B = 2$' denotes the number of bits used to quantize each filter coefficient. While increasing the value of $B = 3, 4, 5, 6, \ldots$, the quantized filter's impulse and frequency response will approach the original filter's impulse and

```
from scipy import signal
import numpy as np
import matplotlib.pyplot as plt
n = 31
n1=np.arange(0,n);
a = signal.firwin(n, cutoff = 0.25, window = "hamming")
B = 16;# Number of Bits
Q = 1/(2**(B))
Qhr=Q*np.round(a/Q)#Rounding
#Obtaining the Frequency response
w,H=signal.freqz(a)
wq,Hq=signal.freqz(Qhr)
#Obtaining the pole-zero plot
z,p,k=signal.tf2zpk(a,1)
zq,pq,kq=signal.tf2zpk(Qhr,1)
#Plotting the responses
plt.figure(1),plt.subplot(2,2,1),plt.stem(a),plt.xlabel('n-->'),plt.ylabel('h[n]')
plt.title('h[n]'),plt.subplot(2,2,2),plt.plot((w/np.pi),20*np.log10(np.abs(H)))
plt.xlabel(r'$\omega$(x$\pi$rad/sample)'),plt.ylabel('Magnitude'),plt.title('|H($\omega$)|')
plt.subplot(2,2,3),plt.plot(np.real(z),np.imag(z),'ro'),plt.plot(np.real(p),np.imag(p),'kx')
theta=np.linspace(0,2*np.pi,100)
plt.plot(np.cos(theta),np.sin(theta)),plt.xlabel('Real part'),plt.ylabel('Imaginary part'),
plt.title('Pole-zero plot'),plt.subplot(2,2,4),plt.plot((w/np.pi),np.unwrap(np.angle(H)))
plt.xlabel(r'$\omega$(x$\pi$rad/sample)'),plt.ylabel('Phase'),plt.title('$\Phi(e^{jw})$')
plt.tight_layout()
plt.figure(2),plt.subplot(2,2,1),plt.stem(Qhr),plt.xlabel('n-->'),plt.ylabel('$h_q[n]$')
plt.title('$h_q[n]$ with N = {} bits'.format(B))
plt.subplot(2,2,2),plt.plot((w/np.pi),20*np.log10(np.abs(Hq)))
plt.xlabel(r'$\omega$(x$\pi$rad/sample)'),plt.ylabel('Magnitude'),plt.title('$|H_q(\omega)|')
plt.subplot(2,2,3),plt.plot(np.real(zq),np.imag(zq),'ro')
plt.plot(np.real(pq),np.imag(pq),'kx')
theta=np.linspace(0,2*np.pi,100)
plt.plot(np.cos(theta),np.sin(theta)),plt.xlabel('Real part'),plt.ylabel('Imaginary part'),
plt.title('Pole-zero plot'),plt.subplot(2,2,4),plt.plot((w/np.pi),np.unwrap(np.angle(Hq)))
plt.xlabel(r'$\omega$(x$\pi$rad/sample)'),plt.ylabel('Phase'),plt.title('$\Phi_q(e^{jw})$')
plt.tight_layout()
```

Fig. 9.7 Python code to analyse coefficient quantization effect of FIR filter

frequency response. The simulation result of the python code given in Fig. 9.7 is displayed in Fig. 9.8.

Inferences

From Fig. 9.8, the following observations can be made:

1. Figure 9.8a displays the impulse response of the original FIR filter components and its magnitude, phase and pole-zero plot.
2. Figure 9.8b shows the number of bits selected as 2 for quantizing FIR filter coefficients and its impulse response, magnitude, phase and pole-zero plot. From this figure, it is possible to infer that the impulse response of the quantized FIR filter looks like a rectangular pulse, which is completely deviated from the original impulse response (i.e.) sinc function. Magnitude and phase responses

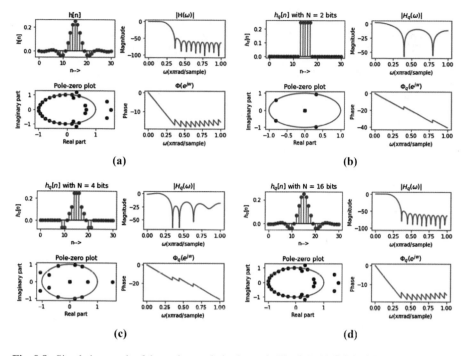

Fig. 9.8 Simulation result of the python code is shown in Fig. 9.7. (**a**) Original (unquantized). (**b**) Quantized with $N = 2$ bits. (**c**) Quantized with $N = 4$ bits. (**d**) Quantized with $N = 16$ bits

differ completely from the original magnitude and phase responses. In the pole-zero plot, some zeros occur at the origin, whereas in the original pole-zero plot, there is no zero at the origin; it indicates that after the filter coefficients, a lot of coefficients become zero due to less number of bits allocated for the representation.

3. From Fig. 9.8c, d, when the number of bits (N) is chosen as 4 and 16, the impulse response of the quantized FIR filter approaches the original impulse response, and the magnitude and phase response approaches the original one.

4. From these figures, it is possible to confirm that allocating the number of bits to represent the filter coefficients plays a significant role in DSP computations.

Experiment 9.5 Verify the Effect of Quantization Using the Two's Complement Truncation Approach of FIR Filters

This experiment discusses the effect of quantization using two's complement truncation method of FIR filter coefficients. The python code to perform the two's complement truncation-based quantization of FIR filter coefficients is shown in Fig. 9.9. In this experiment, we have chosen the order of the FIR filter as 31, the window function is considered 'Hamming', and the cut-off frequency is 0.25π rad/samples. The finite number of bits required to represent each filter coefficient

```
# This program verifies the effect of coefficient quantization of FIR filters
  # Twos complement truncation
from scipy import signal
import numpy as np
import matplotlib.pyplot as plt
n = 31
n1=np.arange(0,n);
a = signal.firwin(n, cutoff = 0.25, window = "hamming")
B = 16;# Number of Bits
Q = 1/(2**(B))
Qhr=Q*np.floor(a/Q)# Twos complement truncation
#Obtaining the Frequency response
w,H=signal.freqz(a)
wq,Hq=signal.freqz(Qhr)
#Obtaining the pole-zero plot
z,p,k=signal.tf2zpk(a,1)
zq,pq,kq=signal.tf2zpk(Qhr,1)
#Plotting the responses
plt.figure(1),plt.subplot(2,2,1),plt.stem(a),plt.xlabel('n-->'),plt.ylabel('h[n]')
plt.title('h[n]'),plt.subplot(2,2,2),plt.plot((w/np.pi),20*np.log10(np.abs(H)))
plt.xlabel(r'$\omega$(x$\pi$rad/sample)'),plt.ylabel('Magnitude'),plt.title('|H($\omega$)|')
plt.subplot(2,2,3),plt.plot(np.real(z),np.imag(z),'ro'),plt.plot(np.real(p),np.imag(p),'kx')
theta=np.linspace(0,2*np.pi,100)
plt.plot(np.cos(theta),np.sin(theta))
plt.xlabel('Real part'),plt.ylabel('Imaginary part'),plt.title('Pole-zero plot')
plt.subplot(2,2,4),plt.plot((w/np.pi),np.unwrap(np.angle(H)))
plt.xlabel(r'$\omega$(x$\pi$rad/sample)'),plt.ylabel('Phase'),plt.title('$\Phi(e^{jw})$')
plt.tight_layout()
plt.figure(2),plt.subplot(2,2,1),plt.stem(Qhr),plt.xlabel('n-->'),plt.ylabel('$h_q[n]$')
plt.title('$h_q[n]$ with N = {} bits'.format(B))
plt.subplot(2,2,2),plt.plot((w/np.pi),20*np.log10(np.abs(Hq)))
plt.xlabel(r'$\omega$(x$\pi$rad/sample)'),plt.ylabel('Magnitude'),plt.title('$|H_q(\omega$)|')
plt.subplot(2,2,3),plt.plot(np.real(zq),np.imag(zq),'ro'),plt.plot(np.real(pq),np.imag(pq),'kx')
theta=np.linspace(0,2*np.pi,100)
plt.plot(np.cos(theta),np.sin(theta)),plt.xlabel('Real part'),plt.ylabel('Imaginary part'),
plt.title('Pole-zero plot'),plt.subplot(2,2,4),plt.plot((w/np.pi),np.unwrap(np.angle(Hq)))
plt.xlabel(r'$\omega$(x$\pi$rad/sample)'),plt.ylabel('Phase'),plt.title('$\Phi_q(e^{jw})$')
plt.tight_layout()
```

Fig. 9.9 Python code for two's complement truncation

is selected as 2, 4 and 16. The simulation result of the python code given in Fig. 9.9 is shown in Fig. 9.10.

Inferences

The following observations can be drawn from Fig. 9.10:

1. When the number of bits $N = 2$, the impulse response of the FIR filter is completely deviated from the original impulse response of the FIR filter. Similarly, the magnitude and phase responses differ from the original one. From the pole-zero plot, the locations of the zeros of quantized filter are dislocated from the original positions.

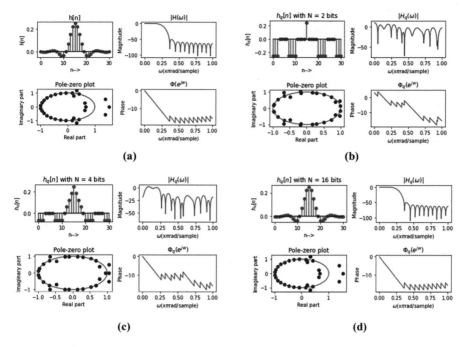

Fig. 9.10 Simulation result of the python code given in Fig. 9.9. (**a**) Original filter. (**b**) Quantized filter with $N = 2$ bits. (**c**) Quantized filter with $N = 4$ bits. (**d**) Quantized filter with $N = 16$ bits

2. When the number of bits is chosen as 4, the impulse response is still not the same as the original one. From this, 4 bits for each filter coefficient are insufficient for the finite precision.
3. When the number of bits is considered as 16. The impulse, magnitude and phase responses are exactly the same as the original one. This is also evident from the pole-zero plot.
4. From this experiment, we must understand that allocating the number of bits to represent the filter coefficients is very important for the FIR filter design. Even though the FIR filter is stable after the quantization, the filter response is not as expected.

Experiment 9.6 Verify the Effect of Quantization Using the Magnitude Truncation Approach of FIR Filters

This experiment explores the effect of quantization using magnitude truncation of FIR filter coefficients. The magnitude truncation is another type of quantization approach that can represent the infinite precision to the finite precision of filter coefficients. The python code to perform the magnitude truncation of the FIR filter coefficients is shown in Fig. 9.11. In this experiment, we have chosen the order of FIR filter is 31, the window function is considered 'Hamming' and the cut-off frequency is 0.25π rad/samples. The finite number of bits required to represent

```
# This program verifies the effect of coefficient quantization of FIR filters
# Magnitude truncation
from scipy import signal
import numpy as np
import matplotlib.pyplot as plt
n = 31
n1=np.arange(0,n);
a = signal.firwin(n, cutoff = 0.25, window = "hamming")
B = 16;# Number of Bits
Q = 1/(2**(B))
Qhmt=np.zeros(len(a))
# Magnitude Truncation
for i in range(len(a)):
    if a[i] > 0:
        Qhmt[i]=Q*np.floor(a[i]/Q)
    else:
        Qhmt[i]=Q*np.ceil(a[i]/Q)
#Obtaining the Frequency response
w,H=signal.freqz(a)
wq,Hq=signal.freqz(Qhmt)
#Obtaining the pole-zero plot
z,p,k=signal.tf2zpk(a,1)
zq,pq,kq=signal.tf2zpk(Qhmt,1)
#Plotting the responses
plt.figure(1),plt.subplot(2,2,1),plt.stem(a),plt.xlabel('n-->'),plt.ylabel('h[n]'),plt.title('h[n]')
plt.subplot(2,2,2),plt.plot((w/np.pi),20*np.log10(np.abs(H)))
plt.xlabel(r'$\omega$(x$\pi$rad/sample)'),plt.ylabel('Magnitude'),plt.title('|H($\omega$)|')
plt.subplot(2,2,3),plt.plot(np.real(z),np.imag(z), 'ro'),plt.plot(np.real(p),np.imag(p), 'kx')
theta=np.linspace(0,2*np.pi,100)
plt.plot(np.cos(theta),np.sin(theta))
plt.xlabel('Real part'),plt.ylabel('Imaginary part'),plt.title('Pole-zero plot')
plt.subplot(2,2,4),plt.plot((w/np.pi),np.unwrap(np.angle(H)))
plt.xlabel(r'$\omega$(x$\pi$rad/sample)'),plt.ylabel('Phase'),plt.title('$\Phi(e^{jw})$')
plt.tight_layout()
plt.figure(2),plt.subplot(2,2,1),plt.stem(Qhmt),plt.xlabel('n-->'),plt.ylabel('$h_q[n]$')
plt.title('$h_q[n]$ with N = {} bits'.format(B))
plt.subplot(2,2,2),plt.plot((w/np.pi),20*np.log10(np.abs(Hq)))
plt.xlabel(r'$\omega$(x$\pi$rad/sample)'),plt.ylabel('Magnitude'),plt.title('$|H_q(\omega$)|')
plt.subplot(2,2,3),plt.plot(np.real(zq),np.imag(zq), 'ro')
plt.plot(np.real(pq),np.imag(pq), 'kx')
theta=np.linspace(0,2*np.pi,100)
plt.plot(np.cos(theta),np.sin(theta)),plt.xlabel('Real part'),plt.ylabel('Imaginary part'),
plt.title('Pole-zero plot'),plt.subplot(2,2,4),plt.plot((w/np.pi),np.unwrap(np.angle(Hq)))
plt.xlabel(r'$\omega$(x$\pi$rad/sample)'),plt.ylabel('Phase'),plt.title('$\Phi_q(e^{jw})$')
plt.tight_layout()
```

Fig. 9.11 Python code for magnitude truncation

each filter coefficient is chosen as 2, 4 and 16. The simulation result of the python code given in Fig. 9.11 is shown in Fig. 9.12.

Inferences

From Fig. 9.12, the following inferences can be made:

1. When the number of bits is chosen as $N = 2$, the impulse response of the quantized filter has only one non-zero element, and all other elements are zero.

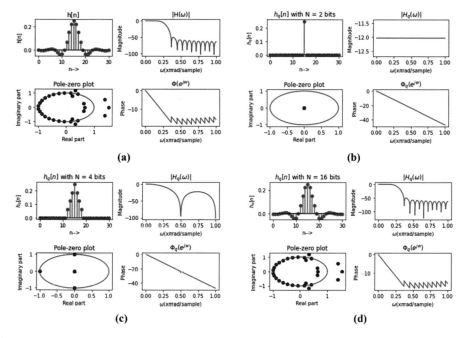

Fig. 9.12 Simulation result of the python code given in Fig. 9.11. (**a**) Original filter. (**b**) Quantized filter with $N = 2$ bits. (**c**) Quantized filter with $N = 4$ bits. (**d**) Quantized filter with $N = 16$ bits

From the pole-zero plot, it is possible to infer that all the zeros of the FIR filters are placed at the origin, and it confirms that improper allocation of bits will lead to error frequency response.

2. While choosing the value N is 4, the impulse response of the quantized filter looks like a triangular function; it is not a Gaussian function. Also, it shows that all the zeros of the FIR filters are squeezed into only four locations.

3. If $N = 16$, the impulse, magnitude, and phase responses of the quantized filter are very close to the responses of the original filter. Also, the pole-zero plot is similar to the original one.

Experiment 9.7 Verify the Effect of Quantization Using the Rounding Approach of Butterworth IIR Filter

This experiment discusses the quantization of Butterworth filter coefficients computed using the bilinear transformation technique (BLT) that has a passband gain of 0 to -3 dB, cut-off frequency of 2 kHz and an attenuation of at least 20 dB for frequencies greater than 5 kHz. Assume the sampling frequency to be 20 kHz. Butterworth filter coefficients are quantized using the rounding approach. The python code to verify the concept of this experiment is shown in Fig. 9.13. Here the number of bits to be allocated to represent each coefficient of the IIR Butterworth filter is chosen as 4, 8, 12 and 16. The simulation results of the python code given in Fig. 9.13 are shown in Fig. 9.14.

```
import numpy as np
import matplotlib.pyplot as plt
from scipy import signal
# Specifications of Filter
fsam, fp, fs, Ap, As, Td=20000, 2000, 5000, 3, 20, 1 # Sampling frequency
wp=2*np.pi*(fp/fsam) # pass band freq in radian per sample
ws=2*np.pi*(fs/fsam) # Stop band freq in radian per sample
# prewarping process
omega_p=(2/Td)*np.tan(wp/2)
omega_s=(2/Td)*np.tan(ws/2)
#Computation of order and normalized cut-off frequency
N, omega_c=signal.buttord(omega_p,omega_s,Ap,As,analog=True)
print('Order of the Filter N =', N),print('Cut-off frequency= {:.4f} rad/s '. format(omega_c))
# Computation of H(s)
b, a=signal.butter(N,omega_c,'low', analog=True)
bz, az=signal.bilinear(b, a, Td)
n=15
n1=np.arange(0,n);
x=(n1==0)
y=signal.lfilter(bz,az,x)
W,H = signal.freqz(bz,az)
#Obtaining the pole-zero plot
z,p,k=signal.tf2zpk(bz,az)
B = 2;# Number of Bits
Q = 1/(2**(B))
Qbr=Q*np.round(bz/Q)#Rounding
Qar=Q*np.round(az/Q)#Rounding
yr=signal.lfilter(Qbr,Qar,x)
Wq,Hq = signal.freqz(Qbr,Qar)
zq,pq,kq=signal.tf2zpk(Qbr,Qar)
#Plotting the responses
plt.figure(1),plt.subplot(2,2,1),plt.stem(n1,y),plt.xlabel('n-->'),plt.ylabel('h[n]'),plt.title('h[n]')
plt.subplot(2,2,2),plt.plot((W/np.pi),20*np.log10(np.abs(H)))
plt.xlabel(r'$\omega$(x$\pi$rad/sample)'),plt.ylabel('Magnitude'),plt.title('|H($\omega$)|')
plt.subplot(2,2,3),plt.plot(np.real(z),np.imag(z),'ro'),plt.plot(np.real(p),np.imag(p),'kx')
theta=np.linspace(0,2*np.pi,100)
plt.plot(np.cos(theta),np.sin(theta))
plt.xlabel('Real part'),plt.ylabel('Imaginary part'),plt.title('Pole-zero plot')
plt.subplot(2,2,4),plt.plot((W/np.pi),np.unwrap(np.angle(H)))
plt.xlabel(r'$\omega$(x$\pi$rad/sample)'),plt.ylabel('Phase'),plt.title('$\Phi(e^{jw})$')
plt.tight_layout()
plt.figure(2),plt.subplot(2,2,1),plt.stem(n1,yr),plt.xlabel('n-->'),plt.ylabel('$h_q[n]$')
plt.title('$h_q[n]$ with N = {} bits'.format(B))
plt.subplot(2,2,2),plt.plot((Wq/np.pi),20*np.log10(np.abs(Hq)))
plt.xlabel(r'$\omega$(x$\pi$rad/sample)'),plt.ylabel('Magnitude'),plt.title('$|H_q(\omega$)|')
plt.subplot(2,2,3),plt.plot(np.real(zq),np.imag(zq),'ro'),plt.plot(np.real(pq),np.imag(pq),'kx')
theta=np.linspace(0,2*np.pi,100)
plt.plot(np.cos(theta),np.sin(theta)),plt.xlabel('Real part'),plt.ylabel('Imaginary part'),
plt.title('Pole-zero plot'),plt.subplot(2,2,4),plt.plot((Wq/np.pi),np.unwrap(np.angle(Hq)))
plt.xlabel(r'$\omega$(x$\pi$rad/sample)'),plt.ylabel('Phase'),plt.title('$\Phi_q(e^{jw})$')
plt.tight_layout()
```

Fig. 9.13 IIR filter coefficients quantization using rounding

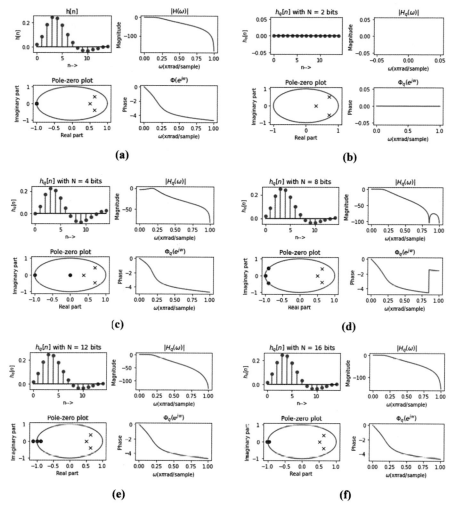

Fig. 9.14 Simulation results of Experiment 9.7. (**a**) Infinite precision. (**b**) Finite precision ($N = 2$ bits). (**c**) Finite precision ($N = 4$ bits). (**d**) Finite precision ($N = 8$ bits). (**e**) Finite precision ($N = 12$ bits). (**f**) Finite precision ($N = 16$ bits)

Inferences

The following inferences can be made from Fig. 9.14:

1. When the number of bits allocated to represent each coefficient of the IIR filter is 2, the impulse response of the quantized filter becomes zero, and the pole-zero plot, magnitude and phase responses confirm it.
2. The number of bits chosen as 4, 8 and 12, the impulse responses of the quantized filter are not the same as the impulse response of the original filter, which reflects in the pole-zero plot, magnitude and phase responses.

3. When the number of bits selected is 16, the impulse response of the quantized filter is exactly the same as the impulse response of the original filter. Also, the pole-zero plot, magnitude and phase responses are on par with the original result.

Task

Repeat the Experiment 9.7 with the filter quantization using two's complement truncation approach.

Experiment 9.8 Chebyshev Type I IIR Filter Coefficients Quantization Effect

Let us consider a Chebyshev Type I IIR filter designed using the bilinear transformation technique (BLT) that has a passband gain of 0 to -3 dB, cut-off frequency of 2 kHz and an attenuation of at least 20 dB for frequencies greater than 5 kHz. Assume the sampling frequency to be 20 kHz. This experiment deals with the quantization effect of IIR filter coefficients. The order and coefficients of Chebyshev Type I filter are computed based on the given specifications. These filter coefficients are quantized by either two's complement or magnitude truncation method. The performance of the infinite precision and finite precision is displayed. The python code to quantize the IIR filter coefficients is given in Fig. 9.15, and the simulation result is shown in Fig. 9.16.

When executing the python code given in Fig. 9.15, first enter the type of quantization '1' for two's complement truncation and '2' for magnitude truncation. The simulation result is shown in Fig. 9.16.

Inference

From Fig. 9.16, it is possible to observe the quantization effect of IIR filter. The impulse and magnitude responses are zero for the low-bit representation of filter coefficients. When the high-bit representation of filter coefficients, the result of impulse and magnitude responses approaches the original responses.

Task

1. Repeat Experiment 9.8 with the filter quantization using two's complement truncation approach.
2. Repeat Experiment 9.8 with the filter quantization using the rounding approach.

9.4   Limit Cycle Oscillations

A digital filter is a non-linear system affected by the quantization of the arithmetic operations. This non-linearity of the digital filter may give stable output under infinite precision arithmetic for specific input. Also, it may give unstable output under finite precision arithmetic for specific input signals. This type of instability usually results in an oscillatory periodic output called 'limit cycle oscillation'. These limit cycle oscillations do not have FIR filters because they do not have a feedback path. But it will exist in IIR filters because it has a feedback path. The limit cycle oscillation is broadly classified into (1) granular and (2) overflow.

```
import numpy as np
import matplotlib.pyplot as plt
from scipy import signal
# Specifications of Filter
fsam, fp, fs, Ap, As, Td=20000, 2000, 5000, 3, 30, 0.9 # Sampling frequency
wp=2*np.pi*(fp/fsam) # pass band freq in radian per sample
ws=2*np.pi*(fs/fsam) # Stop band freq in radian per sample
# prewarping process
omega_p=(2/Td)*np.tan(wp/2)
omega_s=(2/Td)*np.tan(ws/2)
#Computation of order and normalized cut-off frequency
N, omega_c=signal.cheb1ord(omega_p,omega_s,Ap,As,analog=True)
print('Order of the Filter N =', N),print('Cut-off frequency= {:.4f} rad/s '. format(omega_c))
# Computation of H(s)
b, a=signal.cheby1(N, Ap, omega_c,'low', analog=True)
bz, az=signal.bilinear(b, a, Td)
n=15
n1=np.arange(0,n);
x=(n1==0)
y=signal.lfilter(bz,az,x)
W,H = signal.freqz(bz,az)
#Obtaining the pole-zero plot
z,p,k=signal.tf2zpk(bz,az)
B = 8;# Number of Bits
Q = 1/(2**(B))
QT=int(input('Enter the type of Qunatization:1-2s Complement truncation; 2-Magnitude truncation: '))
import sys
if (QT == 1):
   Qbr=Q*np.floor(bz/Q)# Twos complement truncation
   Qar=Q*np.floor(az/Q)# Twos complement truncation
elif (QT == 2):
   Qbr=np.zeros(len(bz)) # Magnitude Truncation
   Qar=np.zeros(len(az)) # Magnitude Truncation
   for i in range(len(bz)):
     if bz[i] > 0:
        Qbr[i]=Q*np.floor(bz[i]/Q)
     else: Qbr[i]=Q*np.ceil(bz[i]/Q)
   for j in range(len(az)):
     if az[j] > 0:
        Qar[j]=Q*np.floor(az[j]/Q)
     else: Qar[j]=Q*np.ceil(az[j]/Q)
else:
   print("'Please select the proper quantization method"');
   sys.exit()
yr=signal.lfilter(Qbr,Qar,x)
Wq,Hq = signal.freqz(Qbr,Qar)
zq,pq,kq=signal.tf2zpk(Qbr,Qar)
#Plotting the responses
plt.figure(1),plt.subplot(2,2,1),plt.stem(n1,y),plt.xlabel('n-->'),plt.ylabel('h[n]')
plt.title('h[n]')
plt.subplot(2,2,2),plt.plot((W/np.pi),20*np.log10(np.abs(H)))
plt.xlabel(r'$\omega$(x$\pi$rad/sample)'),plt.ylabel('Magnitude'),plt.title('|H($\omega$)|')
plt.subplot(2,2,3),plt.plot(np.real(z),np.imag(z),'ro')
plt.plot(np.real(p),np.imag(p),'kx')
theta=np.linspace(0,2*np.pi,100)
```

Fig. 9.15 Python code for IIR filter coefficients quantization

```
plt.plot(np.cos(theta),np.sin(theta))
plt.xlabel('Real part'),plt.ylabel('Imaginary part'),plt.title('Pole-zero plot')
plt.subplot(2,2,4),plt.plot((W/np.pi),np.unwrap(np.angle(H)))
plt.xlabel(r'$\omega$(x$\pi$rad/sample)'),plt.ylabel('Phase'),plt.title('$\Phi(e^{jw})$')
plt.tight_layout()
plt.figure(2)
plt.subplot(2,2,1),plt.stem(n1,yr),plt.xlabel('n-->'),plt.ylabel('$h_q[n]$')
plt.title('$h_q[n]$ with N = {} bits'.format(B))
plt.subplot(2,2,2),plt.plot((Wq/np.pi),20*np.log10(np.abs(Hq)))
plt.xlabel(r'$\omega$(x$\pi$rad/sample)'),plt.ylabel('Magnitude'),plt.title('$|H_q(\omega$)|')
plt.subplot(2,2,3),plt.plot(np.real(zq),np.imag(zq),'ro')
plt.plot(np.real(pq),np.imag(pq),'kx')
theta=np.linspace(0,2*np.pi,100)
plt.plot(np.cos(theta),np.sin(theta)),plt.xlabel('Real part'),plt.ylabel('Imaginary part'),
plt.title('Pole-zero plot'),plt.subplot(2,2,4),plt.plot((Wq/np.pi),np.unwrap(np.angle(Hq)))
plt.xlabel(r'$\omega$(x$\pi$rad/sample)'),plt.ylabel('Phase'),plt.title('$\Phi_q(e^{jw})$')
plt.tight_layout()
```

Fig. 9.15 (continued)

Experiment 9.9 Limit Cycle Oscillation in IIR Filter

This experiment tries to verify that the limit cycle oscillation can be occurred in IIR filter due to coefficient quantization.

Let us consider a second-order recursive system which is given by

$$y[n] - \frac{7}{8}y[n-1] + \frac{5}{8}y[n-2] = \delta[n]$$

The input and output coefficients of the recursive system are quantized using the rounding fixed-point 3-bit quantization approach. The quantized result of the recursive system is written as

$$y_q[n] = Q_r\left\{\delta[n] + \frac{7}{8}y[n-1] - \frac{5}{8}y[n-2]\right\}$$

The python code for this above equation is given in Fig. 9.17. From this figure, it is possible to confirm that the number of bits used in the quantization is 3, and the rounding quantization approach is implemented. After executing the python code given in Fig. 9.17, the obtained result is shown in Fig. 9.18. From this figure, it is possible to infer that the output is oscillated and confirms that the limit cycle oscillation exists in this filter.

Inferences

From this experiment, the following inferences can be made:

1. The finite arithmetic operation in the digital implementation of a recursive system may introduce limit cycle oscillations in the final output. This causes due to the feedback that exists in the recursive system.

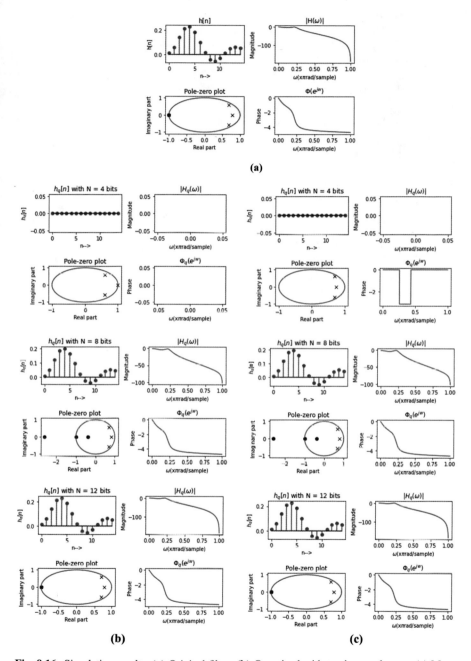

Fig. 9.16 Simulation results. (**a**) Original filter. (**b**) Quantized with two's complement. (**c**) Magnitude truncation

```
import numpy as np
import matplotlib.pyplot as plt
B = 3;# Number of Bits
Q = 1/(2**(B))
N=100
x=np.zeros(N)
y1=np.zeros(N)
y=np.zeros(N)
n=np.arange(0,len(x))
x[0]=1#5/8
y1[-1], y[-1]=0,0
y1[-2], y[-2]=0,0
for i in range(len(x)):
    Y=x[i]+((7/8)*y1[i-1])-(5/8)*y1[i-2]
    y[i]=x[i]+((7/8)*y[i-1])-(5/8)*y[i-2]
    y1[i]=(Q*np.round(Y/Q))
    out=y1
plt.subplot(311),plt.stem(n,x),plt.xlabel('n-->'),plt.ylabel('x[n]'),plt.title('Input')
plt.subplot(312),plt.stem(n,y),plt.xlabel('n-->'),plt.ylabel('y[n]'),plt.title('Infinite precision Output')
plt.subplot(313),plt.stem(n,y1),plt.xlabel('n-->'),plt.ylabel('y1[n]'),plt.title('Finite arithmetic Output')
plt.tight_layout()
```

Fig. 9.17 Python code for limit cycle oscillation

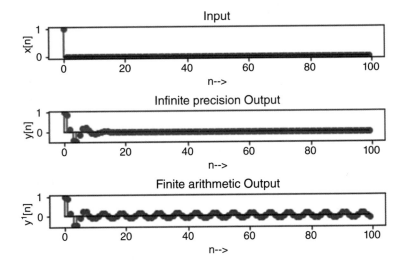

Fig. 9.18 Simulation result

2. After some time, the infinite precision arithmetic operation settles its output as zero, whereas the finite precision result does not become zero.
3. Since the feedback path does not exist in the non-recursive system, the limit cycle oscillation does not occur.

Task

1. Execute the python code given in Fig. 9.17, and determine the minimum value that will eliminate the limit cycle oscillation in Experiment 9.9 (i.e. change the value of B (4, 5, 6, ...)).

9.5 Cascade Form of a Higher Order Filters

The sensitivity reduction to coefficient quantization can be reduced by splitting high-order filters into lower-order filters. This cascade form can split the higher-order filters into multiple lower-order filters.

Experiment 9.10 Verify the Cascade Structure of FIR Filter May Reduce the Quantization Effect of FIR Filter

This experiment explains the concept of the reduction of the quantization effect of FIR filter using cascade structure implementation. The higher-order FIR filter is decomposed into multiple second-order FIR filters, and those filter's coefficients are quantized with finite precision. The result of the multiple second-order filters are combined to get a final output. The final output is always similar to the infinite precision result of the higher-order FIR filter. The python code to verify this experiment is given in Fig. 9.19, and the corresponding simulation result is displayed in Fig. 9.20. From Fig. 9.19, the '*signal.tf2sos*' python command is used here to decompose the higher-order filter into multiple second-order filters.

The command '*signal.sos2tf*' is used here to convert the multiple second-order filters into higher-order ones.

Inferences

The following inferences can be made from Fig. 9.20:

1. The frequency response of the infinite precision FIR filter coefficients and finite precision ($N = 8$ bits) FIR filter coefficients are not the same.
2. However, the frequency response of the cascade realization filter with finite precision looks similar to the original one.
3. From this experiment, it is possible to confirm that the effect of quantization of the FIR filter can be reduced with the help of cascade realization. In the cascade realization, the higher-order filter is decomposed into multiple second-order filters, and these lower-order filters are quantized with finite precision. Finally, all these quantized filter coefficients are combined to get a higher-order filter.
4. Note that the number of bits ($N = 8$) used to represent the higher-order filter is the same as for the lower-order filters.
5. This realization will help when the hardware is limited in the length of the registers. For example, the hardware is an 8-bit register length, and to represent the filter coefficients, it needs more than 8 bits. Then, the cascade realization will help to represent all the filter coefficients with 8-bit precision.

```
from scipy import signal
import numpy as np
import matplotlib.pyplot as plt
n = 15
n1=np.arange(0,n);
a = signal.firwin(n, cutoff = 0.25, window = "hamming")
A = signal.tf2sos(a,1)#Decomposing higher order filter to second order filters
A1=A
B = 8;# Number of Bits
Q = 1/(2**(B))
Qhr=Q*np.round(a/Q)#Rounding
w,H=signal.freqz(a) #Obtaining the Frequency response
wq,Hq=signal.freqz(Qhr) #Obtaining the Frequency response
z,p,k=signal.tf2zpk(a,1) #Obtaining the pole-zero plot
zq,pq,kq=signal.tf2zpk(Qhr,1) #Obtaining the pole-zero plot
for i in range(len(A)):
    A1[i][0:3]=Q*np.round(A[i][0:3]/Q)
cA=signal.sos2tf(A1)
cwq,cHq=signal.freqz(cA[0][0:n]) #Obtaining the Frequency response
czq,cpq,ckq=signal.tf2zpk(cA[0][0:n],1) #Obtaining the pole-zero plot
plt.figure(1),plt.subplot(2,2,1),plt.stem(a),plt.xlabel('n-->'),plt.ylabel('h[n]')
plt.title('h[n]'),plt.subplot(2,2,2),plt.plot((w/np.pi),20*np.log10(np.abs(H)))
plt.xlabel(r'$\omega$(x$\pi$rad/sample)'),plt.ylabel('Magnitude'),plt.title('|H($\omega$)|')
plt.subplot(2,2,3),plt.plot(np.real(z),np.imag(z),'ro'),plt.plot(np.real(p),np.imag(p),'kx')
theta=np.linspace(0,2*np.pi,100)
plt.plot(np.cos(theta),np.sin(theta)),plt.xlabel('Real part'),plt.ylabel('Imaginary part'),
plt.title('Pole-zero plot'),plt.subplot(2,2,4),plt.plot((w/np.pi),np.unwrap(np.angle(H)))
plt.xlabel(r'$\omega$(x$\pi$rad/sample)'),plt.ylabel('Phase'),plt.title('$\Phi(e^{jw})$')
plt.tight_layout()
plt.figure(2),plt.subplot(2,2,1),plt.stem(Qhr),plt.xlabel('n-->'),plt.ylabel('$h_q[n]$')
plt.title('$h_q[n]$ with N = {} bits'.format(B))
plt.subplot(2,2,2),plt.plot((wq/np.pi),20*np.log10(np.abs(Hq)))
plt.xlabel(r'$\omega$(x$\pi$rad/sample)'),plt.ylabel('Magnitude'),
plt.title('$|H_q(\omega)|'),plt.subplot(2,2,3),plt.plot(np.real(zq),np.imag(zq),'ro')
plt.plot(np.real(pq),np.imag(pq),'kx'),plt.plot(np.cos(theta),np.sin(theta))
plt.xlabel('Real part'),plt.ylabel('Imaginary part'),plt.title('Pole-zero plot')
plt.subplot(2,2,4),plt.plot((wq/np.pi),np.unwrap(np.angle(Hq)))
plt.xlabel(r'$\omega$(x$\pi$rad/sample)'),plt.ylabel('Phase'),plt.title('$\Phi_q(e^{jw})$')
plt.tight_layout()
plt.figure(3),plt.subplot(2,2,1),plt.stem(cA[0][0:n]),plt.xlabel('n-->'),plt.ylabel('$h_q[n]$')
plt.title('Cascade $h_q[n]$ with N = {} bits'.format(B))
plt.subplot(2,2,2),plt.plot((cwq/np.pi),20*np.log10(np.abs(cHq)))
plt.xlabel(r'$\omega$(x$\pi$rad/sample)'),plt.ylabel('Magnitude'),
plt.title(' Cascade $|H_q(\omega$)|'),plt.subplot(2,2,3),plt.plot(np.real(czq),np.imag(czq),'ro')
plt.plot(np.real(cpq),np.imag(cpq),'kx'),plt.plot(np.cos(theta),np.sin(theta))
plt.xlabel('Real part'),plt.ylabel('Imaginary part'),plt.title('Pole-zero plot')
plt.subplot(2,2,4),plt.plot((cwq/np.pi),np.unwrap(np.angle(cHq)))
plt.xlabel(r'$\omega$(x$\pi$rad/sample)'),plt.ylabel('Phase'),plt.title(' Cascade $\Phi_q(e^{jw})$')
plt.tight_layout()
```

Fig. 9.19 Python code for cascade realization of FIR filter

Exercises

1. Write a python code to convert decimal numbers to binary with fixed-point representation.

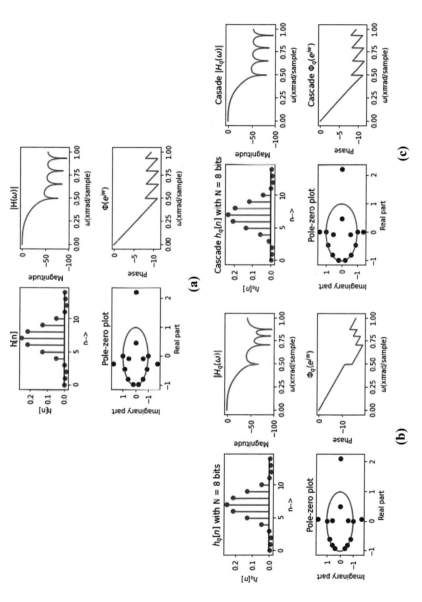

Fig. 9.20 Simulation result. (**a**) Infinite precision output. (**b**) Finite precision output. (**c**) Cascade with finite precision output

2. Write a python code to convert binary representation to decimal representation.
3. Write a python code to verify how the saturation overflow stabilizes the arithmetic operation in digital implementation.
4. Write a python code to plot the quantization characteristics curve of the rounding approach.
5. Write a python code to plot the quantization characteristics curve of two's complement truncation approach.
6. Write a python code to plot the quantization characteristics curve of the truncation approach.
7. Write a python code to verify that the limit cycle oscillation does not occur in FIR filter coefficients quantization.
8. Write a python code to verify the magnitude truncation can inherently suppress the limit cycle oscillation.

Objective Type Questions

1. Input to the python command is a floating-point number 1.45, and the output is 2. Identify the suitable python command.

   A. numpy.round
   B. numpy.floor
   C. numpy.ceil
   D. numpy.array

2. Identify the suitable python command whose input is a floating-point number of 1.45 and the output is 1.

   A. numpy.floor
   B. numpy.ceil
   C. numpy.float
   D. numpy.array

3. The input to the python command 'np.ceil' is 0.9, and the output is

   A. 1.0
   B. 0.5
   C. 0.0
   D. 0.9

4. The python command used to find the order of the Butterworth filter is

   A. signal.buttord
   B. signal.chebord
   C. signal.butter
   D. signal.bilinear

5. The python command used to obtain the coefficients of Butterworth filter is

   A. signal.butter
   B. signal.buttord

C. signal.butter1

D. signal.buttord1

6. The formula to quantize the input value 'x' and step size 'Δ' by rounding approach is

 A. $\Delta \times \text{round}\left(\frac{x}{\Delta}\right)$
 B. $\Delta \times \text{ceil}\left(\frac{x}{\Delta}\right)$
 C. $\Delta \times \text{floor}\left(\frac{x}{\Delta}\right)$
 D. $x \times \text{round}\left(\frac{\Delta}{x}\right)$

7. The formula to quantize the input value 'x' and step size 'Δ' by two's compliment truncation approach is

 A. $\Delta \times \text{round}\left(\frac{x}{\Delta}\right)$
 B. $\Delta \times \text{ceil}\left(\frac{x}{\Delta}\right)$
 C. $\Delta \times \text{floor}\left(\frac{x}{\Delta}\right)$
 D. $x \times \text{round}\left(\frac{\Delta}{x}\right)$

8. The formula to quantize the input value 'x' is greater than '0' and step size 'Δ' by magnitude truncation approach is

 A. $\Delta \times \text{floor}\left(\frac{x}{\Delta}\right)$
 B. $\Delta \times \text{ceil}\left(\frac{x}{\Delta}\right)$
 C. $\Delta \times \text{round}\left(\frac{x}{\Delta}\right)$
 D. $x \times \text{round}\left(\frac{\Delta}{x}\right)$

9. Limit cycle oscillation does not occur in

 A. Recursive system
 B. IIR filter
 C. Stable filter
 D. FIR filter

10. The effect of quantization in a higher-order FIR filter is reduced by using

 A. Parallel realization
 B. Cascade realization
 C. Direct form I realization
 D. Lattice realization

Bibliography

1. Bernard Widrow, Istvan Kollar, "Quantization Noise: Roundoff Error in Digital Computation, Signal Processing, Control and Communications", Cambridge University Press, 2008.

2. N.S. Jayant, Peter Noll, "Digital Coding of Waveforms: Principles and Applications to Speech and Video", Prentice Hall India, 1984.
3. Lawrence R. Rabiner, and Bernard Gold, "Theory and Applications of Digital Signal Processing", Prentice-Hall, 1975.
4. James H. McClellan, Ronald W. Schafer, and Mark A. Yoder, "DSP First", Prentice Hall, 1998.
5. Allen Gersho and Robert M. Gray, "Vector Quantization and Signal Compression", Springer, 1991.

Chapter 10
Multirate Signal Processing

Learning Objectives

After completing this chapter, the reader is expected to

- To perform sampling rate conversion using multirate operators.
- Time-domain and frequency-domain view of multirate operators.
- Demonstrate Type I and Type II polyphase decomposition.
- Signal decomposition using perfect reconstruction filter bank.
- Implementation of crosstalk free two-channel transmultiplexer.

Roadmap of the Chapter

The roadmap of this chapter is given below. From this figure, it is possible to observe that this chapter begins with multirate operators. Downsampling and upsampling operations are discussed in detail. Polyphase decomposition involving downsampling operation is termed as Type I, and polyphase decomposition involving upsampling operation is termed as Type II polyphase decomposition. Subband decomposition enables signals to be divided into different frequency regions. Subband decomposition is done through a filter bank. In this chapter, two-channel, three-channel and tree-structured filter banks are discussed. Finally, this chapter concludes with the design of two-channel crosstalk free transmultiplexer.

© The Author(s), under exclusive license to Springer Nature Singapore Pte Ltd. 2024   403
S. Esakkirajan et al., *Digital Signal Processing*,
https://doi.org/10.1007/978-981-99-6752-0_10

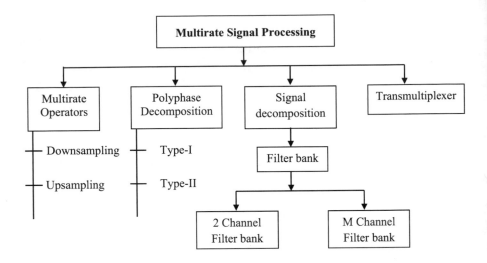

PreLab Questions
1. Mention the need to change the sampling rate of the signal.
2. What is a sampling rate converter?
3. Mention two basic operations in sampling rate conversion or multirate signal processing.
4. What is the need for a filter before the downsampling operation? What is the name of the filter?
5. What is the name given to the filter after upsampling operation? What is the purpose of this filter?
6. Write the time-domain and frequency-domain expression for downsampling by a factor of 'M'.
7. Write the time-domain and frequency-domain expression for upsampling by a factor of 'L'.
8. Mention the three significant properties of the downsampling operation.
9. Why upsampling operation is considered as a linear time-variant operation?
10. What is the condition for interchanging of upsampling by a factor of 'L' and downsampling by a factor of 'M' operation?
11. Why is downsampling by a factor of 'M' followed by upsampling by a factor of 'M' considered an idempotent operation?
12. What is the objective of polyphase decomposition? Mention the types of polyphase decomposition.
13. What is a filter bank? Mention two applications of the filter bank.
14. Mention the threat involved in perfect reconstruction in a two-channel filter bank? Mention the ways to overcome this threat.
15. What is a transmultiplexer? Mention its application.

10.1 Multirate Operators

Multirate operators are used to change the sampling rate of the signal digitally by either by removing (deletion) of samples or inserting zeros between successive samples. Two basic multirate operators are (1) downsampling operator and (2) upsampling operator. The downsampling operation is used to decrease the sampling rate of the signal, whereas upsampling operator is used to increase the sampling rate of the signal.

10.1.1 Downsampling Operation

Downsampling operation reduces the sampling rate by a factor of 'M'. The downsampling operation by a factor of 'M' is shown in Fig. 10.1.

The time-domain expression for downsampling by a factor of 'M' is given by

$$y[n] = x[nM] \tag{10.1}$$

From Eq. (10.1), it is possible to interpret that the output signal consists of every Mth element of the input signal. The transform domain expression for downsampling by a factor of 'M' is given by

$$Y(z) = \frac{1}{M} \sum_{k=0}^{M-1} X\left(z^{\frac{1}{M}} W_M^k\right) \tag{10.2}$$

Downsampling produces expansion in the frequency-domain giving rise to 'aliasing'. Aliasing will occur in the output signal if the input signal $x[n]$ is not bandlimited, which will lead to loss of information. In order to overcome the aliasing problem, a filter is employed before downsampling operation, which is termed as 'anti-aliasing filter'. The combination of downsampler with anti-aliasing filter is termed as 'decimator'. This concept is illustrated in Fig. 10.2. To maintain the bandwidth of the input signal, the cut-off frequency of the filter (anti-aliasing) $H(z)$ is chosen as (π/M) always, which is basically a low pass filter.

Fig. 10.1 Downsampling operation

Fig. 10.2 Decimation by a factor of 'M'

```
#Illustration of downsampling operation
import numpy as np
import matplotlib.pyplot as plt
#Step 1: Generation of input signal x[n]
n=np.linspace(-10,10,21)
x=np.exp(1j*np.pi*n)
#Step 2: Downsampling the input signal
M=2
y=x[::M]
#Step 3: Plotting the results
n1=np.linspace(min(n)/M,max(n)/M,len(y))
plt.subplot(2,1,1),plt.stem(n,x),plt.xticks(range(-10,11))
plt.xlabel('n-->'),plt.ylabel('x[n]'),plt.title('Input Signal (x[n])')
plt.subplot(2,1,2),plt.stem(n1,y),plt.xticks(range(-10,11))
plt.xlabel('n-->'),plt.ylabel('y[n]'),
plt.title('Downsampled Signal (y[n]) by (M={})'.format(M))
plt.tight_layout()
```

Fig. 10.3 Python code which performs downsampling operation

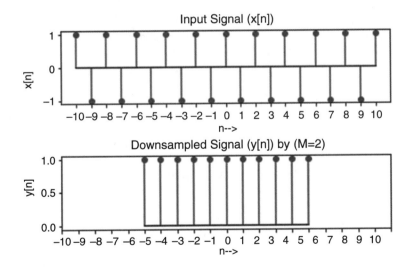

Fig. 10.4 Result of python code shown in Fig. 10.3

Experiment 10.1 Downsampling Operation in Time Domain

The objective of this experiment is to illustrate downsampling operation in time domain. The signal to be downsampled is expressed as $x[n] = e^{j\pi n}, -10 \leq n \leq 10$. The signal is to be downsampled by a factor of 2 to obtain the output signal $y[n]$. The python code, which performs this task, is shown in Fig. 10.3, and the corresponding result is shown in Fig. 10.4.

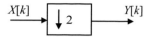

Fig. 10.5 Pictorial representation of problem statement

Inferences

From Fig. 10.3, it is possible to infer the following:

1. The input signal is $x[n] = e^{j\pi n}$ in the range -10 to 10. The input signal toggles between $+1$ and -1. This is the highest frequency in the digital signal.
2. The input signal is downsampled by a factor of 2 to obtain the output signal $y[n]$.

From Fig. 10.4, the following fact can be inferred:

1. The input signal is the highest frequency in digital signal. The signal toggles between -1 and 1 and vice versa.
2. The signal $x[n]$ is downsampled by a factor of 2 to obtain the output signal $y[n]$, which is a DC signal.
3. Downsampling operation has the ability to convert the highest frequency digital signal to a DC signal.
4. The number of samples in the input signal is 21, whereas the number of samples in the output signal is 11. Thus, the downsampling operation reduces the number of samples in the input signal by a factor of 'M'. In this case, the value of 'M' is 2.

Task

1. Change the value of the $M = 4$ in the python code given in Fig. 10.3, and observe the result and comment on it.

Experiment 10.2 Spectrum of Downsampled Signal

The objective of this experiment is to obtain the spectrum of the downsampled signal, and compare it with respect to the spectrum of the input signal. To accomplish this task, a sine wave of 5 Hz signal is generated. This signal is represented as $x[n]$, and its corresponding spectrum is $X[k]$. The signal $x[n]$ is downsampled by a factor of 2 to obtain the output signal $y[n]$, and its corresponding spectrum is $Y[k]$. The objective is to compare these two spectrums. The problem statement is illustrated in Fig. 10.5.

The python code, which performs this task, is shown in Fig. 10.6, and the corresponding output is shown in Fig. 10.7.

Inferences

From Fig. 10.6, the following inferences can be made:

1. The python code generates the sum of sine waves of 0, 2.5, 5 Hz frequency, and it is stored in the variable 'x'.
2. The input sine wave is downsampled by a factor of 2, and the result is stored in the variable 'y'.

```
#Spectrum of downsampled signal
import numpy as np
import matplotlib.pyplot as plt
from scipy.fftpack import fft,fftfreq
#Step 1: Generating the input signal
f1,f2,fs,N=2.5,5,75,512
M,T=2,1/fs
t=np.linspace(0,N*T,N)
x1=0.5*np.ones(len(t))# DC signal
x=x1+np.sin(2*np.pi*f1*t)+np.sin(2*np.pi*f2*t)#
#Step 2: Obtaining the downsampled signal
y=x[::M]
t1=np.linspace(0,N*T/M,len(y))
#Step 3: Obtaining the spectrum of the input signal and downsampled signal
X=fft(x,N)
Y=fft(y,N)
f_axis=fftfreq(N,T)[0:N//M]
#Step4: Plotting the result
plt.subplot(2,2,1),plt.plot(t,x),plt.xlabel('Time'), plt.ylabel('Amplitude'),
plt.title('Input signal'),plt.subplot(2,2,2),plt.plot(t1,y),plt.xlabel('Time'),
plt.ylabel('Amplitude'),plt.title('Downsampled signal')
plt.subplot(2,2,3),plt.plot(f_axis,2/N*np.abs(X[0:N//M])),plt.xlabel('$\omega$-->'),
plt.ylabel('|X($j\omega$)|'),plt.title('Spectrum of input signal')
plt.subplot(2,2,4),plt.plot(f_axis,2/N*np.abs(Y[0:N//M])),plt.xlabel('$\omega$-->'),
plt.ylabel('|Y($j\omega$)|'),plt.title('Spectrum of downsampled signal')
plt.tight_layout()
```

Fig. 10.6 Python code to plot the spectrum of downsampled signal

3. Using the built-in function '*fft*' and '*fftfreq*' in '*scipy*' library, the spectrum of the input signal '*x*' and the output signal '*y*' is obtained and stored in the variable '*X*' and '*Y*', respectively.
4. The magnitude spectrum of the input and output signal is obtained using the built-in function '*abs*', which is available in '*numpy*' library.

From Fig. 10.7, the following inferences can be drawn:

1. From the plot of the input signal and output signal, it is possible to observe that the length of the output signal is lesser than the length of the input signal.
2. The magnitude spectrum of the input signal has peaks at 0, 2.5 and 5 Hz, which shows that the frequencies of the input signal are 0, 2.5 and 5 Hz.
3. The magnitude spectrum of the output signal has peaks at 0, 5 and 10 Hz, which shows that the frequencies of the output signal are 0, 5 and 10 Hz. That is, the bandwidth of the downsampled spectrum increased by 2 because of the downsampling factor chosen as 2.
4. This experiment reveals that compression in the time-domain is equivalent to expansion in the frequency domain.

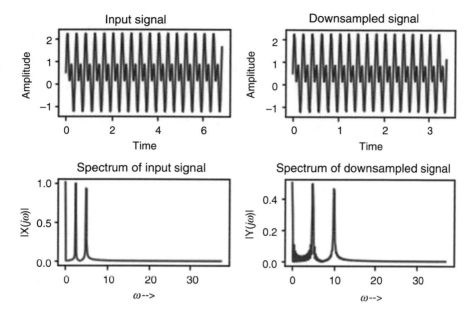

Fig. 10.7 Simulation result

Fig. 10.8 Upsampling
operation

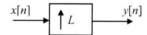

Task

1. Change the value of the $M = 4$ in the python code given in Fig. 10.6, and observe
   the result and comment on it.

10.1.2   Upsampling Operation

Upsampling operation increases the sampling rate by a factor of 'L'. The upsampling
operation by a factor of 'L' is shown in Fig. 10.8.

The time-domain expression for upsampling by a factor of 'L' is given by

$$y[n] = \begin{cases} x\left[\dfrac{n}{L}\right], & n = L, 2L, \ldots \\ 0, & \text{otherwise} \end{cases} \tag{10.3}$$

From Eq. (10.3), it is possible to interpret that upsampling by a factor of 'L' in the
time-domain is accomplished by inserting '$L - 1$' zeros between successive samples
of the input signal $x[n]$. This will increase the length of the input signal; hence,

Fig. 10.9 Interpolation by a factor of '*L*'

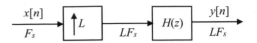

```
#Illustration of upsampling operation
import numpy as np
import matplotlib.pyplot as plt
#Step 1: Generation of input signal x[n]
N=5
n=np.arange(N)
x=np.ones(N)
#Step 2: Upsampling the input signal
L=2
y=np.zeros(L*N)
y[::L]=x
# #Step 3: Plotting the results
n1=np.arange(L*N)
plt.subplot(2,1,1),plt.stem(n,x),plt.xticks(range(0,N))
plt.xlabel('n-->'),plt.ylabel('x[n]'),plt.title('Input Signal (x[n])')
plt.subplot(2,1,2),plt.stem(n1,y),plt.xticks(range(0,L*N+1))
plt.xlabel('n-->'),plt.ylabel('y[n]'),
plt.title('Upsampled Signal (y[n]) by L={}'.format(L))
plt.tight_layout()
```

Fig. 10.10 Python code to perform upsampling operation

upsampling operation can also be termed as 'expansion operation'. The frequency-domain expression for upsampling by a factor of '*L*' is given by

$$Y(z) = X(z^L) \tag{10.4}$$

The above equation can be expressed as

$$Y(e^{j\omega}) = X(e^{j\omega L}) \tag{10.5}$$

The upsampler introduces spectral images. A filter is employed after the upsampler to remove the spectral images. Such type of filter is termed as 'anti-imaging' filter. This is shown in Fig. 10.9.

The cut-off frequency of the filter $H(z)$ is chosen as π/L, which is basically a lowpass filter.

Experiment 10.3 Upsampling Operation in the Time Domain

In this experiment, the input signal $x[n]$ is upsampled by a factor of 2 to obtain the output signal $y[n]$. The python code which performs this task is shown in Fig. 10.10, and the corresponding output is shown in Fig. 10.11.

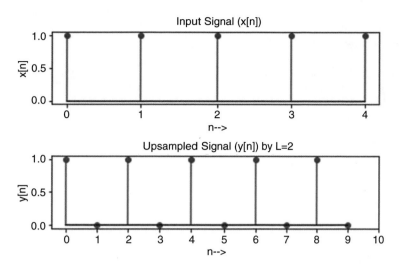

Fig. 10.11 Upsampling operation result

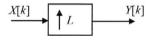

Fig. 10.12 Pictorial representation of problem statement

Inferences
The following inferences can be made from this experiment:

1. From Fig. 10.11, it is possible to observe that the length of the input signal $x[n]$ is 5, whereas the length of the output signal $y[n]$ is 10.
2. Thus, upsampling by a factor of 2 (L) inserts one zero $(L - 1)$ between two successive samples of the input signal $x[n]$.
3. Therefore, upsampling is a 'length stretching operation'.

Experiment 10.4 Spectrum of Upsampled Signal
The objective of this experiment is to obtain the spectrum of upsampled signal and to compare it with the spectrum of input signal. In order to accomplish this task, the following steps are carried out:

- Generate input signal sum of sine waves with the frequency of 0, 4 and 10 Hz.
- Pass this signal through a system that upsamples the input signal by a factor (L) of 2 to obtain the output signal.
- Plot the spectrums of the input and output signal, and comment on the observed result.

The pictorial representation of the problem statement is shown in Fig. 10.12. From Fig. 10.12, $X[k]$ represents the spectrum of the input signal, and $Y[k]$ represents the spectrum of the output signal.

```
#Spectrum of upsampled signal
import numpy as np
import matplotlib.pyplot as plt
from scipy.fftpack import fft,fftfreq
#Step 1: Generating the input signal
f1,f2,fs,N=4,10,50,512
T=1/fs
t=np.linspace(0,N*T,N)
x1=0.5*np.ones(len(t))# DC signal
x=x1+np.sin(2*np.pi*f1*t)+np.sin(2*np.pi*f2*t)#
#x=np.sin(2*np.pi*f*t)
#Step 2: Obtaining the upsampled signal
n = len(x)
L = 2 # Upsample_factor
y = np.zeros(L*n-(L-1))
y[::L] = x
t1=np.linspace(0,N*T*L,len(y))
#Step 3: Obtaining the spectrum of input and downsampled signal
X=fft(x,N)
Y=fft(y,N)
f_axis=fftfreq(N,T)[0:N//(2*L)]
#Step4: Plotting the result
plt.subplot(2,2,1),plt.plot(t,x),plt.xlabel('Time'), plt.ylabel('Amplitude'),
plt.title('Input signal'),plt.subplot(2,2,2),plt.plot(t1,y),plt.xlabel('Time'),
plt.ylabel('Amplitude'),plt.title('Upsampled signal by L={}'.format(L))
plt.subplot(2,2,3),plt.plot(f_axis,2/N*np.abs(X[0:N//(2*L)])),plt.xlabel('$\omega$-->'),
plt.ylabel('|X($j\omega$)|'),plt.title('Spectrum of input signal')
plt.subplot(2,2,4),plt.plot(f_axis,2/N*np.abs(Y[0:N//(2*L)])),plt.xlabel('$\omega$-->'),
plt.ylabel('|Y($j\omega$)|'),plt.title('Spectrum of upsampled signal by L={}'.format(L))
plt.tight_layout()
```

Fig. 10.13 Python code to plot the spectrum of upsampled signal

The python code, which plots the spectrum of the input and upsampled signals, is shown in Fig. 10.13, and the corresponding output is shown in Fig. 10.14.

Inferences
From Fig. 10.14, the following inferences can be drawn:

1. From the plot of the input signal and output signal, it is possible to observe that the length of the output signal is more than the length of the input signal.
2. The magnitude spectrum of input signal has peaks at 0, 4 and 10 Hz, which shows that the frequencies of the input signal are 0, 4 and 10 Hz.
3. The magnitude spectrum of the output signal has peaks at 0, 2 and 5 Hz, which shows that the frequencies of the output signal are 0, 2 and 5 Hz. That is, the bandwidth of the upsampled spectrum is decreased by 2 because of upsampling factor chosen as 2.

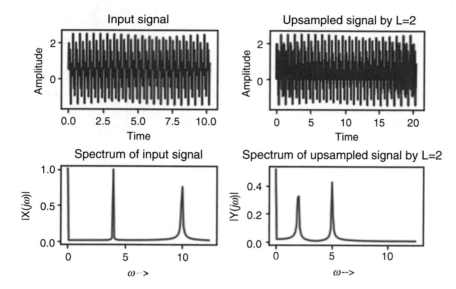

Fig. 10.14 Spectrum of input and upsampled signal

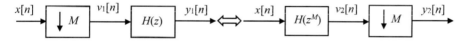

Fig. 10.15 Noble identity for downsampling operation

4. This experiment reveals the fact that expansion in the time-domain is equivalent to compression in the frequency domain.

Task

1. Change the value of the $L = 4$ in the python code given in Fig. 10.13, and comment on the observed spectrum result.

10.2 Noble Identity

Noble identities describe the way to reverse the order of multirate operators and filtering.

10.2.1 Noble Identity for Downsampling Operation

The noble identity of the downsampling operation is depicted in Fig. 10.15.

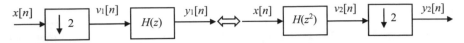

Fig. 10.16 Noble identity for downsampling

```
#Proof of noble identity for downsampling
import numpy as np
import matplotlib.pyplot as plt
from scipy import signal
#Define the signal x[n]
n=np.arange(-10,11)
x=np.ones(len(n))
M=2
#Function to perform downsampling and upsampling operation
def downsample(x,M):
   y=x[::M]
   return(y)
def upsample(x,L):
   n=len(x)
   y=np.zeros(n*L)
   y[::L]=x
   return(y)
#Obtaining the signal v1[n]
v1=downsample(x,M)  #Downsampling of x[n] by a factor of two
h=signal.firwin(5,0.5) #Defining the filter
y1=signal.lfilter(h,1,v1);#Obtaining the signal y1[n]
h1=upsample(h,2);#Equivalent to H(z^2) in time domain
v2=signal.lfilter(h1,1,x)#Obtaining the signal v2[n]
y2=downsample(v2,M)#Obtaining the signal y2[n]
#Plotting the results
plt.subplot(2,1,1),plt.stem(np.abs(y1))
plt.xlabel('n-->'),plt.ylabel('$y_1[n]$'),plt.title('$y_1[n]$')
plt.subplot(2,1,2),plt.stem(np.abs(y2))
plt.xlabel('n-->'),plt.ylabel('$y_2[n]$'),plt.title('$y_2[n]$')
plt.tight_layout()
```

Fig. 10.17 Python illustration regarding noble identity for downsampling operation

Experiment 10.5 Python Illustration of Noble Identity for Downsampling Operation

The block diagram for the noble identity for downsampling operation considered for python illustration is shown in Fig. 10.16.

In this experiment, downsampling factor is chosen as 2. In Fig. 10.16, $H(z^2)$ in time-domain represents upsampling of the filter coefficient $h[n]$ by a factor of 2. The python code, which illustrates the noble identity for downsampling operation, is shown in Fig. 10.17, and the corresponding output is shown in Fig. 10.18.

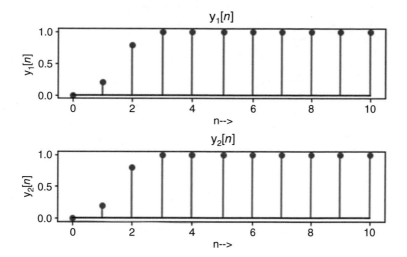

Fig. 10.18 Result of python code shown in Fig. 10.17

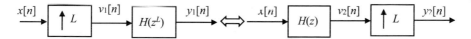

Fig. 10.19 Noble identity for upsampling operation

Inference

From Fig. 10.18, it is possible to observe that the output $y_1[n]$ is equal to the output $y_2[n]$; thus, the noble identity for the downsampling operation is verified.

10.2.2 Noble Identity for Upsampling Operation

The noble identity for upsampling operation is shown in Fig. 10.19

Experiment 10.6 Python Illustration of Noble Identity for Upsampling Operation

This experiment attempts to prove the noble identity for upsampling operation for $L = 2$. The python code which performs this task is shown in Fig. 10.20, and the corresponding output is shown in Fig. 10.21.

Inference

Figure 10.21 shows that the signals $y_1[n]$ and $y_2[n]$ are identical, which means that the noble identity for upsampling operation has been verified for $L = 2$.

```
#Noble identity for upsampling operation
import numpy as np
import matplotlib.pyplot as plt
from scipy import signal
#Step 1: Define the signal x[n]
n=np.arange(-10,11)
x=np.ones(len(n))
L=2
#Step 2: Function to perform upsampling operation
def upsample(x,L):
    n=len(x)
    y=np.zeros(n*L)
    y[::L]=x
    return(y)
v1=upsample(x,L)#Step 3: Obtaining the signal v1[n]
h=signal.firwin(5,0.5)
h1=upsample(h,L)
y1=signal.lfilter(h1,1,v1)#Step 4: Obtaining the signal y1[n]
v2=signal.lfilter(h,1,x)#Step 5: Obtaining the signal v2[n]
y2=upsample(v2,L)#Step 6: Obtaining the signal y2[n]
#Step 7: Plotting the result
plt.subplot(2,1,1),plt.stem(y1)
plt.xlabel('n-->'),plt.ylabel('$y_1[n]$'),plt.title('$y_1[n]$')
plt.subplot(2,1,2),plt.stem(y2)
plt.xlabel('n-->'),plt.ylabel('$y_2[n]$'),plt.title('$y_2[n]$')
plt.tight_layout()
```

Fig. 10.20 Python code to illustrate noble identity for upsampling operation

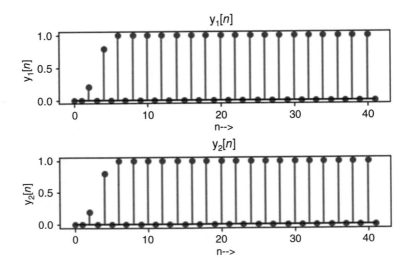

Fig. 10.21 Result of python code shown in Fig. 10.20

10.3 Polyphase Decomposition

Polyphase decomposition refers to the strategy through which the multirate operators can be used to decompose the system function $H(z)$ into its polyphase representation. Polyphase decomposition can be broadly classified into (1) Type I polyphase decomposition and (2) Type II polyphase decomposition.

Experiment 10.7 Python Illustration of Type I Polyphase Decomposition

This python illustration aims to prove the Type I polyphase decomposition illustrated in Fig. 10.22 for the downsampling factor of $M = 2$. In Fig. 10.22, $H(z)$ represents the filter, whereas $E_0(z)$ and $E_1(z)$ represent the polyphase components of $H(z)$. The objective is to prove $y_1[n]$ is equal to $y_2[n]$. The filter chosen in this illustration is a finite impulse response filter designed using the windowing technique.

The python code which implements the Type I polyphase decomposition is shown in Fig. 10.23, and the corresponding output is shown in Fig. 10.24.

Inference

From Fig. 10.24, it is possible to observe that the output $y_1[n]$ is equal to the output $y_2[n]$. Thus, the Type I polyphase decomposition is verified.

Experiment 10.8 Type II Polyphase Decomposition

Type II polyphase decomposition deals with the upsampling operation. Upsampling operation introduces multiple copies of the original signal spectrum, which is termed as 'imaging'. The Type II polyphase decomposition structure is shown in Fig. 10.25.

The python illustration of Type II polyphase decomposition is shown in Fig. 10.26, and the corresponding output is shown in Fig. 10.27.

Inference

From Fig. 10.27, it is possible to observe that the output $y_1[n]$ is equal to $y_2[n]$. This implies that Type II polyphase decomposition is verified.

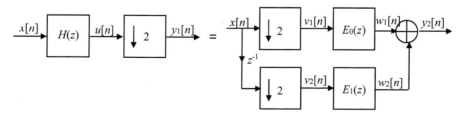

Fig. 10.22 Type I polyphase decomposition

```
#Type-I Polyphase decomposition
import numpy as np
import matplotlib.pyplot as plt
from scipy import signal
#Step 1: Defining the input signal
x=np.ones(8)
h=signal.firwin(8,0.5)
u=signal.lfilter(h,1,x)
y1=u[::2]
#Step 2 Polyphase decomposition
e0=h[0::2]  # Obtaining E0(z)
e1=h[1::2]  # Obtaining E1(z)
x1=np.zeros(len(x)+1)
x1[1:]=x
v1=x[::2]
v2=x1[::2]
w1=signal.lfilter(e0,1,v1)
w2=signal.lfilter(e1,1,v2)
y2=w1+w2[0:len(w1)]
#Step 3: Plotting the result
plt.subplot(2,1,1),plt.stem(y1)
plt.xlabel('n-->'),plt.ylabel('$y_1[n]$'),plt.title('$y_1[n]$')
plt.subplot(2,1,2),plt.stem(y2)
plt.xlabel('n-->'),plt.ylabel('$y_2[n]$'),plt.title('$y_2[n]$')
plt.tight_layout()
```

Fig. 10.23 Python code to illustrate Type I polyphase decomposition

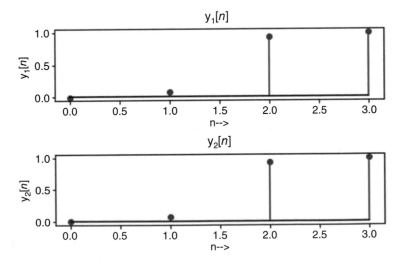

Fig. 10.24 Result of python code shown in Fig. 10.23

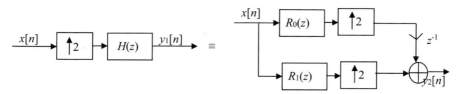

Fig. 10.25 Type II polyphase decomposition structure

```
#Type-II polyphase decomposition
import numpy as np
import matplotlib.pyplot as plt
from scipy import signal
#Step 1: Defining the input signal
x=np.ones(8)
h=signal.firwin(8,0.5)
#Step 2: Polyphase components of H(z)
r0=h[1::2]  # Obtaining R0(z)
r1=h[0::2]  # Obtaining R1(z)
#Step 3: Obtaining the output y1[n]
u=np.zeros(2*len(x))
u[::2]=x
y1=signal.lfilter(h,1,u)
#Step 3: Obtaining the output y2[n]
v1=signal.lfilter(r0,1,x)
v2=signal.lfilter(r1,1,x)
w1=np.zeros(2*len(v1))
w1[::2]=v1
w2=np.zeros(2*len(v2))
w2[::2]=v2
w11=np.zeros(len(w1)+1)
w11[1:]=w1
y2=w2+w11[0:len(w2)]
#Step 4: Plotting the output signals
plt.subplot(2,1,1),plt.stem(y1),plt.xlabel('n-->'),plt.ylabel('$y_1[n]$'),
plt.title('$y_1[n]$'),plt.subplot(2,1,2),plt.stem(y2)
plt.xlabel('n-->'),plt.ylabel('$y_2[n]$'),plt.title('$y_2[n]$')
plt.tight_layout()
```

Fig. 10.26 Python code to demonstrate Type II polyphase decomposition

10.4   Filter Bank

Filter bank is group of filters arranged in a specific fashion. Filter bank is used to split the signal into different frequency bands, which are termed as 'subband coding'. While splitting the signal into various frequency bands, the signal characteristics are

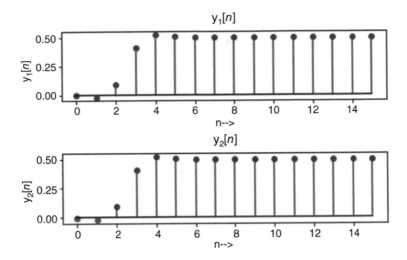

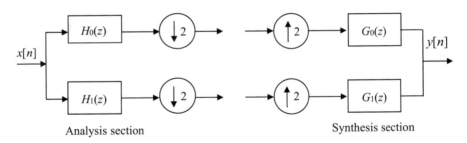

Fig. 10.27 Output of python code shown in Fig. 10.26

Fig. 10.28 Structure of two channel filter bank

different in each band, and different bits can be used for coding the subband signals. This idea is used in speech and image coding. Based on the number of paths available for the input signal, the filter bank can be broadly classified into (1) two-channel filter bank and (2) M-channel filter bank.

10.4.1 Two-Channel Filter Bank

Two-channel filter bank has two sections, namely, (1) analysis section and (2) synthesis section, which is depicted in Fig. 10.28. The input signal fed into the two-channel filter bank is $x[n]$, and the output signal received from the two-channel filter bank is $y[n]$. The channel represents the medium through which the data is transmitted. In Fig. 10.28, the filters in the analysis section are represented as $H_0(z)$ and $H_1(z)$. If $H_0(z)$ represents the lowpass filter, then $H_1(z)$ represents the highpass filter. The corresponding filters in the synthesis section are $G_0(z)$ and $G_1(z)$, respectively.

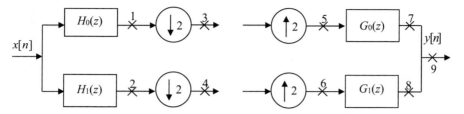

Fig. 10.29 Two-channel filter bank

For perfect reconstruction, the output signal $y[n]$ has to be the delayed version of the input signal $x[n]$. The different threats for perfect reconstruction are (1) aliasing problem due to downsampling operation, (2) amplitude distortion and (3) phase distortion. Proper choice of analysis and synthesis filters will overcome the above-mentioned threats and achieves perfect reconstruction.

Experiment 10.9 Python Implementation of Two-Channel Filter Bank
The structure of the two-channel filter bank which is implemented in this experiment is shown in Fig. 10.29. In Fig. 10.29, different nodes are marked as 1–8.

The input signal $x[n]$ is a sinusoidal signal of 5 Hz frequency. The choice of analysis and synthesis filters are $H_0(z) = \frac{1}{2} + \frac{1}{2}z^{-1}$, $H_1(z) = \frac{1}{2} - \frac{1}{2}z^{-1}$, $G_0(z) = 1 + z^{-1}$ and $G_1(z) = -1 + z^{-1}$. The python code which implements this two-channel filter bank is given in Fig. 10.30, and the corresponding output is shown in Fig. 10.31.

Inferences
1. From Fig. 10.30, it is possible to observe that the filters chosen for the analysis section are $h_0[n] = \{\frac{1}{2}, \frac{1}{2}\}$ and $h_1[n] = \{\frac{1}{2}, -\frac{1}{2}\}$. The filters have only two coefficients. $h_0[n]$ act as lowpass filter, whereas $h_1[n]$ act as high pass filter.
2. The variables chosen in the python code, as shown in Fig. 10.30 like x1, x2, ..., x8, are in line with the nodes shown in Fig. 10.29.
3. From Fig. 10.31, it is possible to observe that the output signal follows the input signal with one sample delay. That is, perfect reconstruction is achieved through the proper choice of filters.

10.4.2 Relationship Between Analysis and Synthesis Filters

Let the analysis filter be expressed as

$$H(z) = H_0(z) \tag{10.6}$$

The expression for $H_1(z)$ in terms of $H(z)$ is given by

```
#Two-channel filter bank
import numpy as np
import matplotlib.pyplot as plt
from scipy import signal
#Functions to perform downsampling and upsampling
def downsample(x,M):
  y=x[::M]
  return(y)
def upsample(x,L):
  y=np.zeros(L*len(x))
  y[::L]=x
  return(y)
#Step 1: Define the filters
h0=np.array([0.5,0.5])
h1=np.array([0.5,-0.5])
g0,g1=2*h0,-2*h1
#Step 2: Generate the input signal
f,fs,N=5,100,256
T=1/fs;
t=np.linspace(0,N*T,N)
x=np.sin(2*np.pi*f*t)
#Step 3: Traversing the path
x1=signal.lfilter(h0,1,x)
x2=signal.lfilter(h1,1,x)
x3=downsample(x1,2)
x4=downsample(x2,2)
x5=upsample(x3,2)
x6=upsample(x4,2)
x7=signal.lfilter(g0,1,x5)
x8=signal.lfilter(g1,1,x6)
y=x7+x8
plt.plot(t,x,'b',t,y,'r--',linewidth=2),plt.legend(['Input','Output'],loc=4)
plt.xlabel('Time'),plt.ylabel('Amplitude'),plt.title('Input and Output signals')
plt.tight_layout()
```

Fig. 10.30 Python code for two-channel filter bank

$$H_1(z) = H(-z) \tag{10.7}$$

The synthesis filters are expressed as

$$G_0(z) = 2H(z) \tag{10.8}$$

$$G_1(z) = -2H(-z) \tag{10.9}$$

From Eqs. (10.6) to (10.9), it is possible to infer the following

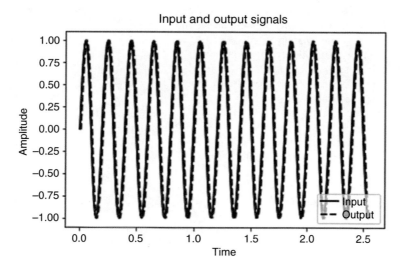

Fig. 10.31 Plot of input and output signals of a two-channel filter bank

- Instead of designing four filters (2 for analysis and 2 for synthesis). It is sufficient to design one prototype filter $H(z)$. All the other filters are obtained as a modified version of the prototype filter.
- If $H(z)$ acts as lowpass filter, then $H(-z)$ acts as highpass filter and vice versa.

Experiment 10.10 Relationship Between Analysis and Synthesis Filters

This experiment tries to obtain the magnitude response of the analysis and synthesis filters. The python code to obtain the magnitude response of the analysis and synthesis filters is given in Fig. 10.32, and the corresponding output is shown in Fig. 10.33.

Inferences

From the magnitude response, the following inferences can be drawn:

1. The analysis filter $h_0[n]$ behaves like a lowpass filter.
2. The analysis filter $h_1[n]$ behaves like a highpass filter.
3. The synthesis filter $g_0[n]$ behaves like a lowpass filter.
4. The synthesis filter $g_1[n]$ behaves like a highpass filter.
5. The filters $h_0[n]$ and $h_1[n]$ are complementary to each other.
6. The filters $g_0[n]$ and $g_1[n]$ are complementary to each other.

Experiment 10.11 Phase Responses of Analysis and Synthesis Filters

This experiment discusses the phase responses of analysis and synthesis filters of two-channel filter bank. The python code to obtain the phase response of the analysis and the synthesis filters are given in Fig. 10.34, and the corresponding output is shown in Fig. 10.35.

```
#Magnitude response of the analysis and synthesis filters
import numpy as np
import matplotlib.pyplot as plt
from scipy import signal
h0=[0.5,0.5]
h0=np.array(h0)
h1=np.array([0.5,-0.5])
g0,g1=2*h0,-2*h1
#Frequency response of four filters
w0, H0 = signal.freqz(h0,1)
w1, H1 = signal.freqz(h1,1)
w2, H2 = signal.freqz(g0,1)
w3, H3 = signal.freqz(g1,1)
#Plotting the result
plt.subplot(2,2,1),plt.plot(w0, 10 * np.log10(abs(H0)))
plt.xlabel('$\omega$ [rad/sample]'),plt.ylabel('Magnitude(dB)')
plt.title('$|H_0(e^{j\omega})|$')
plt.subplot(2,2,2),plt.plot(w1, 10 * np.log10(abs(H1)))
plt.xlabel('$\omega$ [rad/sample]'),plt.ylabel('Magnitude(dB)')
plt.title('$|H_1(e^{j\omega})|$')
plt.subplot(2,2,3),plt.plot(w2, 10 * np.log10(abs(H2)))
plt.xlabel('$\omega$ [rad/sample]'),plt.ylabel('Magnitude(dB)')
plt.title('$|G_0(e^{j\omega})|$')
plt.subplot(2,2,4),plt.plot(w3, 10 * np.log10(abs(H3)))
plt.xlabel('$\omega$ [rad/sample]'),plt.ylabel('Magnitude(dB)')
plt.title('$|G_1(e^{j\omega})|$')
plt.tight_layout()
```

Fig. 10.32 Python code to obtain the magnitude response of analysis and synthesis filters

Inferences
1. From the phase response of the analysis and synthesis filters, it is possible to infer that the filter exhibits linear phase characteristics in the pass band.
2. Because of linear phase characteristics, phase distortion can be avoided.
3. The filters will exhibit constant group delay.

10.4.3   Two-Channel Filter Bank Without Filters

This is a special case of two-channel filter bank in which delay is introduced instead of filters. For example, if $H_0(z) = 1$, $H_1(z) = z^{-1}$, $G_0(z) = z^{-1}$ and $G_1(z) = 1$, then the structure of two-channel filter bank is modified as in Fig. 10.36.

The relationship between the input and output in the frequency-domain is expressed as

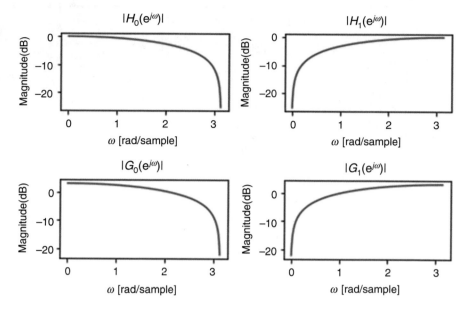

Fig. 10.33 Magnitude responses of analysis and synthesis filters

$$Y(z) = z^{-1}X(z) \tag{10.10}$$

Upon taking inverse Z-transform, the relationship between the input and output is given by

$$y[n] = x[n-1] \tag{10.11}$$

Experiment 10.12 Python Illustration of Two-Channel Filter Bank Without Filters

This experiment deals with the illustration of two-channel filter bank without filters. The python code, which depicts filter bank without filters is shown in Fig. 10.37, and the corresponding output is shown in Fig. 10.38.

Inference

From Fig. 10.38, it is possible to observe that the output signal is a delayed version of the input signal. Hence, perfect reconstruction is achieved. There is one sample delay between the input and output signal.

```
#Phase response of analysis and synthesis filters
import numpy as np
import matplotlib.pyplot as plt
from scipy import signal
h0=[0.5,0.5]
h0=np.array(h0)
h1=np.array([0.5,-0.5])
g0=2*h0
g1=-2*h1
#Frequency response of four filters
w0, H0 = signal.freqz(h0,1)
w1, H1 = signal.freqz(h1,1)
w2, H2 = signal.freqz(g0,1)
w3, H3 = signal.freqz(g1,1)
#Plotting the result
plt.subplot(2,2,1),plt.plot(w0,np.unwrap(np.angle(H0)))
plt.xlabel('$\omega$ [rad/sample]'),plt.ylabel('Degree')
plt.title('$\Phi_{h0}(e^{j\omega})}$')
plt.subplot(2,2,2),plt.plot(w1,np.unwrap(np.angle(H1)))
plt.xlabel('$\omega$ [rad/sample]'),plt.ylabel('Degree')
plt.title('$\Phi_{h1}(e^{j\omega})}$')
plt.subplot(2,2,3),plt.plot(w2,np.unwrap(np.angle(H2)))
plt.xlabel('$\omega$ [rad/sample]'),plt.ylabel('Degree')
plt.title('$\Phi_{g0}(e^{j\omega})}$')
plt.subplot(2,2,4),plt.plot(w3,np.unwrap(np.angle(H3)))
plt.xlabel('$\omega$ [rad/sample]'),plt.ylabel('Degree')
plt.title('$\Phi_{g1}(e^{j\omega})}$')
plt.tight_layout()
```

Fig. 10.34 Python code to obtain the phase response of analysis and synthesis filters

10.4.4   Three-Channel Filter Bank Without Filters

The structure of three-channel filter bank is shown in Fig. 10.39. From this figure, it is possible to infer that there are three channels and each channel contains an analysis and synthesis filters.

If $H_0(z) = 1$, $H_1(z) = z^{-1}$ and $H_2(z) = z^{-2}$, $G_0(z) = z^{-2}$, $G_1(z) = z^{-1}$ and $G_2(z) = 1$, the structure of three-channel filter bank for this choice of filters is given in Fig. 10.40.

The frequency-domain relationship between the input and output signal is given by

$$Y(z) = z^{-2}X(z) \qquad (10.12)$$

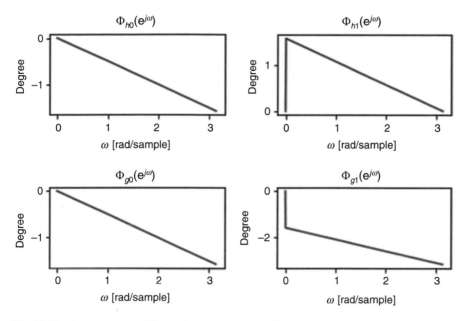

Fig. 10.35 Phase response of the analysis and synthesis filters

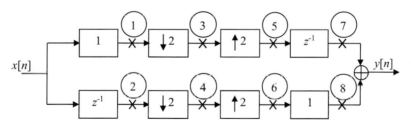

Fig. 10.36 Filter bank without filters

Upon taking inverse Z-transform, the time-domain relationship between the input and output signal is given by

$$y[n] = x[n - 2] \tag{10.13}$$

Experiment 10.13 Illustration of Three-Channel Filter Bank Without Filters
This experiment implements three-channel filter bank without filters using python, and the python code is shown in Fig. 10.41, and the corresponding output is shown in Fig. 10.42.

Inference
1. From Fig. 10.42, it is possible to observe that the output signal is a delayed version of the input signal.

```
#Filterbank without filters
import numpy as np
import matplotlib.pyplot as plt
#Step 1: Generation of sine wave
t=np.linspace(0,1,200)
x=np.sin(2*np.pi*5*t)
#Downsample x
x1=x[::2]
#Upsample x1
x2=np.zeros(2*len(x1))
x2[::2]=x1
#Introduce a delay to get x3
delay=1
x3=np.zeros(len(x)+delay)
x3[delay:]=x
x4=x3[::2]
x5=np.zeros(2*len(x4))
x5[::2]=x4
x6=np.zeros(len(x2)+delay)
x6[delay:]=x2
y=x6+x5[0:len(x6)]
plt.plot(t,x,t,y[0:len(t)]),plt.legend(["Input", "Output"], loc ="upper right"),
plt.xlabel('Time'),plt.ylabel('Amplitude'),plt.title('Input and Output signal')
plt.xlim((0, 1)),plt.ylim((-1, 1))
```

Fig. 10.37 Python code which implements filter bank without filters

2. There is two sample delay between the input and output signal, which is in
   agreement with the theoretical result.

10.5   Tree-Structured Filter Bank

The structure of uniform tree-structured filter bank is given in Fig. 10.43. The
numbers after the block are used to understand the sequence of the process of the
input signal. The same numbers are used as variables in the python code to
understand the sequence of process.

Experiment 10.14 Tree-Structured Filter Bank
This experiment illustrates the concept of tree-structured filter bank using python.
The python code for tree-structured filter bank is shown in Fig. 10.44, and the
corresponding output is shown in Fig. 10.45.

Inference
The following inferences can be made from this experiment:

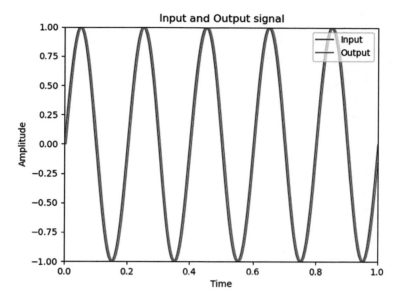

Fig. 10.38 Result of filter bank without filters

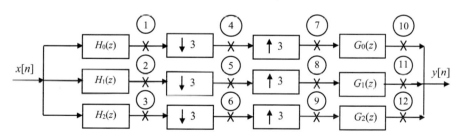

Fig. 10.39 Structure of three-channel filter bank

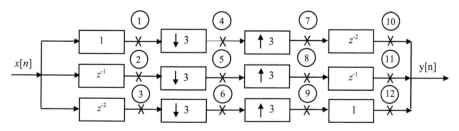

Fig. 10.40 Three-channel filter bank for the choice of filters

```
import numpy as np
import matplotlib.pyplot as plt
from scipy import signal
def downsample(x,M):
   y=x[::M]
   return(y)
def upsample(x,L):
   y=np.zeros(L*len(x))
   y[::L]=x
   return(y)
#Define the filters
h0,h1,h2=[1],[0,1],[0,0,1]
g0,g1,g2=[0,0,1],[0,1],[1]
#Input signal
t=np.linspace(0,1,200)
x=signal.sawtooth(2*np.pi*5*t)
x1=signal.lfilter(h0,1,x)
x2=signal.lfilter(h1,1,x)
x3=signal.lfilter(h2,1,x)
x4=downsample(x1,3)
x5=downsample(x2,3)
x6=downsample(x3,3)
x7=upsample(x4,3)
x8=upsample(x5,3)
x9=upsample(x6,3)
x10=signal.lfilter(g0,1,x7)
x11=signal.lfilter(g1,1,x8)
x12=signal.lfilter(g2,1,x9)
y=x10+x11+x12
plt.plot(t,x,t,y[0:len(t)]),plt.xlabel('Time'),plt.ylabel('Amplitude')
plt.title('Input and Output signal'),
plt.legend(['Input','Output'],loc=1), plt.tight_layout()
```

Fig. 10.41 Three-channel filter bank without filters

1. The input signal to tree-structured filter bank is a sawtooth signal of 5 Hz frequency. The output signal is also a sawtooth signal.
2. The output signal is a delayed version of the input signal. Thus, tree-structured filter bank obeys the perfect reconstruction criterion.
3. Perfect reconstruction is achieved through the proper choice of analysis and synthesis filters.

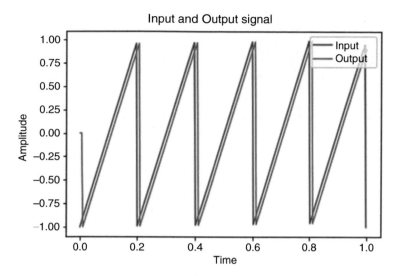

Fig. 10.42 Input and output signals of three-channel filter bank without filters

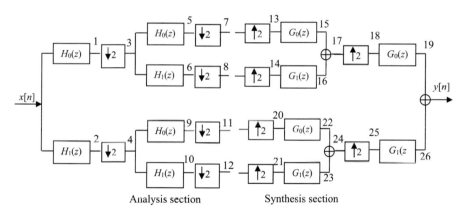

Fig. 10.43 Structure of tree-structured filter bank

10.6   Transmultiplexer

A transmultiplexer converts time division multiplexing (TDM) signals to frequency division multiplexing (FDM) and vice versa. A technique for sending multiple signals through the same physical medium is to use different portions of the available frequency spectrum. Frequency division multiplexing refers to the process of spectral separation to permit the simultaneous transmission of signals from multiple users. In frequency division multiplexing, all the signals operate at the same time with different frequencies. In time-division multiplexing, all the signals operate with the same frequency at different times. The operation of converting from one form of

```python
#Tree structured filter bank
import numpy as np
import matplotlib.pyplot as plt
from scipy import signal
def downsample(x,M): #Function to perform downsampling operation
    y=x[::M]
    return(y)
def upsample(x,L): #Function to perform upsampling operation
    y=np.zeros(L*len(x))
    y[::L]=x
    return(y)
#Step 1: Define the filters
h0=[0.5,0.5]
h0=np.array(h0)
h1=signal.qmf(h0)
g0,=2*h0,-2*h1
#Step 2: Generate the input signal
t=np.linspace(0,1,100)
x=signal.sawtooth(2*np.pi*5*t)
#Step 3: Traversing the path(1: Analysis section)
x1=signal.lfilter(h0,1,x)
x2=signal.lfilter(h1,1,x)
x3=downsample(x1,2)
x4=downsample(x2,2)
x5=signal.lfilter(h0,1,x3)
x6=signal.lfilter(h1,1,x3)
x7=downsample(x5,2)
x8=downsample(x6,2)
x9=signal.lfilter(h0,1,x4)
x10=signal.lfilter(h1,1,x4)
x11=downsample(x9,2)
x12=downsample(x10,2)
#2: Synthesis section
x13=upsample(x7,2)
x14=upsample(x8,2)
x15=signal.lfilter(g0,1,x13)
x16=signal.lfilter(g1,1,x14)
x17=x15+x16
x18=upsample(x17,2)
x19=signal.lfilter(g0,1,x18)
x20=upsample(x11,2)
x21=upsample(x12,2)
x22=signal.lfilter(g0,1,x20)
x23=signal.lfilter(g1,1,x21)
x24=x22+x23
x25=upsample(x24,2)
x26=signal.lfilter(g1,1,x25)
y=x19+x26
plt.plot(t,x,'k',t,y,'r'),plt.legend(['Input','Output'],loc=1),plt.xlabel('Time'),
plt.ylabel('Amplitude'),plt.title('Input-Output waveform')
```

Fig. 10.44 Python code for uniform tree-structured filter bank

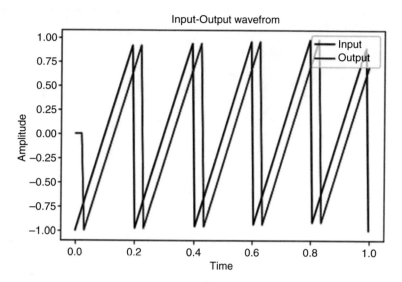

Fig. 10.45 Input-output waveform of tree-structured filter bank

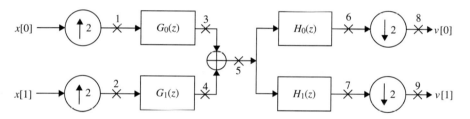

Fig. 10.46 Structure of two-channel transmultiplexer

multiplexing to another is termed as 'transmultiplexing'. The structure of two-channel transmultiplexer is shown in Fig. 10.46.

Proper choice of filters will avoid the problem of crosstalk in two-channel transmultiplexer.

Experiment 10.15 Implementation of Two-Channel Transmultiplexer

This experiment discusses the implementation of two-channel transmultiplexer using python. The python code, which implements a two-channel transmultiplexer, is shown in Fig. 10.47, and the corresponding output is shown in Fig. 10.48.

Figure 10.47 shows that the signal $x[0]$ is a sine wave of 5 Hz frequency, and the signal $x[1]$ is a cosine wave of 5 Hz frequency. The transmitted signals are sine and cosine waves. The variables 'n0 to n9' in the python code shown in Fig. 10.46 align with the nodes depicted in Fig. 10.47. From Fig. 10.48, it is possible to observe that the received signals $y[0]$ is a sine wave similar to the signal $x[0]$, and the signal $y[1]$ is a cosine wave similar to the transmitted signal $x[1]$.

```
#Two-channel transmultiplexer
import numpy as np
import matplotlib.pyplot as plt
from scipy import signal
def downsample(x,M):
    y=x[::M]
    return(y)
def upsample(x,L):
    y=np.zeros(L*len(x))
    y[::L]=x
    return(y)
#Step 1 defining x0 and x1
f,fs,N,N1=5,100,256,128;
T=1/fs
t=np.linspace(0,N*T,N)
x0=np.sin(2*np.pi*f*t)
x1=np.cos(2*np.pi*f*t)
#Step 2: Define the filters
g0,g1,h0,h1=[0, 1, 1], [0, -1, 1], [0.5, 0.5], [0.5, -0.5];
#Step3 Tracing the structure
n1=upsample(x0,2)  #At node 1
n2=upsample(x1,2)  #At node 2
n3=signal.lfilter(g0,1,n1) #At node 3
n4=signal.lfilter(g1,1,n2) #At node 4
n5=n3+n4  #At node 5
n6=signal.lfilter(h0,1,n5) #At node 6
n7=signal.lfilter(h1,1,n5) #At node 7
n8=downsample(n6,2) #At node 8
n9=downsample(n7,2) #At node 9
plt.subplot(2,1,1),plt.plot(t,x0,t,n8),plt.legend(['Transmitted(x0)','Received(y0)'],loc=1)
plt.xlabel('Time'),plt.ylabel('Amplitude')
plt.subplot(2,1,2),plt.plot(t,x1,t,n9),plt.legend(['Transmitted(x1)','Received(y1)'],loc=1)
plt.xlabel('Time'),plt.ylabel('Amplitude')
plt.suptitle('Transmitted and received signals in transmultiplexer')
plt.tight_layout()
```

Fig. 10.47 Python code for two-channel transmultiplexer

Inferences

The following inferences can be drawn from Fig. 10.48:

1. The received signals are similar to the transmitted signal without distortion; thus, crosstalk problem is avoided. The proper choice of synthesis and analysis filter avoids the crosstalk problem.
2. Perfect reconstruction of transmultiplexer achieves complete crosstalk cancellation and is distortion-free.

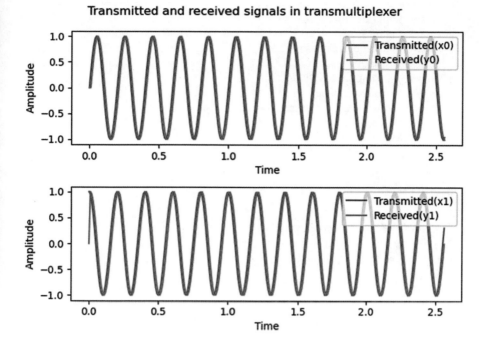

Fig. 10.48   Transmitted and received signals in transmultiplexer

Experiment 10.16 Audio Signal Transmission

This experiment tests the process of two-channel transmultiplexer using audio signal as an input. Instead of transmitting sine wave and cosine wave as the input through transmultiplexer, pass male and female voices through two-channel transmultiplexer, and observe whether the male and female voices can be received in the output without distortion.

The python code which performs this task is shown in Fig. 10.49, and the corresponding output is shown in Fig. 10.50.

Inferences

From Fig. 10.49, it is possible to observe the following:

1. The library 'sounddevice' is used to play the speech signal.
2. The built-in function 'wavfile' from 'scipy' library is used to read the speech signal.
3. The two signals fed to the transmultiplexer are (a) x0 is a male voice and (b) x1 is a female voice.
4. The output (variable name 'n8') represents the received male voice corresponding to the transmitted male voice (x0).
5. The output (variable name 'n9') represents the received female voice corresponding to the transmitted female voice (x1).
6. The transmitted and the received male and female voices are plotted and heard.

```
#Transmultiplexer for speech signal
import numpy as np
import matplotlib.pyplot as plt
from scipy import signal
from scipy.io import wavfile
import sounddevice as sd
#Functions to perform downsampling and upsampling
def downsample(x,M):
    y=x[::M]
    return(y)
def upsample(x,L):
    y=np.zeros(L*len(x))
    y[::L]=x
    return(y)
#Step 1 Reading the speech signals
fs, x0 = wavfile.read('Male.wav')
fs, x1 = wavfile.read('Female.wav')
x0=x0[:,0]
x1=x1[:,1]
#Step 2: Define the filters
g0,g1,h0,h1=[0,1, 1], [0,-1, 1], [0.5, 0.5], [0.5, -0.5];
#Step3 Tracing the structure
n1=upsample(x0,2)
n2=upsample(x1,2)
n3=signal.lfilter(g0,1,n1)
n4=signal.lfilter(g1,1,n2)
n5=n3+n4
n6=signal.lfilter(h0,1,n5)
n7=signal.lfilter(h1,1,n5)
n8=downsample(n6,2)
n9=downsample(n7,2)
plt.subplot(2,2,1),plt.plot(x0),plt.title('Transmitted male voice')
plt.subplot(2,2,2),plt.plot(x1),plt.title('Transmitted female voice')
plt.subplot(2,2,3),plt.plot(n8),plt.title('Received male voice')
plt.subplot(2,2,4),plt.plot(n9),plt.title('Received female voice')
plt.tight_layout()
#Hearing the audio signal
sd.play(x0,fs)   #Transmitted male voice
sd.wait()
sd.play(n8,fs)  #Received male voice
sd.wait()
sd.play(x1,fs)   #Transmitted female voice
sd.wait()
sd.play(n9,fs)  #Received female voice
```

Fig. 10.49 Python code for Experiment 10.16

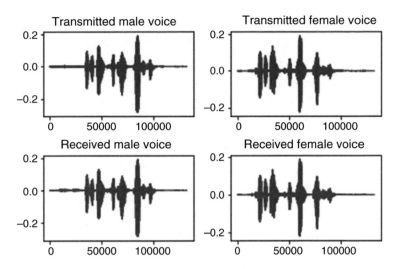

Fig. 10.50 Transmitted and received speech signals through two-channel transmultiplexer

Fig. 10.51 Multirate system

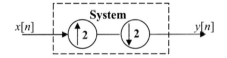

From Fig. 10.50, it is possible to observe that the received speech signal resembles the transmitted speech signal, which confirms that the two-channel transmultiplexer is free from cross-talk.

Exercises

1. Write a python code to simulate the comb signal whose expression is given by

$$x[n] = \frac{1}{M} \sum_{k=0}^{M-1} e^{j\frac{2\pi}{M}kn}, \quad 0 \le n \le 10$$

For $M = 1, 2, 3$ and 4, comment on the observed output.

2. Write a python code to prove that the downsampling operation obeys the superposition principle. The objective is to prove that downsampling operation obeys both additivity and homogeneity properties.

3. Write a python code to prove that upsampling by a factor of 2 is a time-varying operation.

4. Generate a square wave of 5 Hz fundamental frequency. Downsample this signal by a factor of 2. Plot the spectrum of the input and downsampled square waves, and comment on the observed result.

5. Write a python code to prove the fact that the output signal $y[n]$ is identical to the input signal $x[n]$ for the multirate system shown in Fig. 10.51.

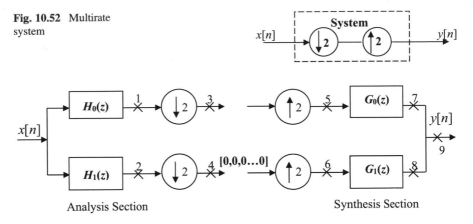

Fig. 10.52 Multirate system

Fig. 10.53 Two-channel filter bank

6. Write a python code to implement the following multirate system depicted in Fig. 10.52.

   Plot the input and output signal and comment on the observed output.

7. Read an audio signal. Downsample it by a factor of 2. Hear the original and downsampled audio signal and comment on the observation.

8. Read an audio signal. Upsample it by a factor of 2. Hear the original and upsampled audio signal and comment on the observation.

9. The input to the two-channel filter bank shown in Fig. 10.53 is a square of 5 Hz fundamental frequency. The impulse response of the analysis and synthesis filters are $h_0[n] = \{0.5, 0.5\}$, $h_1[n] = \{0.5, -0.5\}$, $g_0[n] = \{1, 1\}$ and $g_1[n] = \{-1, 1\}$. The high pass filter section has to be masked. The bitstream from node 4 has to be strings of zeros. The impact of masking the high frequency component has to be analysed.

   Plot the input and output signal and comment on the observed output.

10. Record two voice signals, namely, $x[0]$ and $x[1]$, with a sampling frequency of 8000 Hz. The signal $x[0]$ corresponds to the word '*YES*', and the signal $x[1]$ corresponds to the word '*NO*'. That is, the signal $x[0]$ and $x[1]$ are recorded voice signals with the word '*YES*' and '*NO*', respectively. Pass these two voice signals to crosstalk free transmultiplexer to obtain the output signal $y[0]$ and $y[1]$. Comment on the observed output.

Objective Questions

1. Downsampling by a factor of 'M' is a

   A. Linear, time-invariant operation
   B. Linear, time-variant operation
   C. Non-linear, time-invariant operation
   D. Non-linear, time-variant operation

2. The time-domain expression for downsampling by a factor of 'M' is given by

A. $y[n] = x\left[\frac{n}{M}\right]$

B. $y[n] = x[nM]$

C. $y[n] = x[n]^M$

D. $y[n] = x[n]^{\frac{1}{M}}$

3. The input signal $x[n]$ is upsampled by a factor of 'L'; then, the result is downsampled by the same factor 'L' to obtain the signal $y[n]$. The relationship between $y[n]$ and $x[n]$ is given by

A. $y[n] = x[nL]$

B. $y[n] = x[n]$

C. $y[n] = x\left[\frac{n}{L}\right]$

D. $y[n] = x[n]^L$

4. The input signal $x[n] = \left\{\underset{\uparrow}{1}, 1, 1, 1\right\}$ is passed through downsampling by a factor of 2. The result of downsampling operation is then passed through upsampling by a factor of 2 to obtain the output signal $y[n]$. The expression for the output signal is

A. $y[n] = \left\{\underset{\uparrow}{1}, 1, 1, 1\right\}$

B. $y[n] = \left\{\underset{\uparrow}{1}, 0, 1, 0\right\}$

C. $y[n] = \left\{\underset{\uparrow}{1}, -1, 1, -1\right\}$

D. $y[n] = \{0, 0, 0, 0\}$

5. Which of the following operation is an example of an idempotent operation

   A. Upsampling followed by downsampling
   B. Downsampling followed by upsampling
   C. Downsampling followed by downsampling
   D. Upsampling followed by upsampling

6. Which of the following is an example of an identity operation

   A. Upsampling followed by downsampling
   B. Downsampling followed by upsampling
   C. Downsampling followed by downsampling
   D. Upsampling followed by upsampling

7. Which of the following results in idempotent operation

   A. Downsampling by a factor of 'M' followed by upsampling by a factor of 'M'
   B. Upsampling by a factor of 'M' followed by downsampling by the same factor

C. Downsampling followed by unit delay operation
D. Upsampling followed by unit delay operation

8. If the variable 'x' contain the input signal, the python command y=x[::2] results in

A. Upsampling of the input signal by a factor of 2
B. Downsampling of the input signal by a factor of 2
C. Delaying of input signal by a factor of 2
D. Advance of input signal by a factor of 2

9. A function 'operation' is given below. The function accepts the input signal (x) and gives an output signal (y). What is the relationship between the input and output signal?

```
def operation(x,L):
    y=np.zeros(L*len(x))
    y[::L]=x
    return(y)
```

A. Output signal 'y' is downsampled by a factor of 2
B. Output signal 'y' is upsampled by a factor of 2
C. Output signal 'y' is delayed by a factor of 2
D. Output signal 'y' is advanced by a factor of 2

10. Downsampling is a

A. Linear, time-invariant operation
B. Linear, time-variant operation
C. Non-linear, time-invariant operation
D. Non-linear, time-variant operation

11. Interchanging of upsampling by a factor of 'L' and downsampling by a factor of 'M' is possible if and only if

A. L and M are of same value
B. L and M should be odd number
C. L and M should be even number
D. L and M are relatively prime

Bibliography

1. Ronald E. Crochiere, and Lawrence R. Rabiner, "Multirate Digital Signal Processing", Pearson, 1983.
2. P. P. Vaidyanathan, "Multirate Systems and Filter Banks", Prentice Hall, 1993.

3. N. J. Fliege, "Multiratge Digital Signal Processing: Multirate Systems, Filterbanks, Wavelets", John Wiley and Sons, 1999.
4. Bruce W. Suter, "Multirate and Wavelet Signal Processing", Academic Press, 1997.
5. Vikram M Gadre, and Aditya S. Abhyankar, "Multiresolution and Multirate Signal Processing: Introduction, Principles and Applications", McGraw Hill, 2017.

Chapter 11
Adaptive Signal Processing

Learning Objectives
After reading this chapter, the reader is expected to

- Implement and analyse the Wiener filter.
- Write a python code to implement the LMS algorithm and its variants.
- Perform system identification using the LMS algorithm.
- Perform inverse system modelling using the NLMS algorithm.
- Implement adaptive line enhancer using the LMS algorithm and its variants.
- Implement the RLS algorithm.

Roadmap of the Chapter
The roadmap of this chapter is depicted below. This chapter starts with the Wiener filter, least mean square (LMS) algorithm and its variant approaches for adaptive signal processing applications like system identification and signal denoising. Next, the RLS algorithm is discussed with the suitable python code.

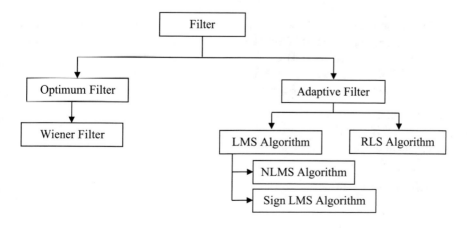

PreLab Questions
1. List out the valid differences between the optimal filter and the adaptive filter.
2. What is an adaptive filter? How it differs from the ordinary filter.
3. Examples of adaptive filter.
4. When are adaptive filters preferred?
5. List out the performance measures of the adaptive filter.
6. What is an LMS algorithm?
7. What do you mean by least square estimation?
8. List out the variants of LMS algorithm.
9. How the step size impacts the LMS algorithm?
10. What is the RLS algorithm, and how it differs from LMS?

11.1   Wiener Filter

Wiener filter is the mean square error (MSE) optimal stationary linear filter for signal corrupted by additive noise. The Wiener filter computation requires the assumption that the signal and noise are in the random process. The general block diagram of the Wiener filter is shown in Fig. 11.1. The main objective of the Wiener filter is to obtain the filter coefficient of the LTI filter, which can provide the final output ($y[n]$) as much as the minimum MSE between the output and the desired signal or target ($d[n]$). In Fig. 11.1, $s[n]$ denotes the original signal, which is a clean signal, and it is corrupted by additive noise $\eta[n]$ to give the signal $x[n]$. The parameters of the filter have to be designated has to be designed in such a way that the output of the filter $y[n]$ should resemble the desired signal $d[n]$ such that the error '$e[n]$' is minimum.

The expression for the optimal Wiener filter is given by

$$h_{\text{opt}} = R^{-1}p \tag{11.1}$$

The above expression is termed as 'Wiener-Hopf' expression, which is named after American-born Norbert Wiener and Austrian-born Eberhard Hopf. The expression for optimal filter depends on the autocorrelation matrix (R) of the observed signal ($x[n]$) and the cross-correlation vector (p) between the observed signal ($x[n]$) and the desired signal ($d[n]$). h_{opt} denotes the optimal filter coefficients.

Experiment 11.1 Wiener Filtering
The aim of this experiment is to implement the Wiener filtering using python. Here the optimal filter coefficients are obtained using the Wiener-Hopf equation given in

Fig. 11.1 Block diagram of Wiener filter

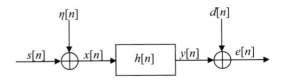

```
#Wiener filter
import numpy as np
from numpy.random import randn
import matplotlib.pyplot as plt
from scipy.linalg import toeplitz
from scipy import signal
#Step 1: Generation of signal s[n]
t=np.linspace(0,1,100)
s=np.sin(2*np.pi*5*t)
Ns=len(s)
#Step 2: Generation of random noise
# n=randn(len(t))*0.1
n=np.random.normal(0,.2,len(s))
#Step 3: Observed signal x[n]
x=s+n
#Step 4: Autocorrelation of observed signal
rxx=np.correlate(x,x,mode='full')
#Step 5: Cross-correlation between desired and observed signal
rsx=np.correlate(s,x,mode='full')
#Step 6: Deciding the length of the filter
Nh=11
#Step 7: Trimming the autocorrelation and cross-correlation values
rxx1=rxx[Ns-1:Ns+Nh-1]
rsx1=rsx[Ns-1:Ns+Nh-1]
#Step 8: Obtaining the autocorrelation matrix
Rx=toeplitz(rxx1)
#Step 9: Inverse of the autocorrelation matrix
Rx1=np.linalg.inv(Rx)
#Step 10: Obtaining the filter coefficient
w1=np.matmul(Rx1,rsx1)
#Step 11: Filtering the noisy signal
y=signal.lfilter(w1,1,x)
plt.subplot(3,1,1),plt.plot(t,s),plt.xlabel('t-->'),plt.ylabel('Amplitude'),
plt.title('Clean signal'),plt.subplot(3,1,2),plt.plot(t,x),plt.xlabel('t-->'),
plt.ylabel('Amplitude'),plt.title('Noisy signal'),plt.subplot(3,1,3), plt.plot(t,y)
plt.xlabel('t-->'),plt.ylabel('Amplitude'),plt.title('Filtered signal'),plt.tight_layout()
```

Fig. 11.2 Python code for Wiener filtering

Eq. (11.1). The python code for Wiener filter is shown in Fig. 11.2. Simulation result of the python code given in Fig. 11.2 is depicted in Fig. 11.3.

The built-in functions used in python code shown in Fig. 11.2 is summarized in Table 11.1.

Inference

From Fig. 11.3, it can be made the following observations:

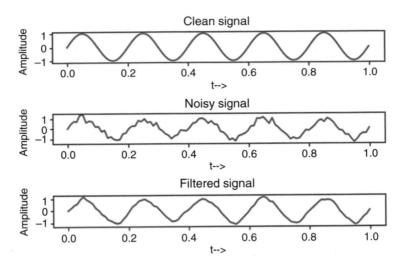

Fig. 11.3 Simulation result of Wiener filter

Table 11.1 Built-in functions used in the python code given in Fig. 11.2

S. No.	Objective	Built-in function	Library
1	To generate a clean sinusoidal signal of 5 Hz frequency	np.sin()	Numpy
2	To add white noise, which follows normal distribution to clean signal	np.random.normal()	Numpy
3	To perform autocorrelation	np.correlate()	Numpy
4	To obtain the inverse of the matrix	np.linalg.inv()	Scipy
5	To perform convolution	signal.lfilter()	Scipy

1. The input or clean signal frequency is 5 Hz, and it is a smooth sine waveform.
2. The additive noise added signal as input to the Wiener filter, and it is a distorted signal.
3. The filtered signal is not a smooth sine waveform. However, this waveform is far better than the noisy signal. Hence, the Wiener filter has a capability to minimize the impact of additive noise in a signal.

Task
1. Change the value of standard deviation in random noise generation python command 'np.random.normal(0,.2,len(s))' given in Fig. 11.2. Execute and make the appropriate changes in this python code to get 'filtered signal' as similar as 'clean signal'.

Experiment 11.2 Wiener Filter Using Built-In Function
This experiment performs the Wiener filtering using built-in function in '*scipy*' library. The built-in function is available in *the* '*scipy*' library '*wiener*' can be used to filter out the noisy components. In this experiment, noise-free sinusoidal signal of 5 Hz frequency is generated. The clean signal is corrupted by adding

Table 11.2   Steps followed and built-in functions

S. No.	Objective	Built-in function	Library
1	To generate a clean sinusoidal signal of 5 Hz frequency	np.sin()	Numpy
2	To add white noise, which follows normal distribution to clean signal	np.random.normal()	Numpy
3	To minimize the impact of noise using Wiener filter	signal.wiener()	Scipy

```
#Wiener filter
import numpy as np
import matplotlib.pyplot as plt
from scipy import signal
#Step 1: Generation of clean signal
t=np.linspace(0,1,100)
s=np.sin(2*np.pi*5*t)
#Step 2: Adding noise
n=np.random.normal(0,.2,len(s))
x=s+n
#Step 3: Wiener filter
y=signal.wiener(x)
#Step 4: Plotting the results
plt.subplot(3,1,1),plt.plot(t,s),
plt.xlabel('Time'),plt.ylabel('Amplitude'),plt.title('Clean signal')
plt.subplot(3,1,2),plt.plot(t,x),plt.xlabel('Time'),plt.ylabel('Amplitude'),
plt.title('Noisy signal'),plt.subplot(3,1,3),plt.plot(t,y)
plt.xlabel('Time'),plt.ylabel('Amplitude'),plt.title('Filtered signal')
plt.tight_layout()
```

Fig. 11.4   Wiener filtering using built-in function

random noise, which follows the normal distribution with zero mean and 0.2 standard deviation. The corrupted signal is then passed through the Wiener filter to minimize the impact of noise. The steps followed along with the built-in functions used in the program are given in Table 11.2.

The python code which performs this task is shown in Fig. 11.4, and the corresponding output is shown in Fig. 11.5.

Inference

From Fig. 11.5, it is possible to infer that the impact of noise is minimized after passing the noisy signal through Wiener filter.

11.1.1   Wiener Filter in Frequency Domain

From Wiener-Hopf equation, the expression for the optimal Wiener filter is given by

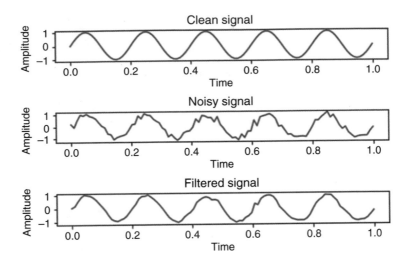

Fig. 11.5 Result of Wiener filtering

$$h_{opt} = R^{-1}p \qquad (11.2)$$

The above equation can be expressed as

$$h_{opt} = \frac{p}{R} \qquad (11.3)$$

In the above expression, 'p' represents the cross-correlation between desired signal and the observed signal, and 'R' represents the autocorrelation of the observed signal. Taking Fourier transform on both sides of Eq. (11.3), we get

$$FT\{h_{opt}\} = \frac{FT\{p\}}{FT\{R\}} \qquad (11.4)$$

According to the Wiener-Khinchin theorem, Fourier transform of autocorrelation function gives power spectral density. Using this theorem, Eq. (11.4) is expressed as

$$H(e^{j\omega}) = \frac{S_{dx}(e^{j\omega})}{S_{xx}(e^{j\omega})} \qquad (11.5)$$

In Eq. (11.5), $H(e^{j\omega})$ represents the frequency response of the Wiener filter, $S_{dx}(e^{j\omega})$ represents the cross-power spectral density estimation between desired and observed signal and $S_{xx}(e^{j\omega})$ represents the power spectral density of the observed signal.

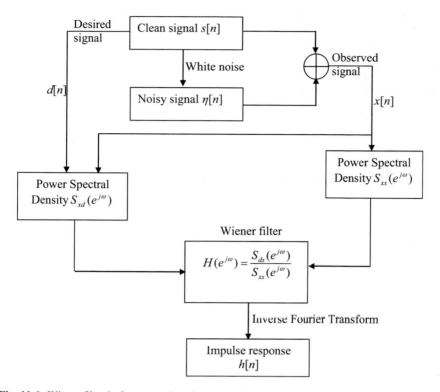

Fig. 11.6 Wiener filter in frequency domain

Experiment 11.3 Wiener Filter in Frequency Domain

The steps followed in the implementation of Wiener filter in frequency domain are given in Fig. 11.6. The noisy signal is obtained by adding white noise, which follows normal distribution to the clean signal. The observed signal is a clean signal with white noise added to it. The power spectral density of the observed signal is represented by $S_{xx}(e^{j\omega})$. The power spectral density between the desired and observed signal is represented by $S_{dx}(e^{j\omega})$. The Wiener filter is obtained in the frequency domain using the relation $H(e^{j\omega}) = \frac{S_{dx}(e^{j\omega})}{S_{xx}(e^{j\omega})}$. Here the desired signal is the clean signal $s[n]$. Upon taking inverse Fourier transform of $H(e^{j\omega})$, the impulse response of the Wiener filter is obtained.

The python code used to implement the Wiener filter in frequency domain is shown in Fig. 11.7, and the corresponding output is in Fig. 11.8.

The built-in functions used in the program and its purpose are given in Table 11.3.

Inference

From Fig. 11.8, the following observations can be made:

```
#Wiener filter in frequency domain
import numpy as np
import matplotlib.pyplot as plt
from scipy import signal
from matplotlib import patches
t=np.linspace(0,1,100)
s=np.sin(2*np.pi*5*t) #Step1: Generation of clean signal s[n]
n=np.random.normal(0,0.1,len(t)) #Step 2: Generation of noise
x=s+n #Step 3: Generation of observed signal x[n]
Nh=25
f,Pxx=signal.csd(x,x,nperseg=Nh) #Step 4: Power spectral density of observed signal
f,Psx=signal.csd(s,x,nperseg=Nh) #Step 5: PSD of desired and observed signal
H=Psx/Pxx #Step 6: Wiener filter in frequency domain
h=np.fft.irfft(H) #Step 7: Wiener filter in time domain
w, H1 = signal.freqz(h, 1)
y=signal.filtfilt(h,1,x) #Step 8: Filtered signal
plot1 = plt. figure(1)
bx=plt.subplot(3,1,1)
bx.plot(t,s),bx.set(title='Clean signal',xlabel='Time',ylabel='Amplitude')
bx=plt.subplot(3,1,2)
bx.plot(t,x),bx.set(title='Noisy signal',xlabel='Time',ylabel='Amplitude')
bx=plt.subplot(3,1,3)
bx.plot(t,y),bx.set(title='Filtered signal',xlabel='Time',ylabel='Amplitude')
plt.tight_layout()
plot2 = plt. figure(2)
#Pole-zero plot of the filter
ax = plt.subplot(2,2,3);
unit_circle = patches.Circle((0,0),radius = 1 , fill = False,color='black',ls='solid',alpha = 0.1)
ax.add_patch(unit_circle),ax.axhline(0,color='black',alpha = 0.5)
ax.axvline(0,color='black',alpha = 0.5)
b,a = h,[1]
z,p,k = signal.tf2zpk(b,a)
ax.plot(np.real(z),np.imag(z),'or',label='zeros')
ax.plot(np.real(p),np.imag(p),'xb',label = 'poles')
ax.set(title='Zeros and poles',xlabel='$\sigma$', ylabel='$j\omega$'),ax.legend(loc = 2),ax.grid()
ax = plt.subplot(2,2,1)
ax.stem(h),ax.set(title='Impulse response',xlabel='n-->',ylabel='Amplitude')
ax = plt.subplot(2,2,2)
ax.plot(w/np.pi,20*np.log10(abs(H1))),
ax.set(title='Magnitude response',xlabel='w',ylabel='Magnitude')
ax=plt.subplot(2, 2, 4)
ax.plot(w/np.pi, 180/np.pi*np.unwrap(np.angle(H1)))
ax.set(title='Phase response',xlabel='w',ylabel='Phase'),plt.tight_layout()
```

Fig. 11.7 Python code to implement Wiener filter in frequency domain

1. The impact of noise is minimized by applying the Wiener filter.
2. The impulse response of the Wiener filter is not symmetric; hence, the phase response of the filter is not a linear curve.
3. From the magnitude response, it is possible to observe that the filter is a lowpass filter, and it performs smoothing actions to minimize the impact of noise.

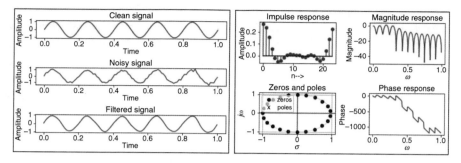

Fig. 11.8 Result and characteristics of Wiener filtering

Table 11.3 Built-in functions used in this experiment

S. No.	Objective	Built-in function	Library
1	To generate clean sinusoidal signal of 5 Hz frequency	np.sin()	Numpy
2	To add white noise which follows normal distribution to clean signal	np.random.normal()	Numpy
3	To compute the power spectral density	signal.csd()	Scipy
4	To compute the impulse response of the filter from the frequency response	np.fft.irfft()	Numpy
5	To obtain the frequency response of the filter	signal.freqz()	Scipy
6	To obtain the poles, zeros and the gain of the filter from the transfer function	signal.tf2zpk()	Scipy

Fig. 11.9 General block diagram of adaptive filtering

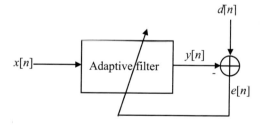

4. From the pole-zero plot, it is possible to observe that poles and zeros lie within the unit circle; hence the filter is stable.

11.2 Adaptive Filter

The adaptive filter is a non-linear filter, which updates the value of the filter coefficients based on some specific criterion. The general block diagram of the adaptive filter is shown in Fig. 11.9. From this figure, it is possible to observe that the filter coefficients are updated based on the error, $e[n]$ between the output of the filter $y[n]$ and reference data $d[n]$. Examples of adaptive filters are LMS filter and RLS filter.

11.2.1 LMS Adaptive Filter

The LMS is a least mean square algorithm that works based on the stochastic gradient descent approach to adapt the estimate based on the current error. The estimate is called the weight or filter coefficient. The weight or filter coefficient update equation of the LMS algorithm is given by.

$$w[n + 1] = w[n] + \mu x[n]e[n] \tag{11.6}$$

where $w[n + 1]$ represents the new weight or updated weight, $w[n]$ denotes the old weight, μ indicates the step size or learning rate, $x[n]$ is the input signal or data and the error signal $e[n] = d[n] - y[n]$. $d[n]$ is the reference data or target data, and $y[n]$ is the actual output of the adaptive filter of the system.

Experiment 11.4 Implementation of LMS Algorithm
This experiment discusses the implementation of LMS algorithm for adaptive filtering using python. The python code to define the LMS algorithm as a function is shown in Fig. 11.10. This code can be called a function in the different applications of the LMS algorithm, which will be discussed in the subsequent experiments. From Fig. 11.10, it is possible to see that the weight updation formula of the LMS algorithm given in Eq. (11.6) exists in it.

Inference
1. From Fig. 11.10, it is possible to observe that the LMS algorithm is written as a function, and it can be called a signal processing application whenever needed.

```
# This python code for LMS algorithm
def LMS_algorithmm(x,mu,N,t):
    # x = input data, mu = step size, t = reference data, N = Filter length
    N1=len(x)
    w = np.zeros(N) # Initial filter
    e = np.zeros(N1-N)
    for n in range(0, N-F):
        xn = x[n+N:n:-1]
        en = t[n+N] - np.dot(xn,w) # Error
        w = w + mu * en * xn # Update filter (LMS algorithm)
        e[n] = en # Record error
    return w,e
```

Fig. 11.10 Python code for LMS algorithm

Fig. 11.11 Block diagram
of system identification

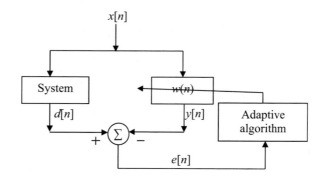

```
import numpy as np
import matplotlib.pyplot as plt
from scipy import signal
N1 = 500 # Size of the Input data
N = 25 # Filter size
n_iter=[10,50,100,150]# it must be less than (N1-N)
x = np.random.randn(N1) # Input to the filter
h = signal.firwin(N, 0.25) # FIR filter to be identified
t = signal.convolve(x, h) # Target/desired signal
t = t + 0.01 * np.random.randn(len(t)) # with added noise
mu = 0.04 # LMS step size
plt.figure(),plt.title('Filter to be Identified'),plt.stem(h),plt.xlabel('n-->'),plt.ylabel('h[n]')
for i in range(0,len(n_iter)):
    [w,e]=LMS_algorithmm(x,mu,N,t,n_iter[i]);
    plt.figure(),plt.title('Error signal at iteration %d' % n_iter[i])
    plt.stem(e),plt.xlabel('n-->'),plt.ylabel('e[n]')
    plt.figure(),plt.title('Identified Filter at iteration %d' % n_iter[i])
    plt.stem(w),plt.xlabel('n-->'),plt.ylabel('w[n]')
```

Fig. 11.12 Python code for unknown system identification

2. The inputs to the LMS function are 'x', 'mu', 'N' and 't'. 'x' denotes the input
   data, 'mu' represents step size, 't' denotes the reference data or target data and 'N'
   indicates the length of the adaptive filter.
3. The outputs from this LMS function are 'w', which denotes the adaptive filter
   coefficients, and 'e' is an error between the estimate and target data.

Experiment 11.5 System Identification Using LMS Algorithm
This experiment deals with unknown system identification using the LMS algorithm.
Let us consider the unknown system as an FIR filter with a length of 25. In this
experiment, the output filter coefficients are obtained by using LMS algorithm with
different number of iterations. The block diagram of the system identification is
shown in Fig. 11.11. The python code to find the unknown system using the LMS
algorithm is given in Fig. 11.12, and its simulation result is shown in Fig. 11.13.

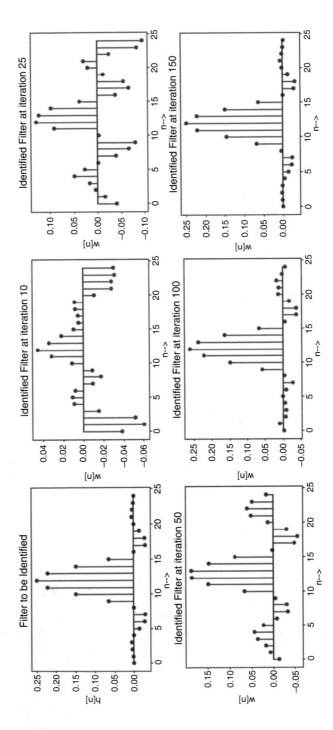

Fig. 11.13 Simulation results of Experiment 11.5

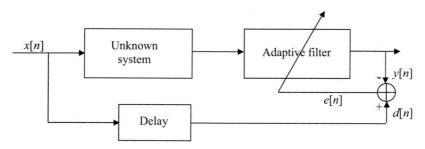

Fig. 11.14 Inverse system modelling using adaptive filter

Figure 11.12 indicates that the number of iterations is considered as 10, 50, 100 and 150, and the length of the unknown FIR filter is chosen as 25. The input to the LMS algorithm is a random signal with a length of 500 samples. The targeted or desired or reference data is obtained by convolving the input random signal with the unknown FIR filter coefficients along with the random noise.

Note that the inputs to the LMS algorithm ($[w,e]=LMS\_algorithmm(x,mu,N,t, n\_iter[i])$)) are random signal ($x$), learning rate ($mu$), length of the filter (N), a reference signal (t) and number or iteration (n_iter). Also, note that the filter coefficients (h) are not given as input to the LMS algorithm. The outputs of the LMS algorithm are error signal (e) and identified filter output (w).

The simulation result of the python code given in Fig. 11.12 is displayed in Fig. 11.13.

Inference
From Fig. 11.13, it is possible to observe that the adaptive filter result approaches the original filter coefficients while increasing the number of iterations.

Task
Increase/decrease the length of the FIR filter and fix the number of iterations is 50. Comment on the observed result.

Experiment 11.6 Inverse System Modelling Using LMS Algorithm
This experiment discusses the inverse system modelling using LMS algorithm. The general block diagram of inverse system modelling using adaptive filter is shown in Fig. 11.14. From this figure, it is possible to understand that the unknown system and the adaptive filter are connected in a cascade form, and the delayed version of the input signal act as a reference signal. The aim of adaptive filtering in this experiment is to obtain the inverse system of the unknown system so that $y[n]$ and $d[n]$ will be similar. If $y[n]$ and $d[n]$ are similar, then the adaptive filter is equal to the inverse of the unknown system.

In communication systems, inverse system modelling is used as channel equalization. In such scenario, the adaptive filter is termed as 'equalizer'. Adaptive equalizer can combat intersymbol interference. Intersymbol interference arises because of the spreading of a transmitted pulse due to the dispersive nature of the channel.

```
import numpy as np
import matplotlib.pyplot as plt
from scipy import signal
from scipy.fft import fft
mu,W=0.04,2.2 # learning rate,Channel Capacity
filt_order,t_samples,delay,trial=7,200,4,1000
noise_var,data_var=0.001,1
for i in range(0,trial):
    inp=np.zeros(filt_order)
    data=np.zeros(filt_order+t_samples)
    v=np.zeros(filt_order+t_samples)
    w=np.zeros(filt_order)
    #Generation of random data and random noise
    for j in range(filt_order-1,t_samples+filt_order):
        data[j]=np.fix(np.random.rand(1)+0.5)*2-1
        v[j]=np.fix(np.random.rand(1)+0.5)*2*np.sqrt(noise_var)-np.sqrt(noise_var)
    # Impusle response of the channel
    h=np.zeros(3)
    for j in range(0,3):
        h[j]=(1/2)*(1+np.cos(2*np.pi/W)*(j-(3-1)))
    C_out=signal.convolve(h,data) # Output from Channel
    Err_square=np.zeros(len(C_out))
    data=np.append(np.zeros(len(h)-1), data)
    v=np.append(np.zeros(len(h)-1), v)
    C_outn=C_out+v;
    [w,e]=LMS_algorithmm(C_outn,mu,filt_order,data,len(C_outn)-filt_order);
    e=np.append(e,np.zeros(filt_order))
    Err_square=Err_square+(e**2)
mse=Err_square/trial
plt.figure,plt.subplot(2,2,1),plt.stem(h),plt.title('Impulse Resp. of Channel filter')
plt.xlabel('n-->'),plt.ylabel('h1[n]'),plt.subplot(2,2,2),plt.stem(w),
plt.title('Impulse Resp. of Inverse filter'),plt.xlabel('n-->'),plt.ylabel('h2[n]')
cas=signal.convolve(w,h);#Cascade operation
mag=fft(cas);#Frequency Response
plt.subplot(2,2,3),plt.stem(cas),plt.title('Impulse Resp. of Cascaded filter'),plt.xlabel('n-->'),
plt.ylabel('h1[n]*h2[n]'),plt.subplot(2,2,4),plt.plot(np.abs(mag)),
plt.title('Mag. Resp. of Cascaded filter'),plt.xlabel('$\omega$-->'), plt.ylabel('|H($\omega$)|'),
plt.ylim(0,10),plt.tight_layout()
```

Fig. 11.15 Python code for Inverse system modelling

The impulse response of the channel is given by

$$h[n] = \begin{cases} \frac{1}{2}\left[1 + \cos\left(\frac{2\pi}{W}(n-2)\right)\right], & n=1,2,3 \\ 0, & \text{otherwise} \end{cases} \tag{11.7}$$

In the above equation, 'W' represents the channel capacity. Higher value of 'W' implies that the channel is more complex.

The python code to obtain the inverse of unknown system using LMS algorithm is given in Fig. 11.15, and its corresponding simulation result is shown in Fig. 11.16.

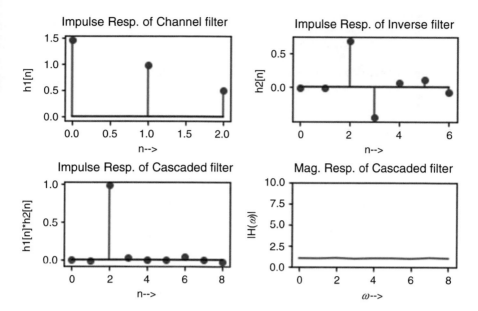

Fig. 11.16 Simulation result of inverse system modelling

Inference

From Fig. 11.16, it is possible to perceive the following facts

1. The impulse response of the cascaded system is an impulse. This implies that the cascade of channel filter and its inverse system results in an identity system.
2. The Fourier transform of an impulse response will result in a flat spectrum. This is obvious by observing the spectrum of the cascaded system.

Task

1. Increase the order of the adaptive filter and obtain the impulse response of the inverse system.

11.2.2 Normalized LMS Algorithm

The weight updation formula for the normalized LMS algorithm is given by

$$w[n+1] = w^T[n] + \frac{\beta}{\|x\|^2 + c} e[n]x[n] \tag{11.8}$$

where 'β' is a positive constant, which controls the convergence speed of the algorithm. 'c' is a small regularization parameter; it is added with the norm of the signal $x[n]$ to avoid the divide by zero error.

```
# This code for NLMS algorithm
def NLMS_algorithmm(x,N,t,beta,c,n_iter):
    # x = input data, N = Filter length t = reference data,
    # beta = Convergence parameter, c = regularization constant,
    # n_iter = number of iteration
    N1=len(x)
    w = np.zeros(N) # Initial filter
    e = np.zeros(N1-N)
    for n in range(0, n_iter):
        xn = x[n+N:n:-1]
        en = t[n+N] - np.dot(xn,w) # Error
        mu=beta/((xn*(np.transpose(xn)))+c)#Learning rate update
        w = w + mu * en * xn # Update filter (NLMS algorithm)
        e[n] = en # Record error
    return w,e
```

Fig. 11.17 Python code for NLMS algorithm

Experiment 11.7 Normalized LMS (NLMS) Algorithm
The python code for the normalized LMS algorithm is given in Fig. 11.17.

Inference
1. From Fig. 11.17, it is possible to observe that it is in the form of a function, and it can be called for the adaptive signal processing applications whenever required.
2. Also, it is possible to know that step size or learning rate is not given as a direct input to the function.
3. The step size is calculated using the input data, β and 'c'.

Experiment 11.8 Inverse System Modelling Using NLMS Algorithm
This experiment is a repetition of the inverse system modelling experiment, which was discussed earlier. Here, Experiment 11.6 is repeated with the same specifications, and NLMS is used for adaptive filtering instead of LMS algorithm. The python code of this experiment is shown in Fig. 11.18, and its corresponding simulation result is displayed in Fig. 11.19.

Inference
The following conclusions can be made from this experiment:

1. From this Fig. 11.19, it is possible to conclude that the cascade of channel and inverse filter gives the impulse response as unit impulse sequence.
2. The magnitude response confirms that the cascaded filter spectrum is a dc.
3. Therefore, the channel filter and the adaptive filter are inverse to each other.

```
import numpy as np
import matplotlib.pyplot as plt
from scipy import signal
from scipy.fft import fft
c,beta,W=1.5,0.25,2.2 # learning rate,Channel Capacity
filt_order,t_samples,delay,trial=7,200,4,1500
noise_var,data_var=0.001,1
for i in range(0,trial):
    inp=np.zeros(filt_order)
    data=np.zeros(filt_order+t_samples)
    v=np.zeros(filt_order+t_samples)
    w=np.zeros(filt_order)
    #Generation of random data and random noise
    for j in range(filt_order-1,t_samples+filt_order):
        data[j]=np.fix(np.random.rand(1)+0.5)*2-1
        v[j]=np.fix(np.random.rand(1)+0.5)*2*np.sqrt(noise_var)-np.sqrt(noise_var)
    # Impusle response of the channel
    h=np.zeros(3)
    for j in range(0,3):
        h[j]=(1/2)*(1+np.cos(2*np.pi/W)*(j-(3-1)))
    C_out=signal.convolve(h,data) # Output from Channel
    Err_square=np.zeros(len(C_out))
    data=np.append(np.zeros(len(h)-1), data)
    v=np.append(np.zeros(len(h)-1), v)
    C_outn=C_out+v;
    [w,e]=NLMS_algorithmm(C_outn,filt_order,data,beta,c,len(C_outn)-filt_order);
    e=np.append(e,np.zeros(filt_order))
    Err_square=Err_square+(e**2)
mse=Err_square/trial
plt.figure,plt.subplot(2,2,1),plt.stem(h),plt.title('Impulse Resp. of Channel filter')
plt.xlabel('n-->'),plt.ylabel('h1[n]'),plt.subplot(2,2,2),plt.stem(w),
plt.title('Impulse Resp. of Inverse filter'),plt.xlabel('n-->'),plt.ylabel('h2[n]')
cas=signal.convolve(w,h);#Cascade operation
mag=fft(cas);#Frequency Response
plt.subplot(2,2,3),plt.stem(cas),plt.title('Impulse Resp. of Cascaded filter')
plt.xlabel('n-->'),plt.ylabel('h1[n]*h2[n]'),plt.subplot(2,2,4),plt.plot(np.abs(mag)),
plt.title('Mag. Resp. of Cascaded filter'),plt.xlabel('$\omega$-->'),
plt.ylabel('|H($\omega$)|'),plt.ylim(0,10),plt.tight_layout()
```

Fig. 11.18 Python code for Experiment 11.8

11.2.3   Sign LMS Algorithm

The weight updation formula for Sign LMS algorithm is given by

$$w[n+1] = w[n] + \mu \ \text{sign}\{e[n]x[n]\} \qquad (11.9)$$

where 'sign' indicates the sign of the number, '$w[n+1]$' represents new weight and '$e[n]$' denotes the error signal between target and estimated signal.

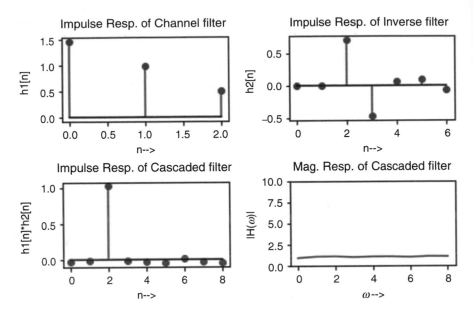

Fig. 11.19 Simulation result of the python code given in Fig. 11.18

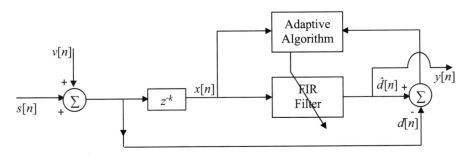

Fig. 11.20 Block diagram of adaptive line enhancer

Experiment 11.9 Adaptive Line Enhancer Using Sign LMS Algorithm

This experiment discusses the python implementation of adaptive line enhancer using sign LMS algorithm. The block diagram of adaptive line enhancer is shown in Fig. 11.20. From this figure, it is possible to observe that input to the FIR filter is a noisy version of the input signal ($x[n]$), and the final output ($y[n]$) is the enhanced input signal or noise-free signal. The aim of this experiment is to remove the noisy components present in the input signal using sign LMS adaptive algorithm. The python code for the "sign LMS algorithm" is given in Fig. 11.21 as a function.

The python code for adaptive line enhancer using sign LMS is given in Fig. 11.22. In this experiment, the input signal has 500, 2000 and 3500 Hz frequencies. The sampling frequency is considered as 8000 Hz. The input signal is added with the external random noise, which is the input to the adaptive filter. The number

```
# This Code for Sign LMS algorithm
def Sign_LMS_algorithmm(x,mu,N,t,n_iter):
    # x = input data, mu = step size, t = reference data, N = Filter length
    # n_iter = number of iteration
    N1=len(x)
    w = np.zeros(N) # Initial filter
    e = np.zeros(N1-N)
    for n in range(0, n_iter):
        xn = x[n+N:n:-1]
        en = t[n+N] - np.dot(xn,w) # Error
        w = w + mu * np.sign(en * xn) # Update filter (LMS algorithm)
        e[n] = en # Record error
    return w,e
```

Fig. 11.21 Python code for Sign LMS algorithm

```
import numpy as np
import matplotlib.pyplot as plt
from scipy import signal
from scipy.fft import fft
f1,f2,f3,Fs=500,2000,3500,8000 # Signal and sampling freq
T=1/Fs
t=np.arange(0,1,T)
noise=np.random.randn(len(t));
d=np.sin(2*np.pi*f1*t)+np.sin(2*np.pi*f2*t)+np.sin(2*np.pi*f3*t)+noise;
delay,N,mu=10,25,0.001 # Delay,Filter length and step size
x=np.append(np.zeros(delay),d);
[w,e]=Sign_LMS_algorithmm(x,mu,N,d,len(t)-N)
y1=signal.convolve(w,x)
mag_x=fft(x)/len(x);#Frequency Response
mag_y=fft(y1)/len(y1);#Frequency Response
plt.figure(),plt.subplot(2,2,1),plt.plot(x),plt.title('Input noisy signal')
plt.xlabel('t-->'),plt.ylabel('x(t)')
plt.subplot(2,2,2),plt.plot(y1),plt.title('Denoised signal')
plt.xlabel('t-->'),plt.ylabel('y(t)')
plt.subplot(2,2,3),plt.plot(np.abs(mag_x[0:4000])),plt.title('Spectrum of noisy signal')
plt.xlabel('$\omega$-->'),plt.ylabel('|X($\omega$)|')
plt.subplot(2,2,4),plt.plot(np.abs(mag_y[0:4000])),plt.title('Spectrum of denoised signal')
plt.xlabel('$\omega$-->'),plt.ylabel('|Y($\omega$)|')
plt.tight_layout()
```

Fig. 11.22 Python code for adaptive line enhancer using sign LMS

of delay is chosen as 10, and length of the adaptive FIR filter is fixed as 25. The main objective of this experiment is to recover or enhance the original signal from the noisy input data using sign LMS algorithm. The simulation result of the python code

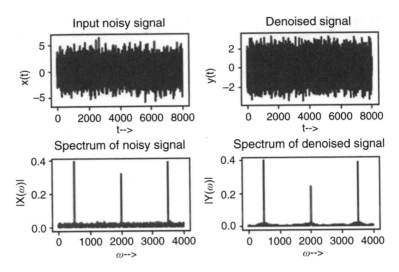

Fig. 11.23 Simulation result of the adaptive line enhancer using sign LMS

given in Fig. 11.22 is shown in Fig. 11.23. From the magnitude spectrum, it is possible to observe that the noise impact is reduced by the sign LMS algorithm.

Inference
From this experiment, the following observations can be drawn:

1. From Fig. 11.23, the magnitude response of the noisy signal indicates that the signal has three unique frequency components and noisy components.
2. The magnitude response of denoised signal has three spikes, and the impact of the noisy components is lesser than the input magnitude response.

Task
1. Do the suitable adjustments in the parameters used in the python code given in Fig. 11.22 to reduce the effect of noise in the denoised or enhanced signal?

11.3   RLS Algorithm

Recursive least square (RLS) is an adaptive algorithm based on the idea of least squares. The block diagram of the adaptive filter based on RLS algorithm is shown in Fig. 11.24. From the figure $x[n]$ is the input to the filter, $d[n]$ is the desired signal and the difference between the desired signal and the output of the filter is the error signal $e[n]$. Forgetting factor is used in RLS algorithm to remove or minimize the influence of old measurements. A small forgetting factor reduces the influence of old samples and increases the weight of new samples; as a result, a better tracking can be realized at the cost of a higher variance of the filter coefficients. A large forgetting factor

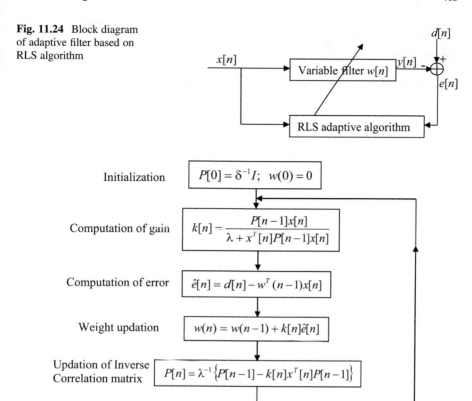

Fig. 11.24 Block diagram of adaptive filter based on RLS algorithm

Fig. 11.25 Flow chart of sequence of steps in RLS algorithm

keeps more information about the old samples and has a lower variance of the filter coefficients, but it takes a longer time to converge.

Let us define the a priori error as $\hat{e}[n] = d[n] - w^T[n-1]x[n]$ and the weight updation formula for the RLS algorithm is given by

$$w[n] = w[n-1] + \frac{P[n-1]x[n]\hat{e}[n]}{\lambda + x^T[n]P[n-1]x[n]} \tag{11.10}$$

If $k[n] = \frac{P[n-1]x[n]}{\lambda + x^T[n]P[n-1]x[n]}$ represents the gain, then the above expression can be written as

$$w[n] = w[n-1] + k[n]\hat{e}[n] \tag{11.11}$$

The flow chart of the sequence of steps followed in RLS algorithm is shown in Fig. 11.25. From the flow chart, it is possible to observe that the algorithm is iterative. Proper initialization of filter coefficients is necessary for convergence.

```
# This Code for RLS algorithm
def RLS_algorithmm(x,lamda,delta,N,t,n_iter):
    # x = input data, lamda = Forgetting factor, delta = Regularization parameter
    # t = reference data, N = Filter length, n_iter = number of iteration
    N1=len(x)
    w = np.zeros(N) # Initial filter
    w=np.transpose(w)
    e = np.zeros(N1-N)
    P=np.eye(N)/delta
    x=np.transpose(x)
    for n in range(0, n_iter):
        xn = x[n+N:n:-1]
        k1=np.dot(P,xn)
        k2=np.dot(np.transpose(xn),P)
        k3=np.dot(k2,xn)
        k =k1/(lamda+k3)
        en = t[n+N] - np.dot(np.transpose(w),xn);# Error
        w = w + np.dot(k,np.conjugate(en)) # Update filter (RLS algorithm)
        P=(1/lamda)*P
        e[n] = en # Record error
    return w,e
```

Fig. 11.26 Python code for RLS algorithm

Experiment 11.10 Implementation of RLS Algorithm

This experiment discusses the implementation of RLS algorithm using python. The python code for RLS algorithm is given in Fig. 11.26, and it is in the form of a function so that this function can be used for different applications.

Experiment 11.11 Adaptive Line Enhancer Using RLS Algorithm

This experiment is a repetition of Experiment 11.9; instead of sign LMS, RLS algorithm is used to filter out the noisy component present in the input signal. The python code for this experiment is given in Fig. 11.27, and its corresponding simulation result is displayed in Fig. 11.28.

Inference

From Fig. 11.28, it is possible to confirm that the magnitude response of the filtered or denoised output is better than the magnitude response of the noisy input. Therefore, RLS algorithm can act as an adaptive line enhancer.

Experiment 11.12 Comparison of System Identification with Different Adaptive Filters

The main objective of this experiment is to compare the simulation result of different adaptive algorithms like LMS, NLMS, Sign LMS and RLS for the system identification process. The python code to compare the simulation results of system identification is given in Fig. 11.29, and its simulation results are shown in Fig. 11.30.

```
import numpy as np
import matplotlib.pyplot as plt
from scipy import signal
from scipy.fft import fft
f1,f2,Fs=500,2000,8000 # Signal and sampling freq
T,lamda,delta=1/Fs,1.9,0.05
t=np.arange(0,1,T)
noise=np.random.randn(len(t));
d=np.sin(2*np.pi*f1*t)+np.sin(2*np.pi*f2*t)+noise;
delay,N=10,50 # Delay,Filter length
x=np.append(np.zeros(delay),d);
[w,e]=RLS_algorithmm(x,lamda,delta,N,d,len(d)-N)
y1=signal.convolve(w,x)
mag_x=fft(x)/len(x);#Frequency Response
mag_y=fft(y1)/len(y1);#Frequency Response
plt.figure(),plt.subplot(2,2,1),plt.plot(x),plt.title('Input noisy signal')
plt.xlabel('t-->'),plt.ylabel('x(t)')
plt.subplot(2,2,2),plt.plot(y1),plt.title('Denoised signal')
plt.xlabel('t-->'),plt.ylabel('y(t)')
plt.subplot(2,2,3),plt.plot(np.abs(mag_x[0:4000])),plt.title('Spectrum of noisy signal')
plt.xlabel('$\omega$-->'),plt.ylabel('|X($\omega$)|')
plt.subplot(2,2,4),plt.plot(np.abs(mag_y[0:4000])),plt.title('Spectrum of denoised signal')
plt.xlabel('$\omega$-->'),plt.ylabel('|Y($\omega$)|')
plt.tight_layout()
```

Fig. 11.27 Python code for adaptive line enhancer using RLS

Inference

From Fig. 11.30, it is possible to observe that proper selection of the adaptive filter parameters like step size or learning rate, forgetting factor and regularization plays a major role in using the adaptive filtering algorithm for the system identification application in signal processing.

Task

Write a python code to compare the simulation result of different adaptive algorithms like LMS, NLMS, sign LMS and RLS for adaptive line enhancement application in signal processing.

Exercises

1. Execute the python code given in Fig. 11.12 and compare the estimated filter 'w' with the original filter coefficients 'h' for different length of the filter. Also, execute the same python code and comment on the convergence of the LMS algorithm with different values of learning rate 'mu', including negative value.
2. Use the python code for the sign LMS algorithm given in Fig. 11.22 to compute the impulse response of the inverse filter and comment on the role of learning rate.
3. Modify the sign LMS algorithm based on the equation of the sign regressor algorithm is given by $w[n + 1] = w[n] + \mu e[n] \text{ sign } \{x[n]\}$, and compute the impulse response of the inverse filter and comment on the simulation result.

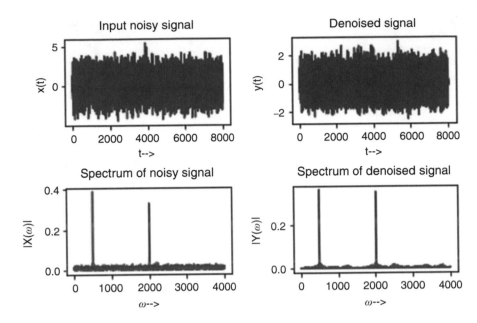

Fig. 11.28 Simulation result of the python code given in Fig. 11.27

```
# Python code for the comparison of adaptive algorithms for system identification
import numpy as np
import matplotlib.pyplot as plt
from scipy import signal
N1 = 1500 # Size of the Input data
N = 25 # Filter size
n=np.arange(0,N,1)
n_iter=200# it must be less than (N1-N)
x = np.random.randn(N1) # Input to the filter
h = signal.firwin(N, 0.25) # FIR filter to be identified
t = signal.convolve(x, h) # Target/desired signal
t = t + 0.01 * np.random.randn(len(t)) # with added noise
mu,mu1,beta,c,lamda,delta = 0.05,0.0005,0.05,1.5,1,0.25 # LMS step size
plt.figure(1),plt.title('Filter to be Identified')
plt.stem(h),plt.xlabel('n-->'),plt.ylabel('h[n]')
[w,e]=LMS_algorithmm(x,mu,N,t,n_iter);
[w1,e1]=NLMS_algorithmm(x,N,t,beta,c,n_iter)
[w2,e2]=Sign_LMS_algorithmm(x,mu1,N,t,n_iter)
[w3,e3]=RLS_algorithmm(x,lamda,delta,N,t,n_iter)
plt.figure(2),plt.subplot(2,2,1),plt.stem(n,w,'g'),plt.xlabel('n-->'),plt.ylabel('w[n]')
plt.title('Identified by LMS'),plt.subplot(2,2,2),plt.stem(n,w1,'k'),plt.xlabel('n-->'),
plt.ylabel('w[n]'),plt.title('Identified by NLMS'),plt.subplot(2,2,3),
plt.stem(n,w2,'r'),plt.xlabel('n-->'),plt.ylabel('w[n]'),plt.title('Identified by Sign LMS')
plt.subplot(2,2,4),plt.stem(n,w3,'b'),plt.xlabel('n-->'),plt.ylabel('w[n]')
plt.title('Identified by RLS'),plt.tight_layout()
```

Fig. 11.29 Python code for unknown system identification

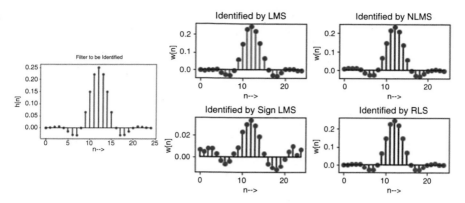

Fig. 11.30 Simulation result of the python code given in Fig. 11.29

4. Modify the sign LMS algorithm based on the equation of sign-sign LMS algorithm is given by $w[n + 1] = w[n] + \mu$ sign $\{e[n]\}$ sign $\{x[n]\}$, and compute the impulse response of the inverse filter and comment on the simulation result.

5. Use the python code for RLS algorithm given in Fig. 11.26 to obtain the inverse filter coefficients and comment on the simulation result. Also, comment on the selection of the forgetting factor and regularization parameter.

Objective Questions

1. The filter which is based on the minimum mean square error criterion, is

   A. Wiener filter
   B. Window-based FIR filter
   C. Frequency sampling-based FIR filter
   D. Savitsky Golay filter

2. If 'R' is the autocorrelation matrix of the observed signal and 'p' represents the cross-correlation between the desired signal and the observed signal, then the expression for the Wiener-Hopf equation is

   A. $w_{opt} = R \times p$
   B. $w_{opt} = R + p$
   C. $w_{opt} = R - p$
   D. $w_{opt} = p/R$

3. The weight update expression of the standard LMS algorithm is

   A. $w(n + 1) = w(n) + \mu x[n]e[n]$
   B. $w(n + 1) = w(n) - \mu x[n]e[n]$
   C. $w(n + 1) = w(n) + \mu x[n]e^2[n]$
   D. $w(n + 1) = w(n) - \mu x[n]e^2[n]$

4. If μ refers to the step size and λ refers to the eigen value of the autocorrelation matrix, then the condition for convergence of LMS algorithm is given by

   A. $0 < \mu < \frac{2}{\lambda_{min}}$

   B. $0 < \mu < \frac{2}{\lambda_{max}}$

   C. $0 < \mu < \frac{2}{\lambda_{max}^2}$

   D. $0 < \mu < \frac{2}{\lambda_{min}^2}$

5. Statement 1: Wiener filter is based on the statistics of the input data.

   Statement 2: Wiener filter is an optimal filter with respect to minimum mean absolute error

   A. Statements 1 and 2 are true
   B. Statement 1 is correct, and Statement 2 is wrong
   C. Statement 1 is wrong, Statement 2 is correct
   D. Statements 1 and 2 are wrong

6. The filter which changes its characteristics in accordance with the environment is termed as

   A. Optimal filter
   B. Non-linear filter
   C. Adaptive filter
   D. Linear filter

Bibliography

1. Simon Haykin, "Adaptive Filter Theory", Pearson, 2008.
2. Bernard Widrow, Samuel D. Stearns, "Adaptive Signal Processing", Pearson, 2002.
3. Dimitris G. Manolakis, Vinay K. Ingle, and Stephen M. Kogon, "Statistical and Adaptive Signal Processing: Spectral Estimation, Signal Modeling, Adaptive Filtering and Array Processing", Artech House Publishers, 2005.
4. Behrouz F. Boroujey, "Adaptive Filters: Theory and Applications", Wiley -Blackwell, 2013.
5. Alexandar D. Poularikas, "Adaptive Filtering", CRC Press, 2015.

Chapter 12
Case Study

Learning Objectives

After completing this chapter, the reader should be familiar with the following

- Applications of signal processing in speech signals
- ECG signal analysis
- Power line signal analysis

Roadmap of the Chapter

The case study discussed in this section focuses on the application of signal processing algorithms in the field of electrical and electronics engineering. Three case studies discussed in this section are (1) speech recognition, (2) QRS detection algorithm in ECG (3) power line signal analysis. Transform domain analysis of speech signal is discussed in the first case study. Analysis of ECG signal is the focus of second case study. Identification of different types of faults in power line signal is done in the third case study.

12.1 Case Study 1: Speech Recognition Using MFCC (Mel-Frequency Cepstral Coefficient)

Speech is the easiest and most widely used way of communication between humans. The interaction between a human and a computer is typical in the current scenario in the communication field. Communication between humans and computers can be made possible only with the help of hardware devices like keyboards, touch screens, mice, etc. However, humans prefer a more natural form of interaction than hardware devices. The speech signal is the most profound means of communication human beings use. For the human to human interaction, voice is the most significant feature, which helps us to recognize the speaker and extract the information from the speaker. The speech recognition system can be used to create documents from

© The Author(s), under exclusive license to Springer Nature Singapore Pte Ltd. 2024
S. Esakkirajan et al., *Digital Signal Processing*,
https://doi.org/10.1007/978-981-99-6752-0_12

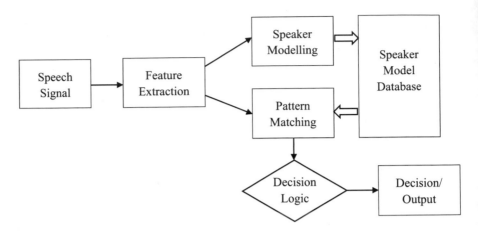

Fig. 12.1 General block diagram of the speech recognition system

speech, saving more time and reducing the burden on a human. In general, voice samples contain more information, including the person's gender and age. We can distinguish whether the voice belongs to a male or female, child or adult, based on the voice samples. Also, sometimes it reflects the state of mind of the speaker. The voice recognition method uses some of the information in the voice and identifies the speaker. Voice recognition is a technique that detects a voice sample from unique properties that may be acoustic or phonetic.

The general block diagram of speech recognition system contains feature extraction, speaker modelling and pattern matching method to identify the speaker, which is illustrated in Fig. 12.1. In the first stage, a speech sample is considered an input to the system. Here the speech sample would be noise-free; to remove the noise components in the speech signal, preprocessing method can be used. The preprocessed speech signal is the input to the feature extraction process in the second stage of the system. Using the feature extraction approaches, some properties of speech, like acoustic or phonetic features, are extracted. Finally, in the third stage, training and testing of the speech recognition model is developed, which will give the final decision of the system (i.e.,) which speech belongs to whom.

12.1.1   Speaker Identification

The process of identifying a speaker's voice from a group of speakers is called speaker identification. In this process, the voice of the input speaker is verified with the voice stored already in the database, and best match can be obtained using a pattern-matching algorithm. If the voice does not match the voices stored in the database, then the voice is a new one. The new voice can be updated in the database. The general block diagram of the speaker identification system is shown in Fig. 12.2.

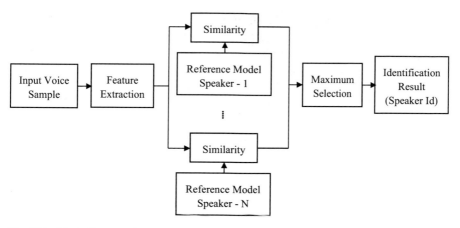

Fig. 12.2 Block diagram of speaker identification system

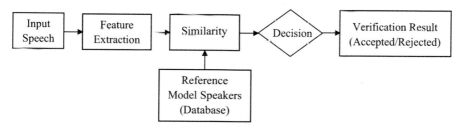

Fig. 12.3 General block diagram of a speaker verification system

12.1.2 Speaker Verification System

The process of authenticating a speaker based on the characteristics of voice samples is called speaker verification. In this system, the final output will be either accepted or rejected. The main applications of this system are military, aircraft and voice-verified authentication areas. The block diagram of the speaker verification system is shown in Fig. 12.3. From this figure, it is possible to confirm that the major blocks in the speaker verification system are feature extraction, similarity identification and decision-maker.

Feature extraction is the common block for both speaker recognition and verification systems. In general, feature extraction approaches help us to extract some good features and characteristics from the voice samples. The mel-frequency cepstral coefficient (MFCC) is the most widely used feature extraction method. More detail about the MFCC is discussed in the next section.

12.1.3   Mel-Frequency Cepstral Coefficient (MFCC) Feature

The step-by-step procedure to compute the MFCC feature is discussed in this section. This procedure contains six major steps involved in the computation of MFCC; they are (1) pre-emphasis, (2) sampling and windowing, (3) fast Fourier transform (FFT), (4) mel filter bank, (5) logarithmic function and (6) discrete cosine transform. The block diagram of MFCC computation is depicted in Fig. 12.4.

12.1.3.1   Pre-emphasis

The voice samples are passed through a highpass filter, which is mathematically expressed as

$$y[n] = x[n] - ax[n-1] \tag{12.1}$$

where $x[n]$ is the input voice samples, 'a' is the filter constant and it takes the value between 0.9 to 1.0 and $y[n]$ denotes the filtered voice samples. In this pre-emphasis process, the input voice samples are passed through a highpass filter, and the filtered output will emphasize the high-frequency component present in the input voice samples.

The python code to read, normalize and display audio files is shown in Fig. 12.5, and its simulation results are shown in Fig. 12.6.

From Fig. 12.6, it is possible to observe that the amplitude of the original audio signal is [−20,000, 20,000], whereas in the normalized audio, the amplitude varies from −1 to 1.

12.1.3.2   Sampling and Windowing

The speech signal is divided into small segments with a duration of 20–30 ms, which are called 'frames'. While splitting the input voice samples may be allowed to overlap between the successive segments. Windowing can be used to avoid discontinuity between consecutive segments. Also, the windowing technique smooths the extreme samples in both starting and ending of the segments. The commonly used windowing function is Hamming or Hanning. The process of windowing the input sequence is mathematically written as

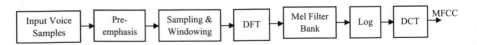

Fig. 12.4 Block diagram of MFCC computation

```
import numpy as np
from scipy.io import wavfile
import scipy.fftpack as fft
from scipy.signal import get_window
import IPython.display as ipd
import matplotlib.pyplot as plt
audio1 = "DSP_UV.wav"
sample_rate, audio = wavfile.read(audio1)
ipd.Audio(audio1)
audio2=audio[50000:100000]
duration = len(audio2)/sample_rate
print(f"Sample rate: {sample_rate}Hz")
print(f"Audio duration: {duration}s")
t = np.linspace(0,duration,len(audio2))
plt.figure(figsize=(15,6)),plt.plot(t,audio2),plt.xlabel("Time (s)"),plt.ylabel("Amplitude")
plt.title("Original Audio in Time domain"),plt.show()
#Normalizing to amplitude ranging between +1 and -1
normalizedAudio = audio2/np.max(np.abs(audio2))
plt.figure(figsize=(15,6)),plt.plot(t,normalizedAudio),
plt.xlabel("Time"),plt.ylabel("Amplitude")
plt.title("Normalized Audio in Time domain"),plt.show()
```

Fig. 12.5 Python code for read, normalize and display audio file

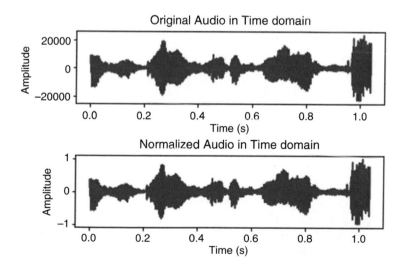

Fig. 12.6 Plot of original and normalized audio signal

```
def frame_audio(normalizedAudio, FFT_size=2048, hop_size=10, sample_rate=8000):
    audio = np.pad(normalizedAudio, int(FFT_size / 2), mode='reflect')
    frame_len = np.round(sample_rate * hop_size / 1000).astype(int)
    frame_num = int((len(audio) - FFT_size) / frame_len) + 1
    frames = np.zeros((frame_num,FFT_size))
    for n in range(frame_num):
        frames[n] = audio[n*frame_len:n*frame_len+FFT_size]
    return frames
hop_size = 15 #ms
FFT_size = 2048
audio_framed = frame_audio(audio, FFT_size=FFT_size, hop_size=hop_size,
sample_rate=sample_rate)
print(f"Framed audio shape: {audio_framed.shape}")
window = get_window("hann", FFT_size, fftbins=True)
plt.figure,plt.subplot(3,1,1),plt.plot(window)
plt.title("Hanning Window"),plt.xlabel("Samples"),plt.ylabel("Amplitude"),plt.grid(True)
audio_win = audio_framed*window
plt.subplot(3,1,2),plt.plot(audio_framed[72])
plt.xlabel("Samples"),plt.ylabel("Amplitude"),plt.title("Before Windowing")
plt.subplot(3,1,3),plt.plot(audio_win[72]),plt.xlabel("Samples"),plt.ylabel("Amplitude")
plt.title("After Windowing"),plt.tight_layout(),plt.show()
```

Fig. 12.7   Python code for framing and windowing audio signal

$$y[n] = x[n] \times w[n] \tag{12.2}$$

where $w[n]$ represents the windowing function.

The python code for framing and windowing of normalized audio signal is shown in Fig. 12.7. From this figure, it is possible to understand that the first part is the framing/segmenting/partitioning of the audio signal. Then Hanning window is used for the windowing operation on the partitioned audio signal in the second part. The simulation result of the python code given in Fig. 12.7 is shown in Fig. 12.8. From this Fig. 12.8, it is evident the importance of the windowing concept. The windowing method is used to smoothen the initial and end of the audio signal frame.

12.1.3.3   Discrete Fourier Transform (DFT)

DFT is a well-known transform to convert the time-domain information of speech signal into frequency-domain information. Also, it extracts useful information/some features of the speech signal without losing the information present in it.

The python code to convert a time-domain audio signal into a frequency-domain magnitude spectrum is shown in Fig. 12.9. 'FFT' library is used here to compute the

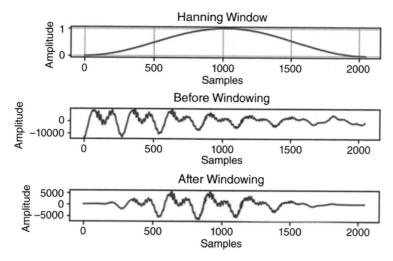

Fig. 12.8 Result of Hanning window and windowing audio signal

```
audio_winT = np.transpose(audio_win)
audio_fft = np.empty((int(1 + FFT_size // 2), audio_winT.shape[1]), dtype=np.complex64, order='F')
for n in range(audio_fft.shape[1]):
    audio_fft[:, n] = fft.fft(audio_winT[:, n], axis=0)[:audio_fft.shape[0]]
audio_fft = np.transpose(audio_fft)
frameNo = 72
f_axis = fft.fftfreq(audio_framed[frameNo].size,1/sample_rate)[0:audio_framed[frameNo].size//2]
plt.figure
plt.plot(f_axis,2/audio_framed[frameNo].size*np.abs(audio_fft[frameNo][0:audio_framed[0].size//2]))
plt.title(f"FFT of Frame:{frameNo}")
plt.xlabel("Frequency"),plt.ylabel("Magnitude")
plt.show()
audio_power = np.square(np.abs(audio_fft))
print(audio_power.shape)
freq_min = 0
freq_high = sample_rate / 2
mel_filter_num = 10
print(f"Minimum frequency: {freq_min}"),print(f"Maximum frequency: {freq_high}")
```

Fig. 12.9 Magnitude spectrum computation using FFT

magnitude spectrum. The output of the python code given in Fig. 12.9 is displayed in Fig. 12.10. Frame number 72 is displayed.

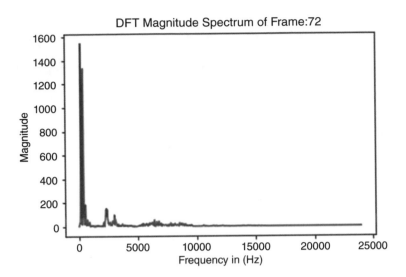

Fig. 12.10 Magnitude spectrum of frame number 72

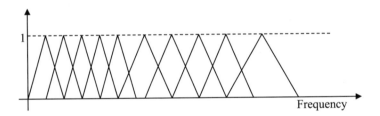

Fig. 12.11 Frequency response of mel-frequency bandpass filter

12.1.3.4 Mel-Frequency Bandpass Filter

The mel-frequency bandpass filter is a triangular-shaped multiple bandpass filter. DFT obtains the magnitude spectrum of the speech signal, and this spectrum is multiplied with the set of triangular bandpass filters to smoothen the magnitude spectrum of speech signal, which is expressed as

$$Y[k] = X[k] \times H[k] \tag{12.3}$$

where $H[k]$ is the magnitude spectrum of the triangular bandpass filter. The sample triangular bandpass filter frequency responses are shown in Fig. 12.11. This figure shows a set of triangular filters that are used to compute weighted sum of filter spectral components so that the output of process approximates to a mel scale. Each filter's magnitude response is triangular in shape, and the magnitude is unity at the centre frequency and decreases linearly to zero at the centre frequency of two adjacent filters. The final output is the sum of its filtered spectral components.

```
def freq_to_mel(freq):
  return 2595.0 * np.log10(1.0 + freq / 700.0)

def met_to_freq(mels):
  return 700.0 * (10.0**(mels / 2595.0) - 1.0)

def get_filter_points(fmin, fmax, mel_filter_num, FFT_size, sample_rate=8000):
  fmin_mel = freq_to_mel(fmin)
  fmax_mel = freq_to_mel(fmax)
  print("MEL min: {0}".format(fmin_mel))
  print("MEL max: {0}".format(fmax_mel))
  mels = np.linspace(fmin_mel, fmax_mel, num=mel_filter_num+2)
  freqs = met_to_freq(mels)
  return np.floor((FFT_size + 1) / sample_rate * freqs).astype(int), freqs
filter_points, mel_freqs = get_filter_points(freq_min, freq_high, mel_filter_num, FFT_size,
sample_rate=sample_rate)
print(f"Filter Points : {filter_points}")

def get_filters(filter_points, FFT_size):
  filters = np.zeros((len(filter_points)-2,int(FFT_size/2+1)))
  for n in range(len(filter_points)-2):
    filters[n, filter_points[n] : filter_points[n + 1]] = np.linspace(0, 1, filter_points[n + 1] -
filter_points[n])
    filters[n, filter_points[n + 1] : filter_points[n + 2]] = np.linspace(1, 0, filter_points[n + 2] -
filter_points[n + 1])
  return filters
filters = get_filters(filter_points, FFT_size)
plt.figure
for filter in filters:
  plt.plot(filter)
plt.title("Mel Filterbank"),plt.xlabel("Frequency"),plt.ylabel("Weights")
plt.show()
enorm = 2.0 / (mel_freqs[2:mel_filter_num+2] - mel_freqs[:mel_filter_num])
filters *= enorm[:, np.newaxis]
plt.figure
for n in range(filters.shape[0]):
  plt.plot(filters[n])
plt.title("Normalized Mel Filterbank"),plt.xlabel("Frequency"),plt.ylabel("Weights")
plt.show()
```

Fig. 12.12 Python code for design of mel-filter bank

The python code to design a mel-filter bank and normalized mel-filter bank is shown in Fig. 12.12. The simulation result of the python code given in Fig. 12.12 is shown in Fig. 12.13.

Figure 12.13 shows that the gain of the filters in the mel-filter bank is unity, whereas, in the normalized mel-filter bank, the gain of the filters is different.

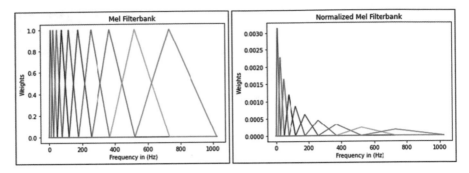

Fig. 12.13 Mel-filter bank and normalized mel-filter bank

```
audio_filtered = np.dot(filters, np.transpose(audio_power))
audio_log = 10.0 * np.log10(audio_filtered)
plt.figure,plt.subplot(2,1,1),plt.plot(audio_filtered),
plt.xlabel("Time in (sec)"),plt.ylabel("Amplitude"),plt.title("Filtered Signal")
plt.subplot(2,1,2),plt.plot(audio_log),
plt.xlabel("Frequency in (Hz)"),plt.ylabel("Magnitude")
plt.title("Log Power Spectrum")
plt.tight_layout()
```

Fig. 12.14 Python code for the computation of cepstral components

12.1.3.5 Log Operation

The logarithmic function is used to compute cepstral components from the filtered acoustic signal. The python code to compute the cepstral components is given in Fig. 12.14, and its simulation result is shown in Fig. 12.15. This figure makes it possible to understand the use of the logarithmic function for the MFCC computation. The amplitude of the filtered signal is too high; the role of the logarithmic function is to reduce the amplitude level.

12.1.3.6 Discrete Cosine Transform (DCT)

The final mel-frequency cepstral coefficient is obtained by taking DCT on the cepstral component, which is the output of the logarithmic function. The mathematical expression of the computation MFCC is given by

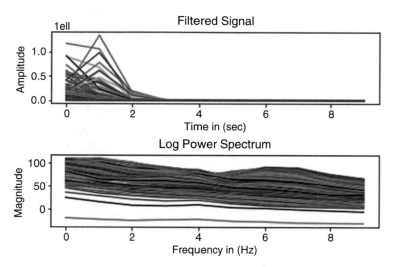

Fig. 12.15 Plot of filtered signal and its cepstral components

$$C[n] = \sum_{k=0}^{L-1} Y[k] \cos\left(\frac{\pi n}{N}\left(k - \frac{1}{2}\right)\right) \tag{12.4}$$

where $n = 0, 1, 2, \ldots, N-1$ and N denote the number of triangular bandpass filters. L represents the number of mel-scale cepstral coefficients. The primary use of DCT is to extract the output of bandpass filters to generate mel scale coefficients, and also it converts the frequency-domain spectrum into a time-domain signal. The outcome of the DCT is called 'mel-scale cepstral coefficients' (MFCC). These MFCC act as a feature of the voice signal, which helps for the different speech signal processing applications.

The python code to compute the DCT of the logarithmic function output is given in Fig. 12.16. The simulation result of the python code, which is shown in Fig. 12.16, is displayed in Fig. 12.17.

These features fed into the different classification methods, like SVM, KNN, random forest, etc., to identify and recognize the speakers.

12.2 Case Study 2: QRS Detection in ECG Signal Using Pan-Tomkins Algorithm

Electrocardiogram (ECG or EKG) is the electrical indication of the contractile process of the heart. Every heart contraction produces an electrical impulse captured by electrodes placed on the skin. ECG gives information about the heart rate, rhythm and morphology. ECG is characterized by a periodic wave sequence of P, QRS, J, T and U wavelets associated with each heartbeat. The QRS complex has a high clinical

```
def dct(dct_filter_num, filter_len):
    basis = np.empty((dct_filter_num,filter_len))
    basis[0, :] = 1.0 / np.sqrt(filter_len)
    samples = np.arange(1, 2 * filter_len, 2) * np.pi / (2.0 * filter_len)
    for i in range(1, dct_filter_num):
        basis[i, :] = np.cos(i * samples) * np.sqrt(2.0 / filter_len)
    return basis
dct_filter_num = 12
dct_filters = dct(dct_filter_num, mel_filter_num)
cepstral_coefficents = np.dot(dct_filters, audio_log)
cepstral_coefficents.shape
print(f"MFCCs : {cepstral_coefficents[:, 1:dct_filter_num+1]}")
plt.figure(figsize=(10,6))
c = plt.imshow(cepstral_coefficents[:,1:dct_filter_num+1], aspect='auto',
origin='lower',cmap='Spectral');
plt.title("Mel Frequency Ceptral Coefficiernt"),plt.ylabel("MFCC"),plt.xlabel("Frames")
plt.tight_layout(),plt.colorbar(c)
```

Fig. 12.16 Python code for the computation of DCT

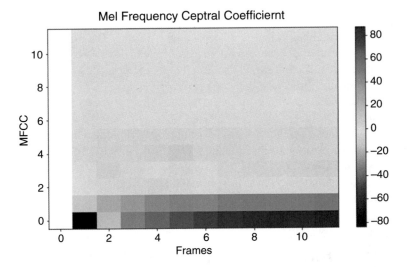

Fig. 12.17 MFCC output

significance, and its detection is the first stage of ECG signal processing. From the
position of the QRS complex, it is possible to obtain the positions of P and T waves.
The normal ECG signal with different intervals of wavelets is shown in Fig. 12.18.
From this figure, it is possible to know that, in particular, QRS complex as compared
to the other waves has the steepest slope, has the highest amplitude in most cases,

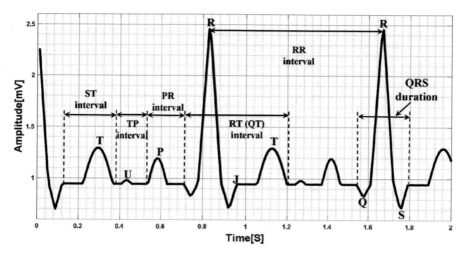

Fig. 12.18 Normal ECG showing different waves

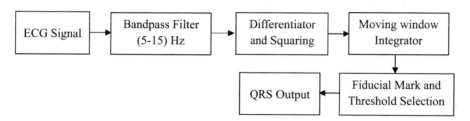

Fig. 12.19 Block diagram QRS Detection

lasts for less than 0.2 s, has a peak at R and is preceded by a P wave and succeeded by a T wave for a normal ECG.

QRS detection consists of three major processing steps; they are (1) linear digital filtering, (2) non-linear transformation and decision rule algorithms. Let us discuss the Pan and Tompkins QRS detection algorithm in this case study. This algorithm starts with the linear process, which includes a bandpass filter, a derivative operation and moving window integration. The second step is the non-linear transformation, which uses amplitude squaring. The final stage is a decision rule algorithm, which includes adaptive thresholds and QRS detection. The block diagram of QRS complex detection in ECG using the Pan and Tompkins algorithm is given in Fig. 12.19.

12.2.1 ECG Signal Preprocessing

The normal ECG signals are time-varying with small amplitude ranging from 10 μV to 5 mV. The typical amplitude of the ECG signal is 1 mV and their frequencies vary from 0.05 to 100 Hz. The ECG signal is mainly concentrated in the 0.05–35 Hz range. For the ECG signal analysis, the system requires a noise-free ECG signal to get an accurate prediction. However, ECG signals are affected by various noises and artifacts practically. The ECG analysis system's first step is to remove its noise by using the filter.

12.2.1.1 Bandpass filter

The bandpass filter is used to reduce the effect of muscle noise, powerline interference, baseline wander and T wave interference. The desirable passband frequency to maximize the QRS energy is approximately 5–15 Hz. In this case study, the Butterworth filter is used with order 3, and passband frequency [0.5, 15] Hz. Instead of bandpass filter, cascaded lowpass and highpass filters may be preferred. Filtering the ECG signal, 'butter' and 'lfilter' python commands are used here. The python code to read the ECG data and noise removal is given in Fig. 12.20, and its corresponding output is shown in Fig. 12.21.

From Fig. 12.21, it is possible to observe that the raw ECG signal has shifted the amplitude to 1 mV, whereas the filtered ECG signal amplitude is between −1 and +1.

```
# QRS peak detection in ECG using Pan-Tomkins algorithm
import numpy as np
import matplotlib.pyplot as plt
from scipy.signal import butter, lfilter
x0=np.loadtxt('ecg_data_1.csv',skiprows=1, delimiter=',')
signal_freq,f_low,f_high,filt_order = 250,0.5,15.0,3
y0=x0[:,1];
qrs_peak_value,noise_peak_value,threshold_value = 0.0,0.0,0.0;
refractory_period = 120  # Change proportionally when adjusting frequency (in samples).
qrs_peak_filtering_factor,noise_peak_filtering_factor,qrs_noise_diff_weight = 0.125,0.125,0.25;
# Detection results.
qrs_peaks_indices = np.array([], dtype=int)
noise_peaks_indices = np.array([], dtype=int)
#Bandpass filtering
Fs = 0.5 * signal_freq
low, high = f_low / Fs, f_high / Fs;
b, a = butter(filt_order, [low, high], btype="band")
y1 = lfilter(b, a, y0)#Band pass filtering
plt.figure(1),plt.subplot(2,1,1),plt.plot(y0),plt.title('Raw ECG Signal')
plt.xlabel('Time-->'),plt.ylabel('Amplitude in mV'),plt.subplot(2,1,2),plt.plot(y1),
plt.title('Filtered ECG Signal'),plt.xlabel('Time-->'),plt.ylabel('Amplitude in mV')
plt.tight_layout()
```

Fig. 12.20 Python code for linear filtering

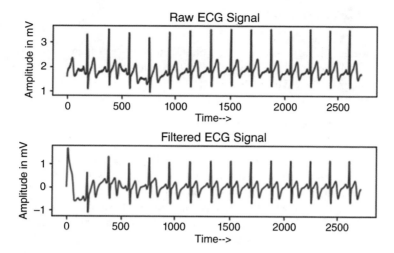

Fig. 12.21 Raw and filtered ECG signals

12.2.1.2   Derivative Process

After bandpass filtering, the ECG signal is differentiated to provide QRS complex slope information. '*np.ediff1d*' python command is used here to obtain the derivative of the filtered ECG signal. The first-order derivative equation can be written as

$$y[n] = \frac{x[n-1] - x[n]}{2} \tag{12.5}$$

where $x[n]$ denotes the input signal and $y[n]$ represents the derivative output signal.

12.2.1.3   Squaring Operation

After the derivative, the resultant signal is squared point by point. The squaring operation can be written as

$$y[n] = x^2[n] \tag{12.6}$$

The squaring operation makes all the data points positive and does non-linear amplification of the derivative output emphasizing the higher frequencies.

12.2.2 Moving Window Integration

Moving window integration helps to obtain the waveform feature information in addition to the slope of the R wave. The mathematical equation to perform moving window integration is given by

$$y[n] = \frac{1}{N} \{x[n - (N - 1)] + x[n - (N - 2)] + \cdots + x[n]\} \qquad (12.7)$$

where N is the number of samples in the width of the integration window. In this case study, the length of the integration window is chosen as 15. The length of the moving window integration (N) is important for QRS wave detection. Generally, the value of N should be approximately the same as the widest possible QRS complex. If the window length (N) is large, the integration waveform will merge the QRS complex and T wave together. If the window length (N) is small, some QRS complexes will produce several peaks in the integration waveform. This may cause difficulty in the subsequent QRS detection. The python code for the derivative process, squaring and moving window integration is shown in Fig. 12.22, and its corresponding output is displayed in Fig. 12.23.

12.2.3 Fiducial Mark

The QRS wave corresponds to the rising edge of the integrated waveform. The time duration of the rising edge is equal to the width of the QRS wave. A fiducial mark for the temporal location of the QRS wave can be obtained from the rising edge, and the desired waveform point is marked as peak of the R wave. The python code for the

```
# Python code for derivative, squaring, and moving window integration
y2=np.ediff1d(y1);#Derivative
y3=y2**2;#Squaring
integral_window=15;
y4=np.convolve(y3, np.ones(integral_window));#Moving Window Integration
plt.figure(2),plt.subplot(3,1,1),plt.plot(y2),plt.title('Derivative ECG output')
plt.xlabel('Time-->'),plt.ylabel('Amp. in mV')
plt.subplot(3,1,2),plt.plot(y3),plt.title('Squared Derivative output')
plt.xlabel('Time-->'),plt.ylabel('Amp. in mV')
plt.subplot(3,1,3),plt.plot(y4),plt.title('Moving window integration output')
plt.xlabel('Time-->'),plt.ylabel('Amp. in mV')
plt.tight_layout()
```

Fig. 12.22 Python code for derivative, squaring and moving window integration

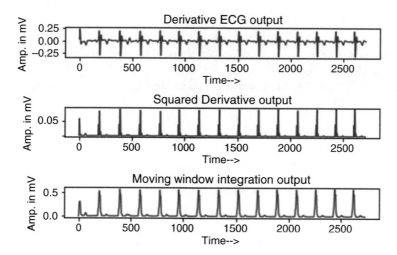

Fig. 12.23 Simulation result of python code is given in Fig. 12.22

```
# Fiducial mark - peak detection on integrated measurements.
spacing=50
kk = y4.size
y5 = np.zeros(kk + 2 * spacing)
y5[:spacing] = y4[0] - 1.e-6
y5[-spacing:] = y4[-1] - 1.e-6
y5[spacing:spacing + kk] = y4
peak_candidate = np.zeros(kk)
peak_candidate[:] = True
for s in range(spacing):
    start = spacing - s - 1
    h_b = y5[start: start + kk]  # before
    start = spacing
    h_c = y5[start: start + kk]  # central
    start = spacing + s + 1
    h_a = y5[start: start + kk]  # after
    peak_candidate = np.logical_and(peak_candidate, np.logical_and(h_c > h_b, h_c > h_a))
ind = np.argwhere(peak_candidate)
ind = ind.reshape(ind.size)#detected_peaks_indices
detected_peaks_vals=y4[ind]
```

Fig. 12.24 Python code for QRS peak detection

peak detection of the integrated ECG measurement is given in Fig. 12.24. From this figure, it is possible to observe that the position of the peak value results in the variable 'ind' and the detected peak value can be obtained in the variable 'detected_peaks_vals'.

12.2.4 Decision Rule Approach

The decision rule consists of adaptive threshold selection. The thresholds are adjusted automatically based on the noise in the ECG signal. The adaptive two thresholds (Th1 and Th2) are calculated using the equation given below.

$$Th1 = NPK + 0.25(SPK - NPK)$$
$$Th2 = 0.5Th1$$
$$(12.8)$$

where NPK represents the running estimate of noise peak and SPK denotes the running estimate of the signal peak, which are computed as

$$SPK = 0.125\,Pk + 0.875\,SPK \quad \text{if Pk is signal peak}$$

$$NPK = 0.125\,Pk + 0.875\,NPK \quad \text{if Pk is noise peak}$$

where Pk denotes peak. A peak is a local maximum determined by observing when the signal changes direction within a predefined time interval. The SPK is a peak the algorithm has already established to be a QRS complex. The NPK is any peak that is not related to the QRS. Here, the thresholds Th1 and Th2 are based on running estimates of SPK and NPK. When a new peak is detected, it must first be classified as a signal peak or noise peak. The peak is a signal peak if the peak exceeds Th1, and the QRS is obtained using Th2. The python code for threshold selection and the final QRS detection is given in Fig. 12.25.

The simulation result of the python code given in Fig. 12.25 is shown in Fig. 12.26. From this figure, it is possible to observe that the R peak of the integrated ECG signal is detected, and the R peak of the filtered ECG is also shown.

Inferences

The following inferences can be made from Figs. 12.21, 12.23 and 12.26.

1. The raw ECG signal is affected by the baseline wander (i.e. the base x-axis of a signal moves up and down rather than straight).
2. The bandpass filtered signal shows that the baseline wander is removed and the x-axis of the signal is in a straight line.
3. The derivated ECG signal output highlights the positive and negative peaks of the ECG signal very clearly. The squared ECG signal displays that all the negative peaks are brought up into the positive peak. These results can be found in Fig. 12.23.
4. The QRS peak marked moving window integrated ECG signal and the final R peak marked filtered ECG signal are displayed, which can be found in Fig. 12.26.

```
for ind, detected_peaks_val in zip(ind, detected_peaks_vals):
  try:
    last_qrs_index = qrs_peaks_indices[-1]
  except IndexError:
    last_qrs_index = 0
  if ind - last_qrs_index > refractory_period or not qrs_peaks_indices.size:
    if detected_peaks_val > threshold_value:
      qrs_peaks_indices = np.append(qrs_peaks_indices, ind)
      qrs_peak_val = qrs_peak_filtering_factor * detected_peaks_val + \
              (1 - qrs_peak_filtering_factor) * qrs_peak_value
    else:
      noise_peaks_indices = np.append(noise_peaks_indices, ind)
      noise_peak_value = noise_peak_filtering_factor * detected_peaks_val + \
              (1 - noise_peak_filtering_factor) * noise_peak_value
    threshold_value = noise_peak_value + \
              qrs_noise_diff_weight * (qrs_peak_value - noise_peak_value)
qrs_peaks_indices_fin=qrs_peaks_indices-8;
plt.figure(3),plt.subplot(2,1,1),plt.plot(y4),plt.title('Integrated ECG with QRS peak marked')
plt.scatter(x=qrs_peaks_indices, y=y4[qrs_peaks_indices], c="red", s=10, zorder=2)
plt.xlabel('Time >'),plt.ylabel('Amplitude in mV')
plt.subplot(2,1,2),plt.plot(y1),plt.title('Filterd ECG with R peak marked')
plt.scatter(x=qrs_peaks_indices_fin, y=y1[qrs_peaks_indices_fin], c="red", s=10, zorder=2)
plt.xlabel('Time-->'),plt.ylabel('Amplitude in mV')
plt.tight_layout()
```

Fig. 12.25 Python code for adaptive threshold selection and QRS detection

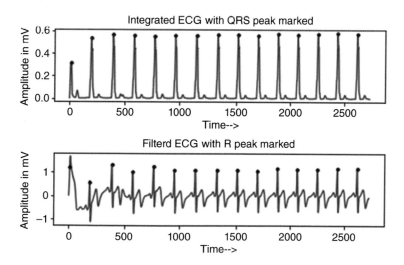

Fig. 12.26 Simulation result of the python code given in Fig. 12.25

12.3   Case Study 3: Power Quality Disturbance Detection

Power quality refers to maintaining a sinusoidal power distribution bus voltage at rated magnitude and frequency. The significant increase in non-linear load and the increased usage of semiconductor devices, lighting controls, solid-state switching devices, inverters and relaying equipment are causing non-linear loads, which lead to power quality disturbances. The basic power quality disturbances (PQD) are voltage sag, voltage swell, voltage interruption, harmonics, flickers, etc. A combination of these disturbances can occur simultaneously. Table 12.1 summarizes the causes and impact of power quality disturbances.

Effective detection and recognition of power quality disturbances are necessary to ensure the reliability of electric power quality. This section focuses on the simulation of the power quality disturbance and analysis of power quality disturbance signals, which is depicted in Fig. 12.27.

The first step in this direction is to generate different types of power quality disturbances. Mathematical models can be developed for different types of power

Table 12.1 Causes and effects of power quality disturbances

S. No.	Power quality disturbance	Causes	Effects
1	Voltage sag	1. Inductive load 2. Switching on and off of large loads	1. Tripping of sensitive equipment 2. Tripping of motors
2	Voltage swell	1. Capacitor switching 2. Switch off large loads	1. Damage to insulation and windings 2. Damage to power supplies
3	Harmonics	1. Non-linear loads 2. Rectifier type equipment	1. Malfunctioning of relays and equipment 2. Capacitor failure
4	Momentary interruption	1. Equipment failure 2. Control malfunction	1. Loss of supply voltage to consumer equipment 2. Shutdown of computers
5	Flicker	1. Machinery with rapid fluctuations in load current or voltage 2. Loads that cause voltage fluctuation include arc welding machines, arc furnaces, etc.	1. Misoperation of relays and contactors 2. Neurological problems in humans

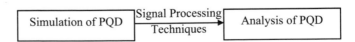

Fig. 12.27 Objectives of this section

Table 12.2 Mathematical model of power quality disturbances

S. No.	PQ disturbance	Mathematical model	Parameters
1	Voltage sag	$A(1 - \alpha(u(t - t_1) - u(t - t_2)))\sin(\omega t)$	$0.1 < \alpha < 0.9$
2	Voltage swell	$A(1 + \alpha(u(t - t_1) - u(t - t_2)))\sin(\omega t)$	$0.1 < \alpha < 0.9$
3	Harmonics	$A\sin(\omega t) + \sum_{n=3}^{7}\alpha_n \sin(n\omega t)$	$0.05 < \alpha_n < 0.15$
4	Momentary interruption	$A(1 - \alpha(u(t - t_1) - u(t - t_2)))\sin(\omega t)$	$0.9 < \alpha < 1$
5	Flicker	$(1 + \lambda \sin(k\omega t)) * \sin(\omega t)$	$0.1 < \lambda < 0.2$; $5 < k < 50$

quality disturbances. The second step is to analyse different types of power quality disturbances using time-frequency and time-scale representation.

The mathematical models of different types of power quality disturbances are given in Table 12.2.

12.3.1 Generation of Power Quality Disturbance

The generation of various power quality disturbances like voltage sag, voltage swell and momentary interruption are simulated, and their results are plotted. The python code which generates various power quality disturbances is shown in Fig. 12.28, and the corresponding output is shown in Fig. 12.29.

The following observations can be drawn from Fig. 12.29:

1. The amplitude of a pure sine wave varies from -1 to $+1$. It is a sine wave of 50 Hz frequency.
2. During the power quality disturbance '*sag*', the amplitude of sine wave decreases for a brief period of time.
3. During the power quality disturbance '*swell*', the amplitude of sine wave increases for a brief period of time.
4. During '*momentary interruption*', the amplitude of sine wave approaches zero value for a brief period of time.
5. During power quality disturbance, the characteristics of the signal (amplitude of the signal) vary with respect to time; hence, the power quality disturbances can be considered as a non-stationary signal.

```
#Power Quality Disturbance
import numpy as np
import matplotlib.pyplot as plt
A,fs,f,N,ph = 1,1000,50,200,0
T = 1/fs
t = np.linspace(0,N*T,N)
#Pure sine wave
pure_sine=np.sin(2*np.pi*f*t+ ph)
#Power quality disturbance
sag = np.sin(2*np.pi*f*t+ ph) - 0.5*np.sin(2*np.pi*f*t+ ph)*((t<0.15)&(t>0.08))
swell = np.sin(2*np.pi*f*t+ ph) + 0.5*np.sin(2*np.pi*f*t+ ph)*((t<0.15)&(t>0.08))
mi = np.sin(2*np.pi*f*t+ ph) - 0.98*np.sin(2*np.pi*f*t+ ph)*((t<0.15)&(t>0.08))
plt.subplot(2,2,1),plt.plot(t,pure_sine),plt.xlabel('Time'), plt.ylabel('Amplitude'),
plt.title('Pure sine wave')
plt.subplot(2,2,2),plt.plot(t,sag),plt.xlabel('Time'), plt.ylabel('Amplitude'),
plt.title('Sag')
plt.subplot(2,2,3),plt.plot(t,swell),plt.xlabel('Time'), plt.ylabel('Amplitude'),
plt.title('Swell')
plt.subplot(2,2,4),plt.plot(t,mi),plt.xlabel('Time'), plt.ylabel('Amplitude'),
plt.title('Momentary interruption')
plt.tight_layout()
```

Fig. 12.28 Python code to simulate power quality disturbances

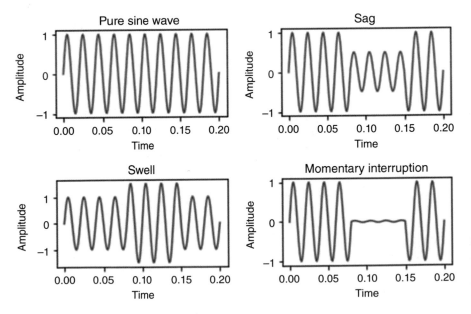

Fig. 12.29 Result of python code shown in Fig. 12.28

```
import numpy as np
import matplotlib.pyplot as plt
A,fs,f,N,ph = 1,1000,50,200,0
T = 1/fs
t = np.linspace(0,N*T,N)
#Pure sine wave
pure_sine=np.sin(2*np.pi*f*t+ ph)
#Power quality disturbance
har = np.sin(2*np.pi*f*t+ ph)+0.2*np.sin(2*np.pi*3*f*t+ ph)+0.3*np.sin(2*np.pi*5*f*t+ ph)
lamda=0.2
k=50
flicker=(1+lamda*np.sin(k*2*np.pi*f*t+ ph)) * np.sin(2*np.pi*f*t+ ph)
#Plotting the results
plt.subplot(3,1,1),plt.plot(t,pure_sine),plt.xlabel('Time'), plt.ylabel('Amplitude'),
plt.title('Pure sine wave')
plt.subplot(3,1,2),plt.plot(t,har),plt.xlabel('Time'), plt.ylabel('Amplitude'),
plt.title('Harmonics')
plt.subplot(3,1,3),plt.plot(t,flicker),plt.xlabel('Time'), plt.ylabel('Amplitude'),
plt.title('Flicker')
plt.tight_layout()
```

Fig. 12.30 Python code to simulate harmonics and flicker

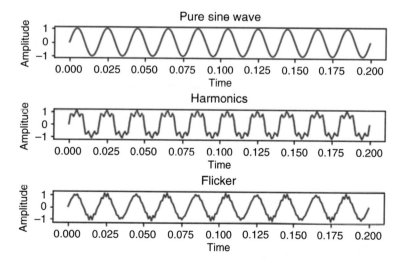

Fig. 12.31 Result of python code shown in Fig. 12.30

12.3.2   Simulation of Power Quality Disturbance

The other power quality disturbances like '*harmonics*' and '*flicker*' are simulated using python. The python code which performs this task is shown in Fig. 12.30, and the corresponding output is shown in Fig. 12.31.

From Fig. 12.31, the following inferences can be drawn

1. A harmonic is an integer multiple of fundamental frequency. The fundamental frequency of the sinusoidal signal is 50 Hz. The odd harmonics of the signal are added to the original sinusoidal signal to obtain the harmonic signal.
2. The term 'flicker' implies the effect the fluctuations in electric voltage have on electrical lighting devices. Loads, like arc furnaces, saw mills, welding machines and high-powered engines with fast stop-and-start cycle, can give rise to the phenomenon of flicker.

12.3.3 Time-Frequency Representation of Power Quality Disturbance

Time-frequency representation is a good tool to analyse non-stationary signal. Spectrogram is a square magnitude of short-time Fourier transform (STFT). STFT gives two-dimensional representation of a one-dimensional signal. In this section, the built-in function '*plt.specgram*' is used here to obtain the time-frequency representation of different types of power quality disturbance. The choice of the window function and the width of the window function is important in obtaining good time-frequency representation so that one obtains good time and frequency resolution. The python code to obtain the time-frequency plot of normal sinusoidal waveform and sinusoidal waveform with sag is shown in Fig. 12.32, and the corresponding output is shown in Fig. 12.33.

```
#Time-Frequency representation of sag
import numpy as np
import matplotlib.pyplot as plt
A,fs,f,N,ph = 1,1000,50,256,0
T = 1/fs
t = np.linspace(0,N*T,N)
#Pure sine wave
pure_sine=np.sin(2*np.pi*f*t+ ph)
#Generation of sag
sag = np.sin(2*np.pi*f*t+ ph) - 0.5*np.sin(2*np.pi*f*t+ ph)*((t<0.15)&(t>0.08))
#Plotting the signal and its STFT
plt.subplot(2,2,1),plt.plot(t,pure_sine),plt.xlabel('Time'), plt.ylabel('Amplitude'),
plt.title('Pure sine wave')
plt.subplot(2,2,2),plt.plot(t,sag),plt.xlabel('Time'), plt.ylabel('Amplitude'),plt.title('Sag')
plt.subplot(2,2,3),plt.specgram(pure_sine, Fs=fs, NFFT=32, noverlap=1,window =None)
plt.xlabel('Time'),plt.ylabel('Frequency'),plt.title('Spectrogram of sine wave')
plt.subplot(2,2,4),plt.specgram(sag, Fs=fs, NFFT=32, noverlap=1,window =None)
plt.xlabel('Time'),plt.ylabel('Frequency'),plt.title('Spectrogram of Sag')
plt.tight_layout()
```

Fig. 12.32 STFT of sinusoidal signal and sinusoidal signal with 'sag'

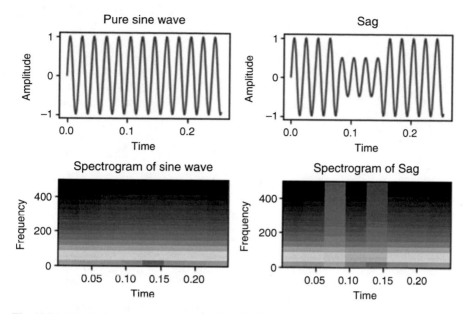

Fig. 12.33 Result of python code shown in Fig. 12.32

From Fig. 12.33, it is possible to infer that the short-time Fourier transform of pure sinusoidal signal shows a horizontal line at 50 Hz, which indicates that the generated sinusoidal signal has a 50 Hz frequency component. The STFT of the sag waveform clearly indicates the starting and ending of the *sag* in the sinusoidal waveform.

The python code to obtain the time-frequency representation of a sinusoidal signal with momentary interruption is given in Fig. 12.34, and its simulation result is depicted in Fig. 12.35.

Figure 12.35 represents the time-frequency representation of momentary interruption in a power line signal. Momentary interruption refers to zeroing of the amplitude of the sinusoidal signal for a brief period of time. From Fig. 12.35, it is possible to observe that the time-frequency representation of momentary interruption is different from the time-frequency representation of the pure sinusoidal waveform. Thus, time-frequency representation clearly distinguishes pure sinusoidal signal from momentary interruption.

12.3.4 Time-Scale Representation of Power Quality Disturbance

Time-scale representation can be obtained using wavelet transform. Wavelet transform has the ability to perform multi-resolution analysis of the signal. In this section,

```
#Time-Frequency representation of MI
import numpy as np
import matplotlib.pyplot as plt
A,fs,f,N,ph = 1,1000,50,256,0
T = 1/fs
t = np.linspace(0,N*T,N)
#Pure sine wave
pure_sine=np.sin(2*np.pi*f*t+ ph)
#Generation of sag
mi = np.sin(2*np.pi*f*t+ ph) - 0.98*np.sin(2*np.pi*f*t+ ph)*((t<0.15)&(t>0.08))
#Plotting the signal and its STFT
plt.subplot(2,2,1),plt.plot(t,pure_sine),plt.xlabel('Time'), plt.ylabel('Amplitude'),
plt.title('Pure sine wave')
plt.subplot(2,2,2),plt.plot(t,mi),plt.xlabel('Time'), plt.ylabel('Amplitude'),
plt.title('Momentary interruption')
plt.subplot(2,2,3),plt.specgram(pure_sine, Fs=fs, NFFT=16, noverlap=1,window =None)
plt.xlabel('Time'),plt.ylabel('Frequency'),plt.title('Spectrogram of sine wave')
plt.subplot(2,2,4),plt.specgram(mi, Fs=fs, NFFT=16, noverlap=1,window =None)
plt.xlabel('Time'),plt.ylabel('Frequency'),plt.title('Spectrogram of MI')
plt.tight_layout()
```

Fig. 12.34 Python code for Time-frequency representation of momentary interruption

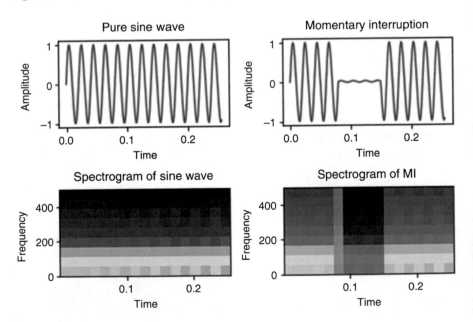

Fig. 12.35 Time-frequency representation of momentary interruption

```
#Scalogram of Momentary Interruption
import numpy as np
import matplotlib.pyplot as plt
import pywt
A,fs,f,N,ph = 1,1000,50,200,0
T = 1/fs
t = np.linspace(0,N*T,N)
#Pure sine wave
pure_sine=np.sin(2*np.pi*f*t+ ph)
#Momentary Interruption (MI)
mi = np.sin(2*np.pi*f*t+ ph) - 0.98*np.sin(2*np.pi*f*t+ ph)*((t<0.15)&(t>0.08))
#Scalogram
scale = [1.,2.]
coef1,freqs1=pywt.cwt(pure_sine,scale,'gaus1')
coef2,freqs2=pywt.cwt(mi,scale,'gaus1')
#Plotting the result
plt.subplot(2,2,1),plt.plot(t,pure_sine),plt.xlabel('Time'), plt.ylabel('Amplitude'),
plt.title('Pure sine wave')
plt.subplot(2,2,2),plt.plot(t,mi),plt.xlabel('Time'), plt.ylabel('Amplitude'),
plt.title('MI'),plt.subplot(2,2,3),
plt.imshow(abs(coef1),extent=[0,200,10,1],interpolation='bilinear',cmap='bone',
        aspect='auto',vmax=abs(coef1).max(),vmin=-abs(coef1).max())
plt.gca().invert_yaxis(),plt.xticks(np.arange(0,201,25))
plt.xlabel('Time'),plt.ylabel('Scale'), plt.title('Scalogram of Sinewave'),plt.subplot(2,2,4),
plt.imshow(abs(coef2),extent=[0,200,10,1],interpolation='bilinear',cmap='bone',
        aspect='auto',vmax=abs(coef2).max(),vmin=-abs(coef2).max())
plt.gca().invert_yaxis(),plt.xticks(np.arange(0,201,25))
plt.xlabel('Time'),plt.ylabel('Scale'), plt.title('Scalogram of MI')
plt.tight_layout()
```

Fig. 12.36 Python code to obtain the scalogram of momentary interruption

time-scale representation of power quality disturbance is obtained using continuous wavelet transform. The library 'pywavelet' is used here to obtain the scalogram of the signal. Scalogram represents the square magnitude to continuous wavelet transform. The choice of the wavelet and the scale are important in obtaining good time-scale representation of the signal. The python code which obtains the time-scale representation of 'momentary interruption (MI)' is shown in Fig. 12.36, and the corresponding output is shown in Fig. 12.37.

From Fig. 12.37, it is possible to observe that the scalogram of signal with momentary interruption is different from the scalogram of a normal sinusoidal signal. This implies that continuous wavelet transform at the proper scale can distinguish power quality disturbances from the normal signal.

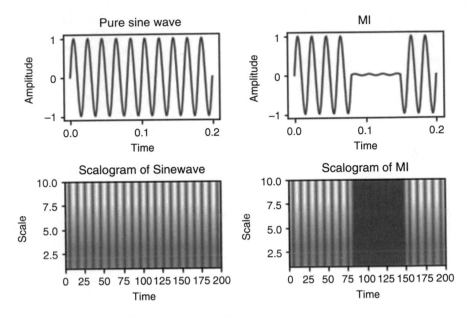

Fig. 12.37 Scalogram of momentary interruption

Bibliography

1. Lawrence Rabiner and Ronald Schafer, "Theory and Applications of Digital Speech Processing", Pearson, 2010.
2. Rangaraj M. Rangayyan, "Biomedical Signal Analysis: A Case-Study Approach", Wiley-Blackwell, 2002.
3. Tompkins Willis J, "Biomedical Digital Signal Processing: C-Language Examples and Laboratory Experiments for the IBMR PC", Prentice Hall India Learning Private Limited, 1998.
4. Waldemar Rebizant, Janusz Szafran, and Andrzej Wiszniewski, "Digital Signal Processing in Power System Protection and Control", Springer, 2011.
5. S. Esakkirajan, T. Veerakumar and Badri N Subudhi, "Digital Signal Processing", McGraw Hill, 2021.

Appendix

Chapter 1: Generation Of Continuous-Time Signals

Answers to PreLab Questions

1. A continuous-time signal can have infinite number of values in a range. Room temperature as a function of time is considered to be continuous-time signal. Speech signal is considered as a continuous-time signal.
2. The built-in functions in the numpy library to create an array of numbers to generate independent variable like time are (a) np.linspace() and (b) np.arange. The linspace is a built-in function available in numpy library to create an evenly spaced sequence of numbers in a specified interval. The syntax of *linspace* is:

    np.linspace(start, stop, num, endpoint, retstep, dtype)

    In the above syntax, 'start' represents the starting value of sequence, 'stop' represents the end value of the sequence and 'num' represents the number of values to generate, and it has to be non-negative. The end point can be either 'true' or 'false'. If it is 'true', the stop is the last sample, it is 'false' then end point value is excluded. The '*retstep*' can be either true or false. If it is true, return (samples, step), where step is the spacing between samples. The '*dtype*' is the data type of the output array. If '*dtype*' is not specified, then it infers the data type from the other input arguments.

    Example: np.linspace(-1,1,5) returns 'array([-1. , -0.5, 0. , 0.5, 1.])' where '-1' is the start value, '1' is the end value. Five sample values are generated, including -1 and +1. The difference between sample values is uniform.

    Note: Similar to '*np.linspace*', we have '*np.logspace*', which is used to create an array of evenly spaced numbers on a log scale value.

    The syntax of *np.arange* built-in function is given by

    np.arange(start, stop ,step, dtype)

    The 'start' and 'stop' represent the beginning and the end value of the interval. The 'step' denotes the step size of the interval. The 'dtype' represent the data type of the output array. The length of the array can be computed using the command ceil((stop-start)/step).

   Example: np.arange(0,1,0.1) generates array of numbers as array([0. , 0.1, 0.2, 0.3, 0.4, 0.5, 0.6, 0.7, 0.8, 0.9]). The length of the array is 10. By default, 'np.arange' command does not allow to include the end point value.

3. Most of the real-world phenomenon like motion of pendulum, under damped spring-mass system can be expressed as sinusoidal signal. The sinusoidal signal is a periodic signal, which varies smoothly with respect to time. According to Fourier series, it is possible to represent periodic signals as sum of sinusoids. Also, sinusoidal signals are eigen functions of linear time-invariant systems.

4. The term 'phase' refers to position of the waveform with respect to the origin. The phase of the signal is measured in degrees or radians.

5. Multidimensional signals require more than one independent variable to represent the signal. Examples of multidimensional signal include (a) grey scale image, (b) colour image and (c) video. Grey scale image is represented as $f(x, y)$, where 'x' and 'y' are termed as spatial variable. The colour image is represented as $f(x, y, \lambda)$, where 'λ' represents colour information. The video signal is basically sequence of image, which is represented as $f(x, y, \lambda, t)$. The video signal is characterized by both spatial and temporal information.

6. (a) The equation of current through the diode is given as

$$I_D = I_s \left\{ e^{\frac{V_D}{\eta V_T}} - 1 \right\}$$

   In the above expression, V_D is the voltage across the diode and I_D is the current through the diode; V_T is the volt-equivalent of temperature, which is 26 mV at room temperature, and η is the ideality factor, which is material dependent. Thus, the current through the diode is modelled as an exponential function. In this case, it is an exponentially growing function.

   (b) The equation for radioactive decay is expressed as

$$A = A_0 e^{-\lambda t}$$

   where 'A' is the ending activity and A_0 is the initial activity, $\lambda = \frac{0.693}{T_{\frac{1}{2}}}$, where $T_{\frac{1}{2}}$ is the half-life period of the element. Thus, the radioactive decay is modelled as exponentially decaying function.

7. Few significant features of complex exponential function are

(a) Complex exponential function is the basis function of Fourier transform.
(b) It is a complex valued signal that simultaneously encapsulates both sine and cosine signal by posting them on the real and imaginary components of the complex signal.
(c) Complex exponentials are Eigen functions of continuous-time linear time-invariant system.

8. Sinc function is mathematically defined as

$$\sin c(t) = \frac{\sin(\pi t)}{\pi t}$$

The sinc function is an even function $\text{sinc}(-t) = \sin c(t)$. Few significant features of sinc function are

(a) Sinc functions are used in the interpolation of signals
(b) Sinc and rectangular functions are dual function. Fourier transform of sinc function results in rectangular function and vice versa.

9. A stationary signal is one whose statistical characteristics do not vary with respect to time. Example is $x(t) = A \sin (2\pi ft + \phi)$. Here the frequency of the signal does not change with respect to time. It is considered as stationary. A non-stationary signal is one whose statistical characteristics change with respect to time. Example of non-stationary signal includes $y(t) = A \sin (2\pi ft^2)$. Here the frequency of the signal changes with respect to time. It is considered as non-stationary. Chirp signal is considered as non-stationary signal.

10. The Gaussian function is characterized by two parameters, which are (a) mean and (b) standard deviation. Few significant features of Gaussian functions are

(a) Gaussian functions are used as smoothing functions. The extent of smoothing is governed by the standard deviation.
(b) Fourier transform of a Gaussian function result in another Gaussian function.

Answers to Objective Questions

Q. No.	1	2	3	4	5	6	7	8	9	10
Key	D	C	A	B	C	B	B	D	B	A

Chapter 2: Sampling and Quantization of Signals

Answers to PreLab Questions

1. The steps involved in converting the analogue signal into a digital signal are (a) sampling, (b) quantization and (c) encoding. Before sampling, it is necessary to ensure that the signal to be sampled is bandlimited. Sampling converts a continuous-time signal into a discrete-time signal. Quantization is basically mapping a large set of values to a smaller set of values. In quantization, the discrete-time signal is converted to a quantized signal. In encoding, the quantized signal is converted to a digital code.

2. The sampling theorem specifies the minimum sampling rate so that the sampled signal can be reconstructed from its samples without aliasing problem. In order to reconstruct the signal from the samples without an aliasing problem, the sampling frequency must be greater than twice the maximum frequency content of the signal. This is expressed as $f_s \geq 2f_{max}$, where f_s represents the sampling frequency and f_{max} represents the maximum frequency content of the signal.

3. Suppose the signal is a periodic and ideal interpolation is employed. In that case, all spectral components less than $f_s/2$ (where f_s represents the sampling frequency) are reconstructed perfectly, but all higher-frequency spectral components are aliased to a frequency less than $f_s/2$.

4. The square wave is not a bandlimited signal. It is not possible to reconstruct the square wave from its samples.

5. Two prominent reasons for aliasing while performing sampling are (a) undersampling and (b) signal which is not a bandlimited signal. Here undersampling implies that the sampling rate $f_s < 2f_{max}$, where f_{max} is the maximum signal frequency.

6. Sampling is basically taking specific instants of the signal. The sampling rate (f_s) is the number of samples per second. Sampling interval (T_s) is the time-interval between two consecutive samples.

7. Nyquist rate $= 2B = 2 \times 5$ kHz $= 10$ kHz.

8. Quantization is mapping a large set of values to a smaller set of values. It will not obey the superposition principle; hence, it is considered as non-linear phenomenon.

9. Quantization maps a large set of values to a smaller set of values. It is not a one-to-one mapping; hence, error is inevitable, and it is considered as irreversible phenomenon. The meaning is that it is difficult to get the original signal exactly after quantization.

10. The process of converting sampled data sequences to a continuous-time signal is termed as signal reconstruction. Different strategies include (a) zero-order hold, (b) first-order hold or linear interpolation and (c) ideal or sinc interpolation.

Answers to Objective Questions

Q. No.	1	2	3	4	5	6	7	8	9	10	11	12	13	14	15
Key	B	A	A	B	D	A	B	C	B	D	B	C	D	B	C

Chapter 3: Generation and Operation on Discrete-Time Sequence

Answers to PreLab Questions

1. Two important steps involved in converting continuous-time signal into a discrete-time signal are (a) sampling and (b) quantization.
2. Different forms of representation of discrete-time signals are (a) graphical form, (b) functional form, (c) sequential form and (d) tabular form.
3. Some standard discrete-time sequences are (a) unit sample sequence ($\delta[n]$), (b) unit step sequence ($u[n]$), (c) unit ramp sequence ($r[n]$) and (d) exponential sequence, which can be broadly classified as real exponential sequence and complex exponential sequence.
4. Some of the salient features of unit sample sequence are:

   (a) Any arbitrary signal $x[n]$ can be expressed in terms of scaled and shifted versions of unit sample sequences. This is expressed as
   $$x[n] = \sum_{k=-\infty}^{\infty} x[k]\delta[n-k].$$
   (b) Convolution of signal $x[n]$ with unit sample signal will result in the signal $x[n]$. This is expressed as $x[n] * \delta[n] = x[n]$.
   (c) If unit sample sequence is applied to linear time-invariant discrete-time system, then the output of the system is termed as the impulse response of the system. This is illustrated in Fig. A.1. From Fig. A.1, it is possible to observe that the input to LTI discrete-time system is unit sample signal; then, the output of the system is termed as the impulse response of the system.

5. It is to be noted that linear time-invariant discrete-time system is characterized by its impulse response. That is, by knowing the impulse response of the system, it is possible to know the properties of the system like causality and stability.

Fig. A.1 Impulse response of the system

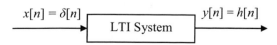

Table A.1 Descriptions of energy and power signals

S. no.	Energy signal	Power signal
1	For a discrete-time energy signal, the energy is finite and non-zero	For a discrete-time power signal, the power is finite and non-zero.
2	Non-periodic signals are energy signals	Periodic signals are power signals
3	Power of energy signal is zero	Energy of power signal is infinite
4	Examples of DT energy signal are $x[n] = a^n u[n]$, $|a| < 1$ $x[n] = u[n] - u[n-1]$	Examples of power signals are Unit step signal $x[n] = A\sin(\omega_o n + \varphi)$

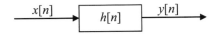

Fig. A.2 Linear time-invariant system

6. Discrete-time signal $x[n]$ is periodic if it obeys the condition $\frac{\omega}{2\pi} = \frac{k}{N}$. Here, '$\omega$' represents the angular frequency, and 'N' represents the fundamental period.

7. The energy and power signal descriptions are summarized in Table A.1.

8. Various mathematical operations that can be performed on DT signal include (a) folding or time reversal; (b) shifting operation, which include delay and advance operation; and (c) scaling operation, which could be time scaling operation, like downsampling and upsampling, and amplitude scaling operation.

9. A DT signal is even if it obeys the condition $x[-n] = x[n]$. A DT signal is odd if $x[-n] = -x[n]$. Example of even signal is cosine signal, whereas example of odd signal is sinusoidal signal. Example of signal which is neither even and nor odd includes unit step signal and unit ramp signal.

10. An energy signal has finite energy, whereas power signals have finite average power. There are certain signals, which are neither energy nor power signal. Example of finite energy signal is $x[n] = (1/2)^n u[n]$. Example of power signal is unit step signal. Example of signal which is neither energy nor power signal is unit ramp signal.

11. Convolution is one of the most important operations in signal processing. The three main mathematical operations involved in convolution are (a) multiplication, (b) addition and (c) shifting operation. Convolution basically performs filtering operation. It is represented in Fig. A.2

    In Figure A.2, the input and output signals are represented as $x[n]$ and $y[n]$ respectively. The impulse response of linear time-invariant system is denoted as $h[n]$. The nature of filtering is decided by $h[n]$. If $h[n] = \{0.5, 0.5\}$, the system behaves like a lowpass filter; on the other hand if $h[n] = \{0.5, -0.5\}$, the system behaves like a highpass filter. The nature of filtering is decided by the impulse response of the system. The expression for the output of the system is given by $y[n] = x[n] * h[n]$, where '$*$' denotes the convolution operation.

    The correlation between two signals $x_1[n]$ and $x_2[n]$ is given by

$$r_{x1x2}(l) = x_1[n] * x_2[-n]$$

Convolving the folded version of sequence $x_2[n]$ with the sequence $x_1[n]$ results in correlation of the signal. The correlation can be broadly classified into (a) autocorrelation and (b) cross-correlation. Autocorrelation is finding the relative similarity of the signal with itself.

Applications of correlation are summarized below:

(a) Correlation is used to find the relative similarity between signals.
(b) Fourier transform of autocorrelation function gives the power spectral density of the signal. This is regarded as Wiener-Khinchin theorem.
(c) Correlation can be used to estimate the pitch of the speech signal.
(d) Correlation can be used for template matching.

Answers to Objective Questions

Q. No.	1	2	3	4	5	6	7	8	9	10	11	12	13	14	15
Key	B	B	C	B	B	C	B	C	D	C	B	B	C	D	B

Q. No.	16	17	18	19	20	21	22	23	24	25
Key	B	B	A	B	A	D	C	B		

Chapter 4: Discrete-Time Systems

Answers to PreLab Questions

1. Different forms of representation of discrete-time systems are (a) block diagram, (b) difference equation, (c) transfer function, (d) impulse response, (e) pole-zero plot and (f) state-space.
2. A discrete-time system is said to be a relaxed system if zero input results in zero output. If $x[n] = 0$, then the corresponding output $y[n]$ should be zero.
3. A discrete-time system is linear if it obeys superposition theorem. Superposition theorem implies (a) homogeneity property and (b) additivity property. According to homogeneity property, scaling of the input results in scaling of the output. According to additivity property, the response of the system to sum of inputs must be equal to sum of the individual responses. Examples of linear

system are as follows: (a) $y[n] = x[-n]$, (b) $y[n] = nx[n]$, (c) $y[n] = Ax[n]$, (d) $y[n] = x[Mn]$ and (e) $y[n] = x[n/L]$.

4. Cascade of two non-linear systems may result in a linear system. For example, consider the cascade of two systems as shown below:

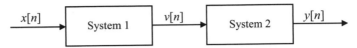

For system 1: The relationship between the input and output is given by $v[n] = \log \{x[n]\}$.

For system 2: The relationship between the input and output is given by $y[n] = \exp \{v[n]\}$.

It can be observed that both System-1 and System-2 are non-linear system, whereas the cascaded system is a linear system. This implies that cascading of two non-linear systems need not be always non-linear.

5. A discrete-time system is causal, if its current output should not depend on the future value of the input. A real-world system cannot react to future input; hence, they are considered as causal systems.

6. A system is memory-less if the current output of the system depends on the current input. All the memoryless systems are causal. A causal system is non-anticipatory. For a causal system, the current output will not depend on the future value of the input. For a causal system, the current output depends on the past input. If the system output depends on the past input, it is a memoryless system. Hence, all memoryless systems are causal, whereas all causal systems are not memoryless. For example, $y[n] = Ax[n]$ is a memory less system, which is also causal. Consider the system $y[n] = Ax[n] + Bx[n - 1]$; the system is causal, whereas it is not memory less.

7. Consider the cascade of two discrete-time time varying system, which is depicted below.

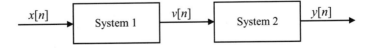

System 1: The system performs upsampling of the input signal by a factor of 2. The relationship between the input and output of the system is given by $v[n] = x[n/2]$. Upsampling by a factor of 2 is a time-varying system.

System 2: System 2 performs downsampling of the input signal by a factor of 2. The relationship between the input and output of the system is given by $y[n] = v[2n]$. Downsampling by a factor of 2 is a time-varying system.

Cascaded system: Cascading of upsampling by a factor of 2 followed by downsampling of two is an identity system. Thus, the cascaded system is a time-invariant system.

Hence, it can be concluded that cascading of two time-varying discrete-time system need not always result in time-varying system.

8. A discrete-time system is said to be invertible if 'distinct input should in distinct output'. Example of invertible system is $y[n] = x[n/2]$, which is upsampling of the input signal by a factor of 2. Example of non-invertible system is $y[n] = x[2n]$, which is downsampling of the input signal by a factor of 2.

9. It is possible to test the causality and stability of linear time-invariant system from its impulse response. A linear time-invariant system is causal if its impulse response is zero for $n < 0$. A linear time-invariant discrete time system is stable if its impulse response is absolutely summable.

10. A discrete-time system is static if current output depends only on current input. A discrete-time system is dynamic if the current output depends on current input, past input and past output. Example of static system is $y[n] = x^2[n]$. Example of dynamic system is $y[n] = x[n] + y[n-1]$.

11. A discrete-time system is said to be non-recursive if the current output of the system depends on the current input and the past input. A discrete-time system is said to be recursive if the current output depends on the previous output of the system.

Example of non-recursive system: $y[n] = 0.5x[n] + 0.5x[n-1]$. For this system, the current output does not depend on the past output, hence it is non-recursive.

Example of recursive system: $y[n] = x[n] + 0.5y[n-1]$. It is a recursive system because the current output is a function of previous output.

12. A zero at $z = 1$ is equivalent to zero at $\omega = 0$. A zero at $\omega = 0$, will block all low frequency component. The system will block DC component of the signal. The system will behave like a highpass filter.

13. A discrete-time system is invertible if distinct input leads to distinct output. Example of a discrete-time system which is invertible is 'accumulator system'. The impulse response of accumulator is $h[n] = u[n]$, whereas the impulse response of the inverse of the accumulator system is given by $h[n] = \delta[n] - \delta[n-1]$. Example of discrete-time system which is non-invertible is downsampler system whose input-output relationship is given by $y[n] = x[2n]$.

14. State-space representation is an application for multiple input and multiple output systems. State-space approach can be used to model non-linear and time-varying systems.

15. From the impulse response of the discrete-time system, it is possible to infer whether the system is causal and stable. A discrete-time system is causal if its impulse response is zero for $n < 0$. A discrete-time system is stable if its impulse response is absolutely summable.

Answers to Objective Questions

Q. No.	1	2	3	4	5	6	7	8	9	10	11	12	13	14	15
Key	D	D	A	A	D	D	C	D	A	A	B	B	C	D	C

Chapter 5: Transforms

Answers to PreLab Questions

1. Spectrum is a compact representation of the frequency content of a signal that is composed of sinusoids.
2. The unilateral or one-sided Z-transform differs from the bilateral or double-sided Z-transform in that the summation is carried out only over non-negative values of time index (n), whereas the bilateral transform includes the both negative and positive values of time index (n). The unilateral Z-transforms can be used to analyse causal systems that are specified by linear constant coefficient difference equations with non-zero initial conditions.
3. The range of variation of 'z' for which Z-transform converges is called region of convergence of Z-transform.
4. The basis function has to be an orthogonal function. The basis function of Fourier transform is complex exponential function.
5. Applying discrete Fourier transform is equivalent to applying discrete Fourier series on a periodically extended finite aperiodic signal. Discrete Fourier series is applied when the signal under analysis is periodic, and discrete Fourier transform is applied when the signal under analysis is aperiodic.
6. DTFT is Z-transform evaluated on a unit circle. The expression for Z-transform is given by

$$X(z) = \sum_{n=-\infty}^{\infty} x[n]z^{-n}$$

Substituting $z = re^{j\omega}$ in the above expression, we get

$$X(e^{j\omega}) = \sum_{n=-\infty}^{\infty} x[n](re^{j\omega})^{-n}$$

For a unit circle, $r = 1$; hence, the above expression can be written as

$$X\left(e^{j\omega}\right) = \sum_{n=-\infty}^{\infty} x[n]e^{-j\omega n}$$

Thus, DTFT is Z-transform evaluated on a unit circle.

7. Double-sided spectrum of a signal composed of sinusoid is expressed as

$$x(t) = X_0 + \sum_{k=1}^{N} \left\{ \frac{X_k}{2} e^{j2\pi f_k t} + \frac{X_k^*}{2} e^{-j2\pi f_k t} \right\}$$

The set of pairs $\left\{ (0, X_0)(f_1, \frac{1}{2}X_1), (-f_1, \frac{1}{2}X_1^*), \ldots, (f_k, \frac{1}{2}X_k), (-f_k, \frac{1}{2}X_k^*) \right\}$ indicates the size and relative phase of sinusoidal component contributing at frequency f_k. This is termed as frequency-domain representation of the signal.

8. Transform is a tool to analyse the signals and systems. Signals are converted from time or spatial domain to frequency domain using transform. Frequency domain is used to describe the signal with respect to frequency. Each frequency has its own amplitude and phase. From the spectrum, it is possible to interpret the frequencies are present in the signal. Thus, the time-domain and the frequency-domain representation of the signal are equivalent. It is possible to transform the signal from time domain to frequency domain and vice versa without any loss of information. Mathematically, transform is taking the inner product of the signal with the basis function. The inner product is one way of quantifying the similarity or the dissimilarity of two signals.

9. The signal $x[n]$ must be conjugate symmetric. This is expressed as $x[n] = x^*[-n]$.

10. The basis function of Fourier transform is complex exponential, which oscillates for all the time. Fourier transform describes the frequency components in the signal averaged over all the time. It is difficult for the Fourier transform to represent signals that are localized in time. Fourier transform is not an effective tool to analyse non-stationary signals.

    Time localization in Fourier transform can be achieved by windowing the signal over which the signal is nearly stationary, which leads to the development of short-time Fourier transform (STFT). It can be represented to non-stationary signal and gives both time and frequency resolutions. However, the time and frequency resolutions are fixed based on the windowing signal.

11. The DCT provides a decomposition of any discrete time signal as a weighted sum of basis functions and these basis functions are cosines. If $x[n]$ is a real for all 'n', then the DCT output $X[k]$ is real for all 'k'. The DCT has excellent energy compaction for many real-world signals (i.e. signals with high correlation among neighbouring samples).

12. Wavelet transform has the ability to perform multi-resolution analysis of the signal, whereas STFT cannot perform multi-resolution analysis. STFT provides time-frequency representation of the signal, whereas wavelet transform provides time-scale representation of the signal.

Answers to Objective Questions

Q. No.	1	2	3	4	5	6	7	8	9	10	11
Key	C	C	B	B	B	C	B	C	D	C	B

Chapter 6: Filter Design Using Pole-Zero Placement Method

Answers to PreLab Questions

1. If $h[n]$ represents the impulse response of the lowpass filter, then the filter whose impulse response is $(-1)^n h[n]$ will act as a highpass filter.
2. A discrete-time system with transfer function$H(z)$ is a minimum phase system if the following conditions are met.

    (a) All the zeros of the system are inside the unit circle centred about the origin.
    (b) All the poles of the system are inside the unit circle centred about the origin.
    (c) The numerator and the denominator of the transfer function $(H(z))$ have equal orders of 'z'.

3. The transfer function of a discrete-time system is given by $H(z) = B(z)/A(z)$. The frequencies for which the values of the denominator and numerator become zero in a transfer function are called poles and zeros. Poles are the roots of the denominator of a transfer function. Similarly, zeros are the roots of the numerator of the transfer function. For a discrete-time system to be stable, the poles of the system should lie within the unit circle.
4. For a discrete-time system to be stable, (a) the poles of the system should lie within the unit circle, and (b) the impulse response should be absolutely summable.
5. The basic principle underlying the pole-zero placement method is to locate poles near points of the unit circle corresponding to frequencies to be emphasized and locate zeros near the frequencies to be deemphasized. All the poles should be placed within the unit circle for the filter to be stable. All complex zeros and poles must occur in complex conjugate pairs for the filter coefficients to be real.
6. All-pass filters can be used as delay equalizer or phase equalizer. When an all-pass filter is placed in cascade with a system that has an undesired phase response, a phase equalizer is designed to compensate for the poor phase characteristics of the system such that the cascaded system will exhibit linear phaseresponse.
7. Notch filter is used to eliminate one particular frequency. It is used to minimize power line interference in biomedical equipment. It can be used in radio receivers to remove unwanted interfering frequencies.

8. Moving average filter basically performs lowpass filtering of the input signal. Lowpass filter converts drastic variation in the signal to a gradual variation. When a square wave is fed as input to the M-point moving average filter, the output will be a triangular wave. Square wave exhibits sharp transition between 'ON' and 'OFF' state. Triangular waveform exhibits gradual variation between 'ON' and 'OFF' state.

9. From the input-output relation, it is possible to observe that the current output is a function of current input and previous input; hence, the given filter is finite impulse response (FIR) filter.

10. Pole-zero plot is a two-dimensional plot with x-axis as the real part and y-axis as the imaginary part. The pole-zero plot shows the unit circle with zeros marked as '0', and poles are indicted with the symbol 'x'. Zeros and poles near the unit circle are expected to have a strong influence on the magnitude response of the filter.

Answers to Objective Questions

Q. No.	1	2	3	4	5	6	7	8	9	10	11	12	13
Key	D	C	C	D	D	B	C	A	D	D	C	D	D

Chapter 7: FIR Filter Design

Answers to PreLab Questions

1. The difference equation relating the input and output of an FIR filter is given by

$$y[n] = \sum_{k=0}^{M} b_k x[n-k]$$

From the difference equation, the following inferences can be drawn:

(a) The current output of FIR filter depends on current input and previous inputs.

(b) The number of previous outputs necessary to compute the current output is termed as the order of FIR filter.

(c) Since the current output is not a function of previous output, FIR filter is considered as a 'non-recursive filter'. It can also be termed as 'all-zero' filter with the poles at the origin.

(d) The filter coefficients are denoted as 'b_k'. The nature of filtering depends on 'b_k'.

2. Based on symmetry and number of coefficients, the FIR filters are classified as Type I, Type II, Type III and Type IV FIR filters

   (a) Type I FIR filter: Even symmetry with odd number of coefficients
   (b) Type II FIR filter: Even symmetry with even number of coefficients
   (c) Type III FIR filter: Odd symmetry with odd number of coefficients
   (d) Type IV FIR filter: Odd symmetry with even number of coefficients

3. An FIR filter with impulse response '$h[n]$' is said to exhibit even symmetry if $h[n] = h[N-1-n]$, where 'N' is the number of coefficients of FIR filter. An FIR filter is said to exhibit odd symmetry if $h[n] = -h[N-1-n]$.

4. A digital filter exhibits linear phase characteristics if its impulse response is either symmetric or anti-symmetric.

5. If a digital filter exhibits linear phase characteristics, then it will not introduce phase distortion. All the frequency components of the input signal will pass through the filter with constant delay so that there will not be any phase distortion.

6. The relationship between group delay (τ_g) and the phase response ($\phi(e^{j\omega})$) of the filter is given by $\tau_g = -\frac{d}{d\omega}(\phi(e^{j\omega}))$.

7. Order of FIR filter (M) is the number of previous input samples necessary to compute the current output. If 'M' denotes the order of the FIR filter and 'N' denotes the number of coefficients of FIR filter, then the relationship between 'M' and 'N' is given by $N = M + 1$.

8. FIR filter is an 'all-zero' filter with the poles occurring at the origin. For a digital system to be stable, the pole should lie within the unit circle. Since the pole of FIR filter occurs at the origin, the FIR filter is an inherently stable filter.

9. The advantages of FIR filter are:

   (a) FIR filter exhibits linear phase characteristics; hence, there will not be phase distortion
   (b) The group delay of FIR filter is constant; hence, all the frequency component of the input signal passes through FIR filter with equal delay.
   (c) The poles of FIR filter lie at the origin; hence, FIR filter is inherently stable filter
   (d) The coefficients of FIR filter are either symmetric or anti-symmetric in nature. Symmetricity of filter coefficient leads to linear phaseresponse of the filter.

10. FIR filters can be designed using (a) windowing method, (b) frequency sampling method and (c) optimal method.

Answers to Objective Questions

Q. No.	1	2	3	4	5	6	7	8	9	10	11	12	13	14	15
Key	A	C	B	A	A	B	A	C	B	D	D	B	B	B	D

Chapter 8: Infinite Impulse Response Filter

Answers to PreLab Questions

1. In a recursive filter, the present output depends on both the inputs and previously calculated outputs.

2. The IIR filter is a recursive filter. The present output of the IIR filter depends on input and past output. Hence, the IIR filter is a recursive filter.

3. Ripple is the fluctuations in the passband or stopband of the filter's frequency response. It is expressed in decibels.

4. Based on the ripples in the frequency response of the IIR filter, the filters are classified as (a) Butterworth filter, (b) Chebyshev filter (Type I), (c) inverse Chebyshev filter (Type II) and (d) elliptic filter.

5. The Butterworth filter's magnitude response is monotonically decreased at all frequencies, and also, there are no local maxima or minima in both the passband and stopband. Hence, it is also called as flat-flat filters.

6. Mapping is a technique used in IIR filter design for converting analogue filter into digital filter. The different types of mapping techniques are (a) backward difference method, (b) impulse invariant technique (IIT), (c) bilinear transformation technique (BLT) and (d) matched Z-transformtechnique.

7. While converting an analogue filter into an equivalent digital filter, it is necessary that stable analogue filter should be mapped to a stable digital filter. The points of the analogue filter in the left half S-plane must be mapped into inside the unit circle in the Z-plane to preserve the stability of the filter. Therefore, all the mapping techniques must preserve the stability of the filter.

8. The steps involved in obtaining the transfer function of a digital filter using the impulse invariant technique are summarized below:

   (a) Obtain the transfer function of an analogue filter $(H(s))$, which has to be converted into an equivalent digital filter $(H(z))$.
   (b) From the analogue transfer function$H(s)$, get the impulse response $h(t)$ using the inverse Laplace transform.
   (c) Apply the sampling process on impulse response $h(t)$ to get $h[nT]$.
   (d) Take Z-transform of the sampled impulse response $h[nT]$ to get the equivalent transfer function$H(z)$.

Table A.2 Comparison of IIR filters

S. no.	Type of IIR filter	Ripple in passband	Ripple in stopband	Transition width	Order of the filter to meet the given specification
1	Butterworth filter	NO	NO	Widest	Highest
2	Chebyshev filter	YES	NO	Narrower	Lower
3	Inverse Chebyshev filter	NO	YES	Narrower	Lower
4	Elliptic filter	YES	YES	Narrowest	Lowest

9. The drawbacks of impulse invariant technique are listed below:

    (a) It is more suitable for all pole filters and does not consider the system's zeros.
    (b) The mapping of analogue frequency 'Ω' to digital frequency 'ω' is 'many-to-one'; hence, aliasing problem exists.
    (c) Due to the presence of aliasing, the impulse invariant method is appropriate for the design of lowpass and bandpass filters only. It is not a suitable technique for the design of highpass and band reject filters.

10. The bilinear transformation is a conformal mapping that maps the '$j\Omega$' axis of the S-plane into the unit circle of the Z-plane only once. Therefore, it can avoid aliasing problems.

11. The bilinear transformation technique maps the analogue frequency and digital frequency in a non-linear fashion. The relationship between the analogue and digital frequency is given by $\omega = 2\tan^{-1}\left(\frac{\Omega T}{2}\right)$. The non-linear relationship between analogue and digital frequency is termed as 'frequency warping'. To overcome the frequency warping problem, 'prewarping' technique is used. Prewarping will preserve the edge frequencies but not the exact shape of the magnitude response.

12. Comparison of different types of IIR filter is given in Table A.2

13. The following steps are involved in the IIR filter design:

    Step 1: Convert the digital filter specifications into an equivalent analogue filter specification.
    Step 2: Convert the analogue filter specifications to normalized lowpass proto-type specifications.
    Step 3: Design a normalized lowpass prototype filter by using any one of the analogue filters: (a) Butterworth filter, (b) Chebyshev Type I filter, (c) Chebyshev Type II filter or (d) elliptic filter.
    Step 4: Use the analogue transformation technique to convert the normalized lowpass prototype filter into the desired analogue filter.

Step 5: Use the mapping technique to convert the desired analogue filter into a desired digital one.

Answers to Objective Questions

Q. No.	1	2	3	4	5	6	7	8	9	10
Key	B	D	C	A	A	A	B	D	C	B

Chapter 9: Quantization Effect of Digital Filter Coefficients

Answers to PreLab Questions

1. The finite word length introduces an error that can affect the performance of the DSP system. The finite word length has limited precision, and it is not sufficient to represent the filter coefficients accurately. This causes errors between the original filter coefficients and finite word length coefficients. The finite number of bits is used in the arithmetic operations in DSP, which is insufficient to give the proper result.

2. In fixed point arithmetic representation, the numbers are represented in a fixed range with a finite number of bits of precision. The numbers beyond the fixed range are either saturated or wrapped around. It is preferred for high speed and low cost.

   In floating-point arithmetic representation, every number is represented in two parts (a) mantissa and (b) exponent. Floating-point representation has a higher dynamic range and no need for scaling. It can be used to perform more complex algorithms in it.

3. The numbers can be represented in binary format, which contains '0s' and '1s'. In sign-magnitude representation, most significant bit (MSB) is used to denote the number as positive or negative. It is called as sign bit, and the remaining bits are used to represent the number, which is called as magnitude bits.

4. Quantization is a process in which a quantity X is approximated into a quantity $Q(X)$. The approximated value will have a distortion between X and $Q(X)$, called quantization error.

5. Rounding is a method to perform quantization operation. It selects the quantized value nearest to the original value. The error between the quantized and original value will not exceed $\pm(\Delta/2)$. Here 'Δ' denotes step size, which is obtained by $\Delta = 2^{-B}$, and B is the number of binary bits.

6. The two's complement truncation is another method to perform the quantization operation. This method always gives a resultant quantized value less than or equal to the original value. The truncation error will to $(-\Delta$ to $0)$.

7. The magnitude truncation is another approach to perform the quantization operation. The result of the quantized value is always less than the original value for $X > 0$, and the quantized result is always greater than the original value for $X < 0$. The advantage of magnitude truncation is that it can inherently suppress the limit cycle oscillation.

8. If the dynamic range of the signals crosses the word length limit, then the overflow exists. The different types of overflows are (a) saturation, (b) zeroing and (c) two's complement.

   Saturation: If the input value crosses the maximum/minimum limit $(X/-X)$, the output will be $X/-X$.

   Zeroing: The output will be zero if the input value exceeds the maximum limit.

   Two's complement: It is a periodic continuation of the 45° straight line. The advantage of this method has the capability of correcting the intermediate overflows automatically.

9. A limit cycle oscillation is a low-level oscillation that can exist in a stable filter due to the non-linearity associated with the quantization operation, like rounding or truncation of the arithmetic calculations in the filtering operation. This limit cycle oscillation is also termed a multiplier round-off limit cycle. These limit cycle oscillations do not occur in non-recursive FIR filters.

10. Infinite precision is needed to represent the filter coefficients. However, the finite number of bits represents the filter coefficients in the real world. Therefore, the representation of the filter coefficients from infinite precision to finite number precision may introduce coefficient quantization. Due to the finite precision representation of filter coefficients, the response of the quantized filter may deviate from the response of the original filter.

Answer to Objective Questions

Q. No.	1	2	3	4	5	6	7	8	9	10
Key	C	A	A	A	A	A	C	A	D	B

Chapter 10: Multirate Signal Processing

Answers to PreLab Questions

1. When two devices operating at different sampling rate are to be interconnected, it is necessary to change the sampling rate of the signal. Often there is a mismatch between the sampling rates of the recording and playback system. The sampling rate of audio signals in compact disc is 44.1 kHz, whereas the sampling rate of audio signals in digital audio tape is 48 kHz. Sampling rate conversion is required for the interconnection of compact disc with digital audio tape.

2. A sampling rate converter is a device or software that accepts digital input signals at one sampling rate and outputs a digital signal at a different sampling rate. This is shown below.

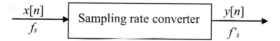

Block diagram of sample rate converter

3. The two basic operations in multirate signal processing are (a) downsampling and (b) upsampling. Downsampling reduces the sampling rate of the input signal, whereas upsampling increases the sampling rate of the input signal.

4. Downsampling in time domain may lead to spectral overlap in the frequency domain, which is termed as 'aliasing'. To overcome aliasing, a lowpass filter is used before downsampling. This filter is termed as 'anti-aliasing filter'. This is depicted below.

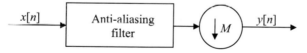

Decimation operation

To overcome the problem of aliasing, the cut-off frequency of the anti-aliasing filter is chosen as f_s/M.

5. Upsampling by a factor of 'L' is the process of inserting '$L - 1$' zeros between successive samples. Upsampling in time domain results in the creation of multiple copies of the original spectrum in the frequency domain. The anti-aliasing filter is a lowpass filter that retains only the original spectrum and removes multiple copies of the original spectrum. The block diagram of anti-imaging filter with upsampling operation is shown below. The sampling frequency of the anti-imaging filter is chosen as f_s/L.

Interpolation operation

6. The time-domain expression for downsampling by a factor of 'M' is given by $y[n] = x[nM]$. The frequency-domain expression for downsampling by a factor of 'M' is given by $Y(z) = \frac{1}{M} \sum_{k=0}^{M-1} X\left(z^{\frac{1}{M}} e^{-j\frac{2\pi}{M}k}\right)$.

7. The time-domain expression for upsampling by a factor of 'L' is given by $y[n] = x[n/L]$. The frequency-domain expression is given by $Y(z) = X(z^L)$.

8. The three significant properties of downsampling operation are

   (a) Downsampling is a linear operation.
   (b) Downsampling is a time-variant operation.
   (c) Downsampling is an irreversible operation.

9. (a) Upsampling is a linear operation, because it obeys additivity and homogeneity properties. Thus, upsampling obeys the superposition principle; hence, it is a linear operation.

   (b) Upsampling has varying responses to the same input at different instants of time; hence, it is considered as time-variant operation.

10. Downsampling by a factor 'M' and upsampling by a factor of 'L' are interchangeable if 'L' and 'M' are relatively prime.

11. Idempotent operations are operations, which can be applied multiple times without changing the result. Downsampling by a factor of 'M' followed by upsampling by the same factor gives the same result if the operation is repeated many times; hence, it is considered as idempotent operation.

12. Polyphase decomposition ensures that filtering operations are performed at the lowest possible sampling rate in the system, which reduces the computational complexity and the overall system's cost. Polyphase decomposition can be classified as (a) Type I polyphase decomposition and (b) Type II polyphase decomposition.

13. Filter bank is a group of filters arranged in a specific fashion. Filter bank is used for subband decomposition of the signal. It is useful for signal denoising and signal compression.

14. The main threats for perfect reconstruction in a two-channel filter bank are (a) aliasing problem, (b) amplitude distortion and (c) phase distortion. Perfect reconstruction can be achieved by proper choice of analysis and synthesis filters.

15. Transmultiplexer is a multiple input-multiple output system (MIMO). It uses multirate operators and filters to combine 'M' signals for transmission across a channel and then recovers the 'M' input signals at the receiver end. The separation of signals should be perfect, and the recovery of each signal should be performed without leakage of signal from one channel to another, which is generally termed as crosstalk. The proper choice of filters can avoid the crosstalk problem.

Answers to Objective Questions

Q. No.	1	2	3	4	5	6	7	8	9	10	11
Key	B	B	B	B	B	A	A	B	B	B	D

Chapter 11: Adaptive Signal Processing

Answers to PreLab Questions

1. In optimal filtering, the input and the desired signals are available for a given time window, and the optimal parameters of the filter are computed only once. In adaptive filtering, the input and the desired signal are provided to the algorithm, and the algorithm computes the parameters of the filter and is updated; hence, it is iterative in nature. Adaptive filter does not require previous knowledge of the signal statistics.

2. An adaptive filter is a filter with filter coefficients that are non-constants. The filter coefficients are adjusted based on some specific criterion defined to optimize the filter's performance. In an ordinary filter, the filter coefficients are constant and do not vary with respect to specific criteria.

3. Some examples of the adaptive filter include Wiener filter, least mean square filter, RLS filter, etc.

4. The adaptive filters are generally preferred in the following contexts:

   (a) The filter characteristics are necessary to be changed or adapted to specific conditions.
   (b) Spectral overlap between the signal and noise
   (c) If the noise present in the signal is unknown or varies with time.

5. The performance measures of the adaptive filter are rate of convergence, misadjustment, tracking, robustness, computational complexity, filter structure, numerical stability and accuracy.

6. The least mean square (LMS) algorithm is an adaptive filter method that uses a gradient-based method of steepest descent to obtain the least mean square error between the output and the reference data. It is an iterative procedure that corrects the weight vector (filter coefficient) in the direction of the negative of the gradient vector, which eventually leads to the minimum mean square error.

7. The cost function of least square estimation is defined as the sum of weighted error squares. The least square estimation is to minimize the error of the filter output to the reference signal. In this process, statistical modelling is not involved directly.

8. The variants of LMS algorithm include (a) normalized LMS, (b) leaky LMS, (c) block LMS and (d) sign LMS.

9. The step size parameter plays a vital role in the LMS algorithm. The larger the step size value, the faster the adaption, increasing residual MSE. Also, it affects the stability of the algorithm. Therefore, the step size selection cannot be arbitrarily large.

10. The recursive least squares is an adaptive filter algorithm, which recursively obtains the filter coefficients that minimize a least squares cost function relating to the input. In this algorithm, input signals are considered deterministic, whereas input signals are considered stochastic in the LMS. As a result, the RLS algorithm converges faster than the LMS algorithm.

Answers to Objective Questions

Q. No.	1	2	3	4	5	6
Key	A	D	A	B	A	C

Index

Printed in the United States
by Baker & Taylor Publisher Services